Essential Mathematics
and
Statistics for Science

Essential Mathematics and Statistics for Science

Graham Currell

Antony Dowman

The University of the West of England, UK

John Wiley & Sons, Ltd

Copyright © 2005   John Wiley & Sons Ltd, The Atrium, Southern Gate, Chichester,
West Sussex PO19 8SQ, England

Telephone (+44) 1243 779777

Email (for orders and customer service enquiries): cs-books@wiley.co.uk
Visit our Home Page on www.wileyeurope.com or www.wiley.com

Reprinted with corrections August 2006
Reprinted November 2006

Other Wiley Editorial Offices

John Wiley & Sons Inc., 111 River Street, Hoboken, NJ 07030, USA

Jossey-Bass, 989 Market Street, San Francisco, CA 94103-1741, USA

Wiley-VCH Verlag GmbH, Boschstr. 12, D-69469 Weinheim, Germany

John Wiley & Sons Australia Ltd, 33 Park Road, Milton, Queensland 4064, Australia

John Wiley & Sons (Asia) Pte Ltd, 2 Clementi Loop #02-01, Jin Xing Distripark, Singapore 129809

John Wiley & Sons Canada Ltd, 22 Worcester Road, Etobicoke, Ontario, Canada M9W 1L1

Wiley also publishes its books in a variety of electronic formats. Some content that appears in print may not be available
in electronic books.

Library of Congress Cataloging-in-Publication Data

British Library Cataloguing in Publication Data

A catalogue record for this book is available from the British Library

ISBN-10: 0-470-02228-0 (H/B)   ISBN-13: 978-0470-02228-3 (H/B)
ISBN-10: 0-470-02229-9 (P/B)   ISBN-13: 978-0470-02229-0 (P/B)

Typeset in 9/11 Times and Century Gothic by TechBooks, New Delhi, India
Printed and bound in Great Britain by Antony Rowe Ltd, Chippenham, Wiltshire
This book is printed on acid-free paper responsibly manufactured from sustainable forestry
in which at least two trees are planted for each one used for paper production.

To
Jenny and Felix
Jan, Ben and Jo.

Contents

Preface   **ix**

1   Mathematics and Statistics in Science   **1**

    1.1   Data and information   1
    1.2   Experimental variation and uncertainty   2
    1.3   Mathematical models in science   4

2   Scientific Data   **7**

    2.1   Scientific numbers   8
    2.2   Scientific quantities   14
    2.3   Angular measurements   26

3   Equations in Science   **35**

    3.1   Basic equation handling   35
    3.2   Introduction to rearranging equations   43
    3.3   Rearranging complicated equations   49
    3.4   Solving equations   53
    3.5   Simultaneous equations   60

4   Linear Relationships   **65**

    4.1   Straight-line graph   66
    4.2   Linear regression   75
    4.3   Linearization   81

5   Logarithmic and Exponential Functions   **89**

    5.1   Mathematics of e, ln and log   89
    5.2   Exponential growth and decay   97

6   Rates of Change   **109**

    6.1   Rate of change   109
    6.2   Differentiation   116

7   Statistics and Information   **121**

    7.1   Describing and inferring   122
    7.2   Frequency statistics   131

7.3   Probability   142
7.4   Bayesian odds   152
7.5   Factorials, permutations and combinations   160

8   Distributions and Uncertainty   167

8.1   Normal distribution   168
8.2   Uncertainties in measurement   179
8.3   Experimental uncertainty   190
8.4   Binomial distribution   196
8.5   Poisson distribution   205

9   Scientific Investigation   209

9.1   Scientific systems   209
9.2   The 'scientific method'   210
9.3   Hypothesis testing   212

10   Parametric Tests   219

10.1 Sample variances: F-test   220
10.2 Sample means: Student's t-test   225
10.3 Linear correlation   236
10.4 Test for proportion   241

11   Chi-squared Tests   245

11.1 Test for frequencies   246
11.2 Contingency tests   252
11.3 Goodness of fit   255

12   Non-parametric Tests   261

12.1 Wilcoxon tests   262
12.2 Mann–Whitney test   265
12.3 Kruskal–Wallis and Friedman tests   267
12.4 Sign test   273
12.5 Spearman's rank correlation   274

13   Experimental Design and Analysis   277

13.1 Data variance   278
13.2 Analysis of variance (ANOVA)   292
13.3 Experiment design   304

Appendixes   313

Review Questions   321

Short Answers to In-text Questions   335

Index   343

Preface

This book has been designed principally as a study text for students on a range of undergraduate science programmes: biological, environmental, chemical, forensic and sports sciences. It covers the majority of mathematical and statistical topics introduced in the first two years of such programmes, but also provides important aspects of experimental design and data analysis that students require when carrying out extended project work in the later years of their degree programmes.

A comprehensive web site is provided that actively supports the content of the book. This also extends the coverage into wider scientific applications and develops experience of relevant data handling software. The book can be used independently of the web site, but the close integration between them provides a greater range and depth of study possibilities. The web site can be accessed at:

wileyeurope.com/go/currellmaths

The introductory level of the book assumes that the reader will have studied mathematics with moderate success to year 11 of normal schooling. Currently in the UK, this is equivalent to a Grade C at Mathematics in the General Certificate of Secondary Education (GCSE).

There are 'revision' notes available on the associated web site for those readers who need to refresh their memories on relevant topics of basic mathematics – fractions, percentages, areas and volumes, etc.

The first eight chapters in the book introduce the *basic* mathematics and statistics that are required for the modelling of many different scientific systems. The remaining chapters are then primarily related to *experimental processes* in science, and introduce the statistical techniques that underpin more complex scientific investigations.

Over 200 *worked Examples* in the text are used to develop the various topics. The calculations for most of these *examples* are also performed using the Microsoft^{TM} spreadsheet program, Excel, and are available through the web site.

The reader can test his/her understanding as each topic develops by working through over 200 *In-text Questions*. The numeric answers are given at the end of the book, but full *worked* answers are also available through the web site in 'pdf' format and, where appropriate, through the use of Excel or other statistics software.

The web site also provides additional *application questions*, which the reader can use to develop an understanding of the wider applications of the relevant mathematics and statistics within the various disciplines.

Throughout the book, the reader has the opportunity of learning how to perform most calculations using Excel files on the web site. This strong integration of paper-based and computer-based calculations supports an understanding of the mathematics and statistics involved, but also supports the development of experience with the use of spreadsheets for data handling and analysis.

Some additional study topics are also introduced via the web site. These topics extend the subject coverage into closely related areas, without making the physical size of the book too large. It is anticipated that the range of additional topics may increase with regular updating of the web site.

At the end of the book there are a number of *review questions* for each chapter that are suitable for use as tutorial questions. The worked answers for these are only available through the Tutor Resources section (password-protected) of the web site. Other restricted material, available through Tutor Resources, includes worked answers to a set of *computer workshops* that can be used to conduct practical, computer-based, workshops in the various topic areas.

Scientific context

The diverse uses of mathematics and statistics in the various disciplines of science place different emphases on the various topics. However, there is a core of mathematical and statistical techniques that is essentially common to all branches of experimental science, and it is this material that forms the basis of this book. We believe that we have developed a coherent approach and consistent nomenclature, which will make the material appropriate across the various disciplines.

Specific groups of students may wish to add emphasis to different aspects of the material. For example, chemists may wish to develop the mathematics, and ecologists may wish to develop the statistics. We endeavour to accommodate some of these advanced specialisms by linking to additional material via the web site.

When developing questions and examples at an introductory level, it is important to achieve a balance between treating each topic as pure mathematics or embedding it deeply in a scientific 'context'. A lack of 'context' can lead to a loss of student interest, but too much can confuse the understanding of the mathematics. The optimum balance varies with topic and level.

In-text Questions and Examples in the book concentrate on clarity in developing the mathematics step-by-step through each chapter. These questions enable students to test their own incremental developments. Where possible we have included a scientific context that is understandable to students from a range of different disciplines. The numerical answers to In-text Questions (where appropriate) are given at the end of the book, but full worked answers are available through the web site.

Application Questions with relevance to a range of specific scientific disciplines are available via the web site. These are given with full worked answers, including use of Excel spreadsheets and other statistics software where appropriate.

Experimental design

The process of good experimental planning and design is a topic that is often much neglected in an overall undergraduate course. Although the topic pervades all aspects of experimental science, it does not have a clear focus in any one particular branch of the science, and is rarely treated coherently in its own right.

Good experiment design is dependent on the availability of suitable mathematical and statistical techniques to analyse the resulting data. A wide range of such methods are introduced in this book:

- Regression analysis (Chapter 4) for relationships that are inherently linear or can be linearized (Unit 4.3).

- Logarithmic and/or exponential functions (Chapter 5) for systems involving natural growth and decay, or for systems where the response increases as a ratio (e.g. loudness, *pH*).

- Modelling with Excel (Chapter 6) for rates of change.

- Probability and Bayesian statistics (Chapter 7) to interpret risk and likelihood.

- Statistical distributions (Chapter 8) for modelling random behaviour in complex systems.

- Parametric and non-parametric tests (Chapters 10, 11 and 12) for differences in variety of different possible systems.

- Analysis of variance (Chapter 13) for hypothesis testing of more complex experimental systems.

An 'Overview' at the beginning of each chapter highlights the relationships between experiment design and the various analytical techniques developed in that chapter.

Chapter 1 provides an introduction to the way in which scientific processes in the real world can be described and modelled by the mathematical and statistical techniques that are then developed in Chapters 2 to 8. In particular, Chapter 4 introduces the technique of linear regression that has become an essential element in the analysis of many different types of experiment.

Chapter 9 looks more closely at the way in which science progresses by experimentation, and develops the 'scientific method' concept of hypothesis testing and deduction. This then links closely with Chapters 10, 11 and 12, which provide the statistical methods for carrying out a range of hypothesis tests.

Chapter 13 develops an understanding of more complex scientific experimentation, where the effects of several factors combine to influence the final results. This chapter demonstrates the strong links between the possible design structures for the experiment and the statistical methods available for its analysis.

Computing software

There are various software packages available that can help scientists in implementing mathematics and statistics. Some university departments have strong preferences for one or the other.

Microsoft^{TM} Excel spreadsheets can be used effectively for a variety of purposes:

- basic data handling – sorting and manipulating data

- data presentation using graphs, charts, tables

- preparing data and graphs for export to other packages, Word, MINITAB, SPSS

- performing a range of mathematical calculations

- performing a range of statistical calculations

Many of the worked examples and in-text questions in this book and web site are presented with Excel calculations available on the web.

Most students find that the statistical functions in Excel are a helpful *introduction* to using statistics, but for particular problems, it is more useful to turn to the packages designed specifically for statistical analysis, e.g. MINITAB, SPSS. Nevertheless, it is usually convenient to use Excel for organizing data into an appropriate layout before exporting to the specialized package.

Apart from Excel, the book does not relate to any specific statistics software package. However, the web site provides examples of the use of other packages in addressing the statistics developed in the book.

We will use the icon 🖳 within the book to make specific reference to relevant computing resources in the web site. The use of the '>' symbol shows the links that should be followed within each web page.

Web site

The web site has six main 'areas' available from the *Homepage*:

Web site Overview:
  Overview of the main features of the web site.

Answers & Examples:
  Worked answers to all the *In-Text Questions* and most *Examples*, including working in Excel and other statistics software.

Computer Techniques:
  Tutorials in Excel, including data transfer to other software.
  Introduction to the use of other *statistics software*.

Study Resources:
  Study Guide
  General resources including *Revision Notes* and *Statistical Tables*.
  Computer Workshops.

Tutor Resources (password-protected):
  Worked answers to *Review Questions* and *Computer Workshops*.

Link to Authors' Web Site:
  Book revisions and up-date.
  Additional material.
  Application questions.
  Statistics software
  Links to other sites

The icon 🖳, at points of particular relevance in the book, is used to prompt the reader to a specific aspect of the web site, and to identify those Examples that have a worked answer available on the web site.

The web site will be developed and expanded, and the reader is advised, when studying a new topic, to check what additional material – notes, questions, exercises – may be available on the web site.

The links identified by the web icons in the book will remain valid even though the information within the web site continues to be updated.

Book structure

Each chapter in the book is subdivided into Units as listed in the Contents. The Units are subdivided into numbered sections, e.g., section 3.1.2 is the second section in Unit 3.1.

Examples in the book are introduced in a clear box with an 'E' number, e.g., E9.2 is the second Example in Chapter 9.

Most Examples have worked answers using software on the web site under *Answers & Examples*.

In-text Questions in the book are presented in a box with a *shaded stripe* down the left-hand side and with a 'Q' number, e.g., Q9.3 is the third Question in Chapter 9. Worked answers are available on the web site under *Answers & Examples*. These are given in 'pdf' format for all questions and using appropriate software for most questions.

Figures and tables are given a sequential number within each chapter, e.g. Figure 5.2 and Table 7.1 respectively.

Equation numbers are given in square brackets, e.g. [9.6], and the equations that are referred to most frequently are emphasized by placing them within a box.

Graham Currell
Tony Dowman

1

Mathematics and Statistics in Science

Overview

A science student encounters mathematics and statistics in three main areas:

- understanding and using theory,
- carrying out experiments and analysing results,
- presenting data in laboratory reports and essays.

Unfortunately, many students do not fully appreciate the need for understanding mathematics and/or statistics until it suddenly confronts them in a lecture or in the write-up of an experiment. There is indeed a 'chicken and the egg' aspect to the problem:

> Some science students have little enthusiasm to study mathematics until it appears in a lecture or tutorial – by which time it is too late! Without the mathematics, they cannot *fully* understand the science that is being presented, and they drift into a habit of accepting a 'second-best' science *without* mathematics. The end result will probably be a drop of at least one grade in their final degree qualification.

All science is based on a *quantitative* understanding of the world around us – an understanding described ultimately by *measurable* values. Mathematics and statistics are merely the processes by which we handle these quantitative values in an effective and logical way.

Mathematics and statistics provide the network of links that tie together the details of our understanding, and create a sound basis for a fundamental appreciation of science as a whole. Without these quantifiable links, the ability of science to predict and move forward into new areas of understanding would be totally undermined.

In recent years, the data-handling capability of Information Technology has made mathematical and statistical calculations far easier to perform, and has transformed the day-to-day work in many areas of science. In particular, a good spreadsheet program, like Excel, enables both scientists and students to carry out extensive calculations quickly, and present results and reports in a clear and accurate manner.

1.1 Data and information

Real world information is expressed in the mathematical world through **data**.

In science, some data values are believed to be fixed in nature. We refer to values that are fixed as **constants**, e.g. the constant, c, is often used to represent the speed of light in a vacuum, $c = 3.00 \times 10^8$ m s^{-1}.

However, most measured values are subject to change. We refer to these values as **variables**, e.g. T for temperature, pH for acidity.

The term, **parameter**, refers to a variable that can be used to describe a relevant characteristic of a scientific system or statistical population (see 7.1.2), e.g. the actual pH of a buffer solution, or the average (mean) age of the whole UK population. The term, **statistic**, refers to a variable that is used to describe a relevant characteristic of a *sampled* set of data (see 7.1.2) e.g. the standard deviation of five repeated measurements of the pH of a buffer solution, or the average (mean) age of 1000 members of the UK population.

Within this book we use the convention of printing letters and symbols that represent quantities (constants and variables) in italics, e.g. c, T and pH.

The letters that represent units are presented in normal form, e.g., m s^{-1} gives the units of speed in metres per second.

There is an important relationship between data and information, which appears when analysing more complex data sets. It is a basic rule that:

> It is impossible to get more 'bits' of information from a calculation than the number of 'bits' of data that are put into the calculation.

For example, if a chemical mixture contains three separate compounds, then it is necessary to make at least three separate measurements on that mixture before it is possible to calculate the concentration of each separate compound.

In mathematics and statistics, the *number* of bits of information that are available in a data set is called the **degrees of freedom**, df, of that data set. This figure appears in many statistical calculations, and it is usually easy to calculate the number of degrees of freedom appropriate to any given situation.

1.2 Experimental variation and uncertainty

The uncertainty inherent in scientific information is an important theme that appears throughout the book.

The **true value** of a variable is the value that we would measure if our measurement process were 'perfect'. However, because no process is perfect, the 'true value' is not normally known.

The **observed value** is the actual value that we produce as our *best estimate* of the true value.

The **error** in the measurement is the difference between the true value and the observed value:

$$\text{Error} = \text{Observed value} - \text{True value} \qquad [1.1]$$

As we do not normally know the 'true value', we cannot therefore know the actual error in any particular measurement. However, it is important that we have some idea of how large the error might be.

The **uncertainty** in the measurement is our *best estimate* of the magnitude of possible errors. The magnitude of the uncertainty must be derived on the basis of a proper understanding of the measurement process involved and the system being measured. The statistical interpretation of uncertainty is derived in Unit 8.2.

The uncertainty in experimental measurements can be divided into two main categories:

1   **Measurement uncertainty**. Variations in the actual process of measurement will give some differences when the same measurement is repeated under exactly the same conditions. For example, repeating a measurement of alcohol level in the same blood sample may give results that differ by a few milligrams in each 100 millilitres of blood.

2   **Subject uncertainty**. A subject is a representative example of the system (Unit 9.1) being measured, but many of the systems in the real world have inherent variability in their responses. For example, in

testing the effectiveness of a new drug, every person (subject) will have a slightly different reaction to that drug, and it would be necessary to carry out the test on a wide range of people before being confident about the 'average' response.

Whatever the source of uncertainty, it is important that any experiment must be designed to counteract the effects of uncertainty, and also to quantify the magnitude of that uncertainty.

Within each of the two types of uncertainty, *measurement* and *subject*, it is possible to identify two further categories:

1   **Random error**. Each subsequent measurement has a random error, leading to *imprecision* in the result. A measurement with a low random error is said to be a *precise* measurement.

2   **Systematic error**. Each subsequent measurement has the same recurring error. A systematic error shows that the measurement is *biased*, e.g., when setting the liquid level in a burette, a particular student may always set the meniscus of the liquid a little too low.

The **precision** of a measurement is the best estimate for the purely *random error* in a measurement.

The **trueness** of a measurement is the best estimate for the *bias* in a measurement.

The **accuracy** of a measurement is the best estimate for the *overall error* in the final result, and includes both the effects of a lack of precision (due to random errors) and bias (due to systematic errors).

E1.1

Four groups of students each measure the *pH* (acidity) of a sample of soil, with each group preparing five replicate samples for testing. The results are given in Figure 1.1.

What can be said about the *accuracy* of their results?

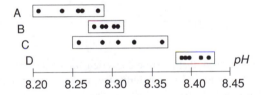

Figure 1.1. Precision and bias in experimental data.

With the information given, very little can be said about the accuracy of the measurements; the 'true' value is not known, and there is no information about possible bias in any of the results.

It is possible to say that the results from groups A and C show greater random uncertainty (less precision) than groups B and D. This could be due to such factors as: a lack of care in preparing the five samples for testing, or some electronic instability in the pH-meter being used.

Groups B and D show greater precision, but at least one of B or D must have some bias in their measurements, i.e. poor 'trueness'. The bias could be due to an error in setting the pH-meter with a buffer solution, which would then make every one of the five measurements in the set wrong by the same amount.

Without knowing the true value of the pH, it is also possible that groups A and/or C are also biased, in addition to being imprecise. For example *if the true value were pH* = 8.40, this would mean that groups A, B and C were all biased, groups A and C also had greater imprecision, with the most *accurate* measurement being group D.

The effect of random errors can be managed and quantified using suitable statistical methods (Units 8.2, 8.3 and 13.2). The presentation of uncertainty as *error bars* on graphs is developed in an Excel Tutorial.

⌨ Computer Techniques > Excel Tutorial: XY-Graph

Systematic errors are more difficult to manage in an experiment, but good experiment design (Unit 13.3) aims to counteract their effect as much as possible.

1.3 Mathematical models in science

A fundamental building block of both science and mathematics is the *equation*.

Science uses the equation as a *mathematical model* to define the *relationship* between one or more factors in the real world (see 3.1.3). It may then be possible to use mathematics to investigate how that equation may lead to *new conclusions* about the world.

Perhaps the most famous equation, arising from the general theory of relativity is:

$$E = mc^2$$

which relates the amount of energy, E (J), that would be released if a mass, m (kg), of matter were to be converted into energy (e.g. in a nuclear reactor). E and m are both variables and the constant c ($= 3.00 \times 10^8$ m s^{-1}) is the speed of light.

E1.2

Calculate the amount of matter, m, that must be converted *completely* into energy, if the amount of energy, E, is equivalent to that produced by a medium-sized power station in one year: $E = 1.8 \times 10^{13}$ J. Rearranging the equation $E = mc^2$ gives:

$$m = \frac{E}{c^2}$$

Substituting values into the equation:

$$m = \frac{1.8 \times 10^{13}}{(3.00 \times 10^8)^2} = 0.000\,20\,\text{kg} = 0.20\,\text{g}$$

This equation tells us that if only 0.20 g of matter is converted into energy, it will produce an energy output equivalent to a power station operating for a year! This is why the idea of nuclear power was initially so very attractive.

Example E1.2 indicates some of the common mathematical processes used in handling equations in science: rearranging the equation, using scientific notation, changing of units, and 'solving' the equation to derive the value of an unknown variable.

Equations are used to represent many different types of scientific processes, and often employ a variety of *mathematical functions* to create suitable models.

In particular, many scientific systems behave in a manner that is best described using an exponential or logarithmic function, e.g. drug elimination in the human body, *pH* values. Example E1.3 shows how both the growth and decay of a bacteria population can be described, in part, by exponential functions.

Another aspect of real systems is that they often have significant inherent *variability*, e.g., similar members of a plant crop grow at different rates, or repeated measurements of the refractive index of glass may give different results. In these situations, we need to develop *statistical models* that we can use to describe the underlying behaviour of the system as a whole.

E1.3

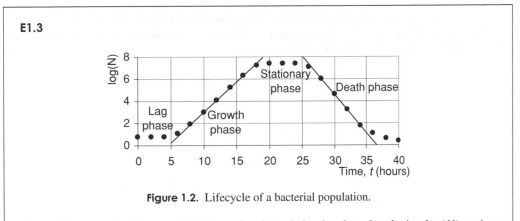

Figure 1.2. Lifecycle of a bacterial population.

Figure 1.2 gives a plot of growth and decay in a bacteria batch colony, by plotting $\log(N)$ against time, t, where N is the number of cells per millilitre.

The 'straight line' sections of the graph in the 'growth' and 'death' phases are two sections of the lifecycle that can be described by *exponential* functions (see Unit 5.2).

The particular statistical model that best fits the observed data is often a good guide to the scientific processes that govern the system being measured. Example E1.4 shows the Poisson distribution that could be expected if plants were distributed *randomly* with an average of 3.13 plants per unit area.

E1.4

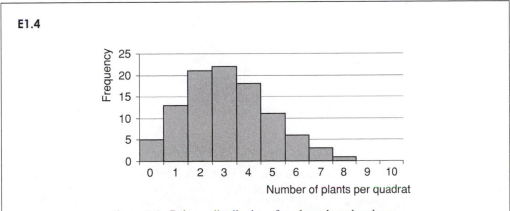

Figure 1.3. Poisson distribution of random plant abundance.

Figure 1.3 shows that, in sampling a *random* distribution of plants with 100 quadrats of unit area, we found that 22 quadrats had 2 plants, 11 quadrats had 5 plants, etc.

If the distribution of plants were affected by clumping or by competition for survival, then we would expect the *shape of the distribution* to be different.

Excel spreadsheets have become particularly useful for implementing mathematical models of very complex scientific systems. Throughout this book we continue to develop mathematics and statistics in conjunction with their practical applications through Excel.

2

Scientific Data

Overview

Data in science appears in a variety of forms. However, there is a broad classification of data into *two* main categories:

1 **Quantitative data**. The numeric value of quantitative data is recorded as a *'measurable (or parametric) variable'*, e.g. time, pH, temperature, etc.

2 **Qualitative (or categorical) data**. Qualitative data is *grouped* into different classes, and the *'names'* of the classes serve only to distinguish, or rank, the different classes, and have no other quantitative value, for example, grouping people according to their nationality, eye colour, etc.

Quantitative data can be further divided into:

Discrete data. Only *specific values* are used, for example, counting the number of students in a class will only give *integer* values.

Continuous data. Using values specified to any accuracy as required, for example, defining time, using seconds, to any number of decimal places as appropriate (e.g. 75.852 06 s).

Quantitative data can be further subdivided into:

Ratio data. The 'zero' of a ratio scale has a true 'zero' value in science, and the ratios of data values also have scientific meaning. For example, the zero, 0 K, of the *absolute temperature scale* in thermodynamics, is a *true* 'absolute zero' (there is nothing colder!), and 100 K is twice the absolute 'temperature' of 50 K.

Interval data. The 'zero' of an interval scale does *not* have a true 'zero' value in science, and the ratios of data values do not have scientific meaning. Nevertheless the data *intervals* are still significant. For example, the zero, 0°C, of the *Celsius temperature scale*, is just the temperature of melting ice and not a true 'zero', and 100°C is not 'twice as hot' as 50°C. However, the degree intervals are the same in both the 'absolute' and Celsius scales.

Qualitative data can be further divided into:

Ordinal data. The classes have a sense of *progression* from one class to the next, e.g. degree classifications (1st, 2.1, 2.2, 3rd), opinion ratings in a questionnaire (excellent, good, satisfactory, poor, bad).

Nominal (named) data. There is *no* sense of progression between classes, e.g. animal species, nationality.

This chapter is concerned mainly with calculations involving *continuous quantitative* data, although examples of other types of data appear elsewhere within the book. The topics included relate to some of the most common calculations that are performed in science:

- using scientific (or standard) notation,
- displaying data to an appropriate precision,
- handling units, and performing the conversions between them,
- performing routine calculations involving quantities, and
- working with angular measurements in both degrees and radians.

2.1 Scientific numbers

2.1.1 Introduction

This Unit describes some of the very common arithmetical calculations that any scientist needs to perform when working with numerical data. Students wishing to refresh their memory of basic mathematics can also refer to the 'revision' notes available on the web site.

⌨ Study Resources > Revision Mathematics

Although much of the data handling is now done with computers (e.g. using Excel), it is still *very* necessary for scientists (including students) to be familiar with the basic 'science' of numbers. In particular, students should ensure that they can use hand-held calculators effectively for scientific notation and functions – see 2.1.10.

2.1.2 Scientific (standard) notation

Scientific notation is also called *standard notation* or *exponential notation*.

In **scientific notation**, the digits of the number are written with the most significant figure before the decimal point, and all other digits after the decimal point. This 'number' is then multiplied by the correct 'power of ten' to make it equal to the desired value, for example,

$$230 = 2.30 \times 10^2$$
$$0.00230 = 2.30 \times 10^{-3}$$
$$2.30 = 2.30 \times 10^0 = 2.30 \times 1 = 2.30$$

In Excel, other software, and some calculators, the 'power of ten' is preceded by the letter 'E', for example, the number -3.56×10^{-11} would appear as $-3.56E-11$.

In a calculator, the 'power of ten' is entered by first pressing the EXP button (*without* entering a '×'). Many calculator manuals use the word 'exponent' (see 5.1.2) to describe the 'power of ten', and some calculators have 'EE' on the relevant button.

⌨ Computing Techniques > Excel Tutorial: Skills

Q2.1 Express the following numbers in scientific notation:

    (i)   42600   (v) 0.045×10^4

    (ii) 0.00362   (vi) 26.6×10^3

    (iii) 10000   (vii) 3.2E3

    (iv) 0.0001   (viii) 4.5E-6

2.1.3 Multiplying (dividing) in scientific notation

When multiplying (or dividing) in scientific notation, it is possible to *separately* multiply (or divide) the numbers and add (or subtract) the 'powers of ten', for example:

E2.1 🖳

Multiplication in scientific notation:

$$(4.2 \times 10^3) \times (2.0 \times 10^4) = (4.2 \times 2.0) \times (10^3 \times 10^4) \quad = 8.4 \times 10^{3+4} \quad = 8.4 \times 10^7$$

                           Separating numbers and powers   Adding powers

Division in scientific notation:

$$(4.2 \times 10^3) \div (2.0 \times 10^4) = (4.2 \div 2.0) \times (10^3 \div 10^4) \quad = 2.1 \times 10^{3-4} \quad = 2.1 \times 10^{-1}$$

                           Separating numbers and powers   Subtracting powers

---

The 'division' equation in E2.1 can be written in fraction form:

$$\frac{4.2 \times 10^3}{2.0 \times 10^4} = \frac{4.2}{2.0} \times \frac{10^3}{10^4} = 2.1 \times 10^{3-4} = 2.1 \times 10^{-1} = 0.21$$

It is often necessary to 'adjust' the position of the decimal point (and 'power of ten') to return the final number to true scientific notation, as in the final step in Example E2.2:

E2.2 🖳

A simple multiplication gives

$$(4.0 \times 10^5) \times (3.5 \times 10^{-3}) = (4.0 \times 3.5) \times (10^5 \times 10^{-3}) = 14 \times 10^{5+(-3)} = 14 \times 10^2$$

However, the result is not in *scientific notation*, and should be adjusted to give:

$$14 \times 10^2 = 1.4 \times 10^3$$

Q2.2 Evaluate the following, giving the answer in scientific notation (calculate 'by hand' and then check answers on a calculator):

(i) 120000×0.003 (iv) $4500 \div 0.09$

(ii) $(1.2 \times 10^5) \times (3.0 \times 10^{-3})$ (v) $0.0056 \times 4.0 \times 10^3$

(iii) $\dfrac{1.2 \times 10^5}{3.0 \times 10^3}$ (vi) $\dfrac{1.2 \times 10^5}{3.0 \times 10^{-3}}$

2.1.4 Adding (subtracting) in scientific notation

Before adding or subtracting scientific numbers it is important to get both numbers to the same '*power of ten*'. It is then possible to simply add (or subtract) the numbers:

E2.3 🖥

To add 3.46×10^3 to 2.120×10^4 we first change 3.46×10^3 to 0.346×10^4
so that both numbers have the multiplier '$\times 10^4$'.
We can then write:

$$3.46 \times 10^3 + 2.120 \times 10^4 = 0.346 \times 10^4 + 2.120 \times 10^4 = (0.346 + 2.120) \times 10^4 = 2.466 \times 10^4$$

Similarly, to subtract 2.67×10^{-2} from 3.0×10^{-3} we first change $2.67 \times 10^{-2} = 26.7 \times 10^{-3}$ so that both numbers have the multiplier '$\times 10^{-3}$', and we can then write:

$$3.0 \times 10^{-3} - 2.67 \times 10^{-2} = 3.0 \times 10^{-3} - 26.7 \times 10^{-3} = (3.0 - 26.7) \times 10^{-3} = -23.7 \times 10^{-3}$$
$$= -2.37 \times 10^{-2}$$

Note that the answer should be left in correct scientific notation form.

Q2.3 Evaluate the following, giving the answer in scientific notation:
(calculate 'by hand' and then check answers on a calculator)

(i) $42.1 - 1.2463 \times 10^3$ (ii) $\dfrac{7.2463 \times 10^6 - 1.15 \times 10^5}{3.0 \times 10^{-3}}$

2.1.5 Significant figures (sf)

The **most significant figure** (or digit) in a number is the first non-zero number reading from the left, e.g., '4' in each of these numbers, 456 and 0.047.

The **least significant figure** (or digit) is the last digit to the right whose value is considered to carry valid information.

E2.4

According to the 1951 census, the population of Greater London was 8 346 137. If I state that the population was 8 350 000, correct to three *significant figures* (*sf*), then I am claiming (correctly) that the population was closer to 8.35 million than to either 8.36 million or 8.34 million.

The figure '8' is the *most* significant figure, and the '5' is the *least* significant figure. The '0's are included to indicate the appropriate 'power of ten'.

After the decimal point, a final '0' should only be included if it is *significant*, for example,

3.800 to 4 sf would be written as 3.800
3.800 to 3 sf would be written as 3.80
3.800 to 2 sf would be written as 3.8

The number of significant figures chosen will depend on the precision or accuracy with which the value is known.

2.1.6 Decimal places (dp)

The **format** of numbers can be specified by defining how many **decimal places (dp)** are included after the decimal point. For example, 9.81 m s^{-2} is the acceleration due to gravity written to two decimal places.

💻 Computer Techniques > Excel Tutorial: Skills

2.1.7 Rounding numbers

It is important, when information is presented in the form of data, that the data is an accurate representation of the information. There is uncertainty in all scientific 'information' (Unit 1.2), and the number of significant figures used in displaying the data should not imply a greater precision than is actually the case. For example, it would not be correct to quote an answer as 1.145917288 simply because the calculator displayed that many digits – it is exceedingly rare for any scientific measurement to be that precise (±0.000000001).

To get the right number of significant figures (sf) or decimal places (dp), it is sometimes necessary to 'round off' the number to the nearest value.

When rounding numbers to specific interval values, any number that is *more than halfway* between values will round *up* to the next value, and any number *less than halfway* will round *down*.

E2.5 💻

Rounding:

(i) 70860 to 3 sf gives 70900

(ii) 70849 to 3 sf gives 70800

(iii) 5.6268×10^{-3} to 4 sf gives 5.627×10^{-3}

(iv) 3.194 to 2 dp gives 3.19

(v) 3.195 to 2 dp gives 3.20

If the number is *exactly halfway* between values, then the number *should* round so that the *last digit is even*. However, it is common practice (including rounding in Excel) that the *halfway* value *always* rounds upwards.

E2.6 🖥

(i) Rounding 70550 to 3 sf gives 70600   (iii) Rounding 0.275 to 2 dp gives 0.28

(ii) Rounding 70850 to 3 sf *should* give 70800   (iv) Rounding 3.185 to 2 dp *should* give 3.18

For (ii) and (iv), 70900 and 3.19 would also be commonly accepted answers – rounding *all* 'halfway' values up!

Q2.4 Round the following numbers to the required numbers of significant figures (sf) as stated:

(i) 0.04651 to 2 sf   (v) 26962 to 3 sf

(ii) 0.04649 to 2 sf   (vi) 11.250 to 3 sf

(iii) 13.97 to 3 sf   (vii) 11.150 to 3 sf

(iv) 7.3548×10^3 to 3 sf   (viii) 5.6450×10^{-3} to 3 sf

Q2.5 Round the following numbers to the required numbers of decimal places (dp) as stated:

(i) 0.04651 to 3 dp   (iii) 426.891 to 2 dp

(ii) 7.9999 to 2 dp   (iv) 1.3450 to 2 dp

When presenting a final calculated value, the number of significant figures or decimal places should reflect the accuracy (see 8.2.5) of the result. Simply performing a mathematical calculation cannot improve the overall accuracy or precision of the original information.

Q2.6 Add the following masses and give the result to an *appropriate* number of decimal places. (Hint: in this case the total value cannot have more decimal places than the *least precise* of all the separate masses.) 0.643 g, 3.10 g, 0.144 g, 0.0021 g

---

It is also important that the 'rounding' process should not be applied until the *end of the calculation*. If the data is 'rounded' too early, then it is quite possible that the small inaccuracies created will be magnified by subsequent calculations. This may result in a final error that is much greater than any uncertainty in the real information.

2.1.8 Order of magnitude

If a value increases by one 'order of magnitude', then it increases by (very) *approximately 10 times*:

- one order of magnitude is an increase of *ten* times,
- two orders of magnitude is an increase of 100 times, and
- an increase of 100 000 times is five orders of magnitude.

E2.7

What are the differences in 'orders of magnitude' between the following pairs of numbers?

(i) 46 800 and 45   (ii) 5.6 mm and 3.4 km

Answers:

 (i) 46 800 is three *orders of magnitude* greater than 45.

(ii) 5.6 mm is six *orders of magnitude* less than 3.4 km.

2.1.9 Estimations

It is often useful to check complicated calculations by carrying out simple calculations 'by hand' using values approximated to one (or two) significant figures.

E2.8

If my calculation suggests that 0.4378×256.2 gives the answer 1121.6436, I can check the result as follows:

Replace the numbers by approximate values 0.4 and 300, and multiply them 'by hand' to get $0.4 \times 300 = 120$. I then find out that my calculated answer is one order of magnitude out – I have put the decimal point in the wrong place, and the correct answer should be 112.164 36.

Q2.7 Estimate, without using a calculator, the approximate speed (in miles per hour) of an aeroplane that takes 4 hours and 50 minutes to fly a distance of 2527 miles.

Is the answer likely to be too high or too low?

2.1.10 Using a calculator

The most appropriate hand calculator for the science student should be inexpensive, easy to use, and have a basic scientific capability. This capability should include logarithms, the exponential function (e), trigonometric functions, the use of brackets, and basic statistical calculations (mean, standard deviation, etc.). The more expensive and sophisticated calculators (e.g. with graphics) should be avoided unless the student is confident in how to use them.

Students who need to improve their familiarity with a calculator should first refer to its *instruction book*! Some limited help is also available on the web site:

💻 Study Resources > Calculator Tutorial

Q2.8 Use a calculator to evaluate the following expressions:

(i) $1/(2.5 \times 10^4)$ use the reciprocal key, '1/x' or 'x^{-1}'

(ii) $(-0.0025) \div (-1.2 \times 10^{-6})$ use the EXP key (sometimes the EE key) for 10^{-6}

(iii) $3.2^{-1.6}$ use the key 'x^y' or '$^\wedge$'

(iv) 3.487^2 use the key 'x^2'

(v)   square root of 0.067   use the square root key '$\sqrt{\ }$'

2.2 Scientific quantities

2.2.1 Introduction

Quantitative measurements are made in relation to agreed 'units' of quantity. For example, the distances for Olympic races are expressed as multiples of an agreed 'unit' of distance (the *metre*): 100 *metres*, 400 *metres*, 1500 *metres*, etc.

The handling of 'units' should be a simple process. However, some students try to work out the conversion of units 'in their heads', and get confused with multiple multiplications and divisions. The answer is to break the problem into a number of very simple steps, writing down each step in turn.

2.2.2 Presenting mixed units

Most people are very familiar with common 'mixed' units such as miles per hour for speed, or dollars per month for wages. However, writing out such units in full using the word 'per' takes up a lot of space, and in science it is more convenient to use abbreviated forms. For example, speed is calculated by *dividing* distance by time, and consequently the units become metres *divided* by seconds: m/s or m s^{-1}. However, the format using the oblique '/' for 'per' (e.g. 'm/s') *should not be used* for units, and should be replaced by formats with negative powers, e.g. 'm s^{-1}'.

The units of a mixed variable represent the *process* used to calculate the value of that variable. Some examples of equivalent forms are given below:

Variable	Units	Unit format:
Speed	metres per second	m s^{-1}
Density	kilograms per cubic metre	kg m^{-3}
Pressure	newtons per square metre	N m^{-2}

By convention, units are shown in normal (not italic) font with a space between each sub-unit. Where a unit is derived from a person's name, the first letter of the unit's name is given in lower case, although the unit symbol is give a capital letter, for example, one newton is written as 1 N.

2.2.3 SI units

The SI (Système Internationale) system of units derives from an international agreement to use a common framework of units, which is based on a set of seven fundamental units – Table 2.1.

Table 2.1. Fundamental SI units.

SI Unit	Symbol	Measures	Defined using:
kilogram	kg	Mass	Standard platinum-iridium mass
second	s	Time	Oscillations of a caesium-137 atom
metre	m	Length	Distance travelled by light in fixed time
kelvin	K	Temperature	Temperature of triple point of water
mole	mol	Amount	Comparison with 0.012 kg of carbon-12
ampere	A	Electric current	Force generated between currents
candela	cd	Light output	Intensity of a light source

Other units are derived as combinations of the fundamental units. Some examples are given in Table 2.2.

Table 2.2. Derived SI units.

SI unit	Symbol	Measures	Equivalence to fundamental units
newton	N	Force	$1\,N = 1\,kg\,m\,s^{-2}$
joule	J	Energy	$1\,J = 1\,N\,m$
watt	W	Power	$1\,W = 1\,J\,s^{-1}$
pascal	Pa	Pressure	$1\,Pa = 1\,N\,m^{-2}$
hertz	Hz	Frequency	$1\,Hz = 1\,\text{cycle per second} = 1\,s^{-1}$

Various prefixes are used to magnify or reduce the size of a particular unit according to the 'power of ten' ratios in Table 2.3.

Table 2.3. Powers of ten in SI units

Power of ten:	10^{12}	10^{9}	10^{6}	10^{3}	10^{1}	10^{-1}
Name:	tera-	giga-	mega-	kilo-	deca-	deci-
Prefix:	T	G	M	k	da	d

Power of ten:	10^{-2}	10^{-3}	10^{-6}	10^{-9}	10^{-12}	10^{-15}
Name:	centi-	milli-	micro-	nano-	pico-	femto-
Prefix:	c	m	μ	n	p	f

Q2.9 By how many 'orders of magnitude' is 4.7 km (kilometres) larger than 6.2 nm (nanometres)?

2.2.4 Conversion of units

Examples E2.9 to E2.14 show that:

- Conversion should be performed in *easy* stages, making *simple* changes at each stage.

- It often helps to write out any complex units as statements in *words*.

For example: if x units of A   are equivalent to y units of B

dividing both sides by 'x':   1 unit of A   is equivalent to $\dfrac{y}{x}$ units of B

multiplying both sides by 'z': z units of A   are equivalent to $z \times \dfrac{y}{x}$ units of B

E2.9

If 5.000 miles are equal to 8.045 km, convert 16.3 miles into kilometres.

Starting with: 5.000 miles are equal to   8.045 km

divide both sides by 5:   1.000 mile is equal to $\dfrac{8.045}{5.000} = 1.609$ km

multiply both sides by 16.3:   16.3 miles are equal to $16.3 \times \dfrac{8.045}{5.000} = 26.2$ km

---

It is important to be able to calculate **'reciprocal'** conversions:

E2.10

Examples of taking the reciprocals of unit conversions:

1.0 m is equivalent to 100 cm $= 1.00 \times 10^2$ cm:

hence: 1.0 cm is equivalent to $\dfrac{1}{100}$ m $= 0.01$ m $= 1.00 \times 10^{-2}$ m

1.000 mile is equivalent to 1.609 km:

hence: 1.000 km is equivalent to $\dfrac{1}{1.609}$ miles $= 0.622$ miles

1.000 L is equivalent to 1000 mL $= 1.000 \times 10^3$ mL:

hence: 1.0 mL is equivalent to $\dfrac{1}{1000}$ L $= 0.001$ L $= 1.0 \times 10^{-3}$ L

and   10 mL is equivalent to $\dfrac{10}{1000}$ L $= 0.01$ L $= 1.0 \times 10^{-2}$ L

Units are often used in a 'power' form, e.g. the units of area are m^2. Conversion factors will also be raised to the **same power as the unit**.

E2.11

Examples using *powers* of units:

Distance: $1.0 \text{ m} = 100 \text{ cm} = 1.0 \times 10^2 \text{ cm}$

Area: $1.0 \text{ m}^2 = 100 \times 100 \text{ cm}^2 = 10\,000 \text{ cm}^2 = 1.0 \times 10^2 \times 10^2 \text{ cm}^2 = 1.0 \times 10^4 \text{ cm}^2$

Volumes: $1.0 \text{ m}^3 = 100 \times 100 \times 100 \text{ cm}^3 = 1\,000\,000 \text{ cm}^3 = 1.0 \times 10^2 \times 10^2 \times 10^2 \text{ cm}^3$

$1.0 \text{ m}^3 = 1.0 \times 10^6 \text{ cm}^3$

$1.0 \text{ cm}^3 = \dfrac{1}{1.0 \times 10^6} \text{ m}^3 = 1.0 \times 10^{-6} \text{ m}^3$

Distance: $1.000 \text{ mile} = 1.609 \text{ km}$

Areas: $1.000 \text{ mile}^2 = 1.609 \times 1.609 \text{ km}^2 = 1.609^2 \text{ km}^2 = 2.589 \text{ km}^2$

$1 \text{ km}^2 = \dfrac{1}{2.589} \text{ mile}^2 = 0.386 \text{ mile}^2$

E2.12 🖳

Express a volume of 30 mm^3 in units of m^3.

The first step is to start from what is known: $1.0 \text{ m} = 1000 \text{ mm}$.

• Then a cubic metre is the volume of a cube with each side of length 1000 mm

• *Hence the volume of a cubic metre*, $1 \text{ m}^3 = 1000 \times 1000 \times 1000 \text{ mm}^3 = 1.0 \times 10^9 \text{ mm}^3$

• Then: $1 \text{ mm}^3 = \dfrac{1}{1.0 \times 10^9} \text{ m}^3 = 1.0 \times 10^{-9} \text{ m}^3$

• Then: $30 \text{ mm}^3 = 30 \times 1.0 \times 10^{-9} \text{ m}^3 = 30 \times 10^{-9} \text{ m}^3 = 3.0 \times 10^{-8} \text{ m}^3$

With mixed units, **convert each unit separately** in a step-by-step conversion.

E2.13 🖳

Express a speed of 9.2 mph in units of m s^{-1}, given that $1.0 \text{ km} = 0.6215$ miles.

Start from what is known: 1.0 km is equal to 0.6215 miles.

• The reciprocal conversion: $1.0 \text{ mile} = \dfrac{1}{0.6215} \text{ km} = 1.609 \text{ km} = 1609 \text{ m}$

• $9.20 \text{ miles} = 9.20 \times 1609 \text{ m} = 14\,803 \text{ m}$

• $1 \text{ hour} = 60 \times 60 = 3600 \text{ s}$

- *A speed of 9.2 miles per hour means 9.2 miles are travelled in 1 hour*

- This is the same as 14 803 m travelled in 3600 s

- Speed = Distance travelled in each second = $\dfrac{14\,803}{3600}$ m s^{-1}= 4.11 m s^{-1}

In a complex conversion, it is useful to convert equivalence equations to **unit values** before calculating a new value. This is illustrated in E2.14:

E2.14 ⌨

A lysozyme solution has 15 enzyme units of activity in 22 mL of solution. Calculate the number of enzyme units in 100 mL.

Start from what is known:

$$22 \text{ mL of solution contain 15 enzyme units}$$

Convert to *unit value* of 1 mL:

$$1 \text{ mL of solution contain } \frac{15}{22} \text{ enzyme units}$$

Taking the new value of 100 mL:

$$100 \text{ mL of solution contain } 100 \times \frac{15}{22} = 68.182 \text{ enzyme units}$$

E2.15 ⌨

A lysozyme solution has 15 enzyme units of activity in 22 mL of solution. If the concentration of protein in the solution is 0.5 grams per 100 mL, calculate the specific lysozyme activity of the solution in enzyme units per mg of protein.

In this case it would be useful to convert to a *common volume* of 100 mL.

From E2.14 we know that 100 mL of solution contain 68.182 enzyme units.

We know also that 100 mL of solution contains 0.5 g protein.

Hence 0.5 g protein is equivalent to an activity of 68.182 units.

1.0 g protein will be equivalent to an activity of $\dfrac{68.182}{0.5} = 136.4$ units.

Activity (per gram of protein) = 136.4 units g^{-1}.

Activity (per mg of protein) = $\dfrac{136.4}{1000} = 0.1364$ units mg^{-1}.

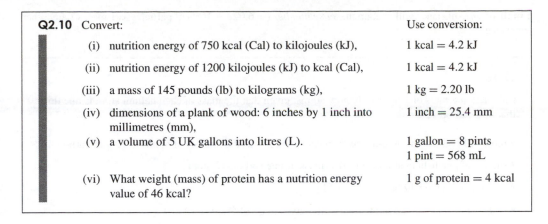

Q2.10 Convert:	Use conversion:
(i) nutrition energy of 750 kcal (Cal) to kilojoules (kJ),	1 kcal = 4.2 kJ
(ii) nutrition energy of 1200 kilojoules (kJ) to kcal (Cal),	1 kcal = 4.2 kJ
(iii) a mass of 145 pounds (lb) to kilograms (kg),	1 kg = 2.20 lb
(iv) dimensions of a plank of wood: 6 inches by 1 inch into millimetres (mm),	1 inch = 25.4 mm
(v) a volume of 5 UK gallons into litres (L).	1 gallon = 8 pints 1 pint = 568 mL
(vi) What weight (mass) of protein has a nutrition energy value of 46 kcal?	1 g of protein = 4 kcal

Q2.11
(i) How many hectares are there in 1.0 km^2? ($1 \text{ hectare} = 1 \times 10^4 \text{ m}^2$)

(ii) If the density of iron is 7.9 g cm^{-3} calculate the density in units of kg m^{-3}.

(iii) A fertilizer is to be spread at the rate of 0.015 g cm^{-2}. What is the spreading rate in kg m^{-2}?

(iv) What is a petrol consumption of 40 miles per gallon (mpg) in litres per 100 kilometres? (Assume $1 \text{ mile} = 1.61 \text{ km}$ and $1 \text{ gallon} = 4.55 \text{ litres}$)

2.2.5 Amount (grams and moles)

There are TWO primary ways of recording the *amount of a substance*. These are *based on*:

1 **Mass** of substance, measured in **grams (g), or kilograms (kg)**.

2 **Number of molecules**, written as *a ratio compared to the number of atoms* in 12 g of carbon-12.
The unit for this ratio is the **mole (mol)**.
For example, 2 *moles* of *any* compound will contain *twice as many* molecules as there are atoms in 12 g of carbon-12.
The mole can also be used to describe the amount of a substance that exists in atomic or sub-molecular form, for example, 3 moles of neon contain three times as many *atoms* as there are atoms in 12 g of carbon-12.

1 mole of carbon-12 has a mass of 12.00 g	[2.1]

Avogadro's number, N_A, is the actual number of atoms in 1 *mole* of carbon-12:

$$N_A = 6.022 \times 10^{23} \qquad [2.2]$$

1 mole of *any* substance will contain the *same number* ($= 6.022 \times 10^{23}$) of particles as 1 mole of any other substance.

E2.16

Calculate the mass of 1 mole of helium atoms, given that the mass of each helium atom is one third that of a carbon-12 atom.

1 mole of helium atoms has the *same number* of atoms as the number of atoms in 12 g of carbon-12.

- Mass of one helium atom $= (1/3) \times$ mass of one carbon-12 atom

- 1 mole of carbon-12 has a mass of 12 g

- The *same number* of helium atoms will have a mass of $(1/3) \times 12 = 4$ g

---

2.2.6 Molar mass

There are a number of 'units' used by scientists in relation to chemical quantities.

> **Molar mass**, M_m, of a substance is the mass, *in grams*, of 1 *mole* of that substance. Units are g mol^{-1}.

Although some units are now considered to be 'obsolete', many are still in frequent use, and it is important to be aware of their meaning:

> **Relative molecular mass**, *RMM*, or **relative molar mass**, M_r (also called **molecular weight**) of a substance is equal to the *ratio* of the (average) mass of 1 molecule of that substance to $1/12$ of the mass of the carbon-12 atom. The term 'relative *atomic* mass' would be used for the mass of an element.
> (Note that 'average' mass is used to allow for the natural mixture of isotopes of different masses.)

In practice, the *RMM* or M_r, for a molecule is calculated by adding the relative atomic masses of its various atoms.

1 mole of a substance has a mass in grams (molar mass) numerically equal to the value of its *RMM* or M_r.

E2.17

Calculate the relative molecular mass, molar mass, and mass of 1 mole of calcium carbonate, $CaCO_3$, given relative atomic masses Ca $= 40.08$, C $= 12.01$, O $= 16.00$.

- $RMM = 40.08 + 12.01 + 3 \times 16.00 = 100.09$ (a pure number)

- Molar mass $= 100.09$ g mol^{-1} (equals the *RMM* in grams per mole).

- Mass of 1 mole $= 100.09$ g (numerically equals the molar mass in grams).

Q2.12 Using relative atomic masses C = 12.01, H = 1.01, O = 16.00, calculate the following values for aspirin, $C_9H_8O_4$, giving the relevant units:

  (i)   relative molecular mass (RMM or M_r),

  (ii)   molar mass, M_m,

  (iii)   mass of 1 mole.

E2.18

Calculate the following for sodium hydroxide, NaOH (RMM = 40):

  (i)   mass (in g) of 1 mol of NaOH

  (ii)   mass (in g) of 0.4 mol of NaOH

  (iii)   number of moles of NaOH that has a mass of 1.0 g

  (iv)   number of moles of NaOH that has a mass of 8.0 g

Answers:

  (i)   1 mol of NaOH has a mass of 40 g (from the definition of a 'mole')

  (ii)   0.4 mol of NaOH has a mass of $0.4 \times 40 = 16$ g

  (iii)   40 g of NaOH is equivalent to 1 mol

    1 g of NaOH is equivalent to $\dfrac{1.0}{40} = 0.025$ mol

  (iv)   8.0 g of NaOH is equivalent to $8.0 \times 0.025 = 0.2$ mol

---

Q2.13 Calculate the following for sodium carbonate, Na_2CO_3 (RMM = 106):

     (i)   mass (in g) of 1 mol of Na_2CO_3

     (ii)   mass (in g) of 0.15 mol of Na_2CO_3

     (iii)   number of moles of Na_2CO_3 that has a mass of 3.5 g

---

Q2.14 A sample of benzoic acid with a mass of 2.2 g was found, by titration, to be an amount equal to 0.018 moles. Calculate

  (i)   molar mass

  (ii)   relative molecular mass

2.2.7 Concentration

The concentration of a solution is the amount of solute per unit volume of solution. The basic unit of volume, m^3, is a large unit, and it is common to use the smaller

- litre, L (which equals a cubic decimetre, 'dm^3') or
- cm^3 (sometimes written as cubic centimetres, 'cc'):

$$1\,L = 1\,dm^3 = 1000\,cm^3 = 1 \times 10^{-3}\,m^3 \qquad [2.3]$$

There are TWO primary ways of recording the concentration of a solution:

1 **Mass of solute per unit volume of solution**, g L^{-1}

$$\text{Concentration (g L}^{-1}) = \frac{\text{Mass (g)}}{\text{Volume (L)}} \qquad [2.4]$$

2 **Molar concentration** (also called **molarity**), **M**, which is the number of moles per litre. 1.0 M is equivalent to 1.0 mol L^{-1} (or mol dm^{-3}).

$$\text{Molar concentration (mol L}^{-1}) = \frac{\text{Number of moles (mol)}}{\text{Volume (L)}} \qquad [2.5]$$

Q2.15 A solution has been prepared such that 100 mL of the solution contains 0.02 mol of sodium hydroxide (NaOH).
Calculate the concentration in moles per litre

E2.19

0.500 L of solution contains 4.00 g of sodium chloride, NaCl (*RMM* = 58.4).
Calculate:

(i) concentration of the solution in g L^{-1}

(ii) molar concentration in mol L^{-1}

Answer:

(i) Using [2.4], concentration = 4.00 / 0.500 = 8.00 g L^{-1}

(ii) 58.4 g of NaCl are equivalent to 1 mol

1 g of NaCl is equivalent to $\dfrac{1.0}{58.4} = 0.01712$ mol

4.00 g of NaCl are equivalent to $= 4.00 \times 0.01712 = 0.0685$ mol

Using [2.5], molar concentration of 0.5 L of solution $= \dfrac{0.0685}{0.500}$ mol L^{-1}

$= 0.137$ mol L$^{-1} = 0.137$ M $= 137$ mM

Q2.16 Calculate the concentration (in mol L^{-1}) of a solution that contains 5.6 g of sodium hydroxide, NaOH (*RMM* = 40.0), in 75 mL of solution.

E2.20

Calculate the mass of sodium hydroxide, NaOH (*RMM*= 40), that must be dissolved in 100 mL of solution to obtain a molar concentration of 0.50 mol L^{-1}.

Convert the volume, 100 mL to litres: 100 mL $= 0.10$ L.

Substitute in [2.5], and let x be the number of moles:

$$0.50 = \frac{x}{0.10}$$

Rearranging the equation gives: $x = 0.50 \times 0.10 = 0.05$ mol.

We know that 1 mol of NaOH is equivalent to 40 g.

Hence the required amount of 0.05 mol is equivalent to $0.05 \times 40 = 2.0$ g.

Q2.17 What mass of sodium chloride, NaCl (*RMM* = 58.4), when dissolved in water to give 100 mL of solution, will give a concentration of 0.10 mol L^{-1} ?

Q2.18 Calculate the mass of hydrated copper sulphate, CuSO$_4$.5H$_2$O (*RMM* = 249.7), that must be dissolved into a final volume of 50 mL of solution to obtain a concentration of 0.50 M.

Other common terminologies relating to *concentration* include:

- Millimoles, mmol: 1 mmol is equivalent to 1.0×10^{-3} mol

- Millimolar, mM: 1 mM is equivalent to 1.0×10^{-3} M $= 1.0 \times 10^{-3}$ mol L^{-1}

- Parts per million, ppm: 1 ppm is equivalent to 1 mg L^{-1}

- Parts per billion, ppb: 1 ppb is equivalent to 1 μg L^{-1}

- Percentage concentrations:
  - weight per volume, w/v
  - volume per volume, v/v
  - weight per weight, w/w
  where the 'weights' are given in grams and 'volumes' in millilitres.

Note that 1.0 mL of water has a mass ('weight') of 1.0 g, and hence we can write that 1.0 mL of a *dilute* solution also has a mass ('weight') of 1.0 g, and 1.0 L of a *dilute* solution has a mass ('weight') of 1.0 kg,

---

E2.21

Examples of typical calculations of equivalence:

10 mL = 0.01 L, 2 mL = 0.002 L, etc.

0.025 mmol in 10 mL is equivalent to $\dfrac{0.025}{0.01}$ mmol L^{-1} = 2.5 mM.

0.023 mg in 10 mL is equivalent to $\dfrac{0.023}{0.01}$ = 2.3 mg L^{-1} = 2.3 ppm.

6.70×10^{-7} g in 2 mL is equivalent to $\dfrac{6.70 \times 10^{-7}}{0.002}$ = 0.000 335 g L^{-1} = 335 μg L^{-1}.

335 μg L^{-1} is equivalent to 335 ppb.

1.2 g of solute in 50 mL of solution has a concentration of $\dfrac{1.2}{50} \times 100$ = 2.4% (w/v).

10 mL of solute diluted to 200 mL has a concentration of $\dfrac{10}{200} \times 100$ = 5% (v/v).

0.88 g of solute in a total of 40 g has a concentration of $\dfrac{0.88}{40} \times 100$ = 2.2% (w/w).

2.2.8 Dilutions

In most dilutions, the *amount of the solute stays the same*.
  In this case it is useful to use the **dilution equation**:

$$V_i \times C_i = V_f \times C_f \qquad [2.6]$$

where V_i and C_i are the initial volumes and concentrations and V_f and C_f are the final values.
  The concentrations can be measured as molarities *or* mass/volume, but the *same units* must be used on each side of the equation.

E2.22 🖳

20 mL of a solution of concentration 0.3 M is transferred to a 100 mL graduated flask, and solvent is added up to the 100 mL mark. Calculate the concentration, C_f, of the final solution.

The amount of the solute is the same in the initial 20 mL as in the final 100 mL, so we can use the dilution equation:

$$20 \times 0.3 = 100 \times C_f$$

$$C_f = \frac{20 \times 0.3}{100} = 0.06\,\text{M}$$

E2.23 🖳

It is necessary to produce 200 mL of 30 mM saline solution (sodium chloride, NaCl, in solution). Calculate the volume of a 35 g L^{-1} stock solution of saline that would be required to be made up to a final volume of 200 mL. (*RMM* of NaCl is 58.4.)

In this question, the concentrations of the two solutions are initially in different forms – moles and grams. We choose to convert 35 g L^{-1} to a molar concentration:

From 2.2.7, 58.4 g (1 mol) of NaCl in 1.00 L gives a concentration of 1.00 M

- 1.00 g of NaCl in 1.00 L gives a concentration $= \dfrac{1.00}{58.4}\,\text{M}$

- 35.0 g of NaCl in 1.00 L gives a concentration $= 35.0 \times \dfrac{1.00}{58.4} = 0.599\,\text{M} = 599\,\text{mM}$

We can now use [2.6] to find the initial volume, V_i, of 0.599 M ($= 599$ mM) saline that must be diluted to give 200 mL of 30 mM saline.

$$V_i \times 599 = 200 \times 30$$

$$V_i = \frac{200 \times 30}{599} = 10.02\,\text{mL}$$

Q2.19 5.0 mL of a solution of concentration 2.0 mol L^{-1} is put into a 100 mL graduated flask and pure water is added, bringing the total volume in the flask to exactly 100 mL.

Calculate the concentration of the new solution.

Q2.20 A volume, V, of a solution of concentration 0.8 mol L^{-1} is put into a 100 mL graduated flask, and pure solvent is added to bring the volume up to 100 mL.

If the concentration of the final solution is 40 mM, what was the initial volume, V?

> **Q2.21** Calculate the volume of a 0.15 M solution of the amino acid alanine that would be needed to make up to a final volume of 100 mL in order to produce 100 mL of 30 mM alanine?

2.3 Angular measurements

2.3.1 Introduction

In many aspects of undergraduate science the students rarely encounter the need to measure angles or solve problems involving rotations. However, angular measurements do occur routinely in a variety of practical situations. The mathematics is not difficult, and, in most cases, it is only necessary to refresh the ideas of simple trigonometry or to revisit Pythagoras!

2.3.2 Degrees and radians

There are 360° (**degrees**) in a full circle.

E2.24

Why are there 360 degrees in a circle?

The choice of 360 was made when 'fractions' were used in calculations far more frequently then they are now. The number 360 was particularly good because it can be divided by many different factors: 2, 3, 4, 5, 6, 8, 9, 10, 12, 15, 18, 20, 24, 30, 36, 40, 45, 60, 72, 90, 120, 180.

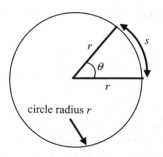

Figure 2.1. Angle in radians.

The **radian** is an alternative measure that is often used in calculations involving *rotations* (Figure 2.1).

The angle, θ, in *radians* is defined as the **arc length**, s, divided by the **radius**, r, of the arc. The angle in radians is given by the simple *ratio*:

$$\theta = \frac{s}{r} \qquad [2.7]$$

$$s = r \times \theta \qquad [2.8]$$

In a *complete circle*, the *arc length*, s, will equal the *circumference* of the circle $= 2\pi r$.

Hence, the angle (360°) of a complete circle $= \dfrac{2\pi r}{r}$ radians $= 2\pi$ radians

$$360° = 2\pi \text{ radians}$$
$$180° = \pi \text{ radians}$$
$$90° = \pi/2 \text{ radians}$$

$$1 \text{ radian} = \frac{180}{\pi} \text{ degrees} = 57.3 \ldots \text{ degrees} \qquad [2.9]$$

2.3.3 Conversion between degrees and radians

$$\text{'}x\text{' in radians becomes } x \times 180/\pi \text{ degrees} \qquad [2.10]$$

$$\text{'}\theta\text{' in degrees becomes } \theta \times \pi/180 \text{ radians} \qquad [2.11]$$

In Excel, to convert an angle

- *from* radians *to* degrees, use the function DEGREES, and

- *from* degrees *to* radians use the function RADIANS.

💻 Computer Techniques > Excel Tutorial: Skills

Q2.22 Convert the following angles from degrees to radians or vice versa:

(i) 360 degrees into radians	(iv) 1.0 radian into degrees
(ii) 90 degrees into radians	(v) 2.1 radians into degrees
(iii) 170 degrees into radians	(vi) 3.5π radians into degrees

---

E2.25

The towns of Nairobi and Singapore both lie approximately on the equator of the Earth at longitudes 36.9°E and 103.8°E respectively. The radius of the Earth at the equator is 6.40×10^3 km. Calculate the distance between Nairobi and Singapore *along the surface* of the Earth.

The equator of the Earth is the circumference of a circle with radius, $r = 6.40 \times 10^3$ km. Nairobi and Singapore are points on the circumference of this circle separated by an angle,

$$\theta = 103.8° - 36.9° = 66.9°$$

Converting this angle to radians: $\theta = 66.9 \times \pi /180$ radians $= 1.168$ radians

The distance on the ground between Nairobi and Singapore will be given by the arc length, s, between them. Using [2.8]:

$$s = r \times \theta = 6.40 \times 10^3 \times 1.168 \, \text{km} = 7.47 \times 10^3 \, \text{km}$$

Q2.23 In Figure 2.2, calculate the distance that the mass rises when the drum rotates by 40°. The radius of the drum is 10 cm.

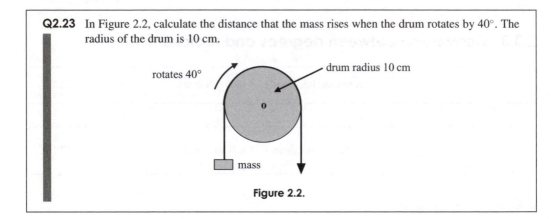

Figure 2.2.

2.3.4 Trigonometric functions

Figure 2.3. Sides of a right-angled triangle.

In a *right-angled* triangle, the longest side is the **hypotenuse**, *H*.

In the triangle shown in Figure 2.3, the angle, θ, is on the left side as shown.

The side opposite the angle is called the **opposite** side, *O*.

The side next to the angle (but not the hypotenuse) is called the **adjacent** side, *A*.

The three main trigonometric functions, sine, cosine and tangent, can be calculated by taking the ratios of sides as in equations [2.12]. Many students use a simple mnemonic to remember the correct ratios: *SOHCAHTOA*.

$$\sin \theta = \frac{O}{H}$$

$$\cos \theta = \frac{A}{H}$$ [2.12]

$$\tan \theta = \frac{O}{A}$$

E2.26 🖳

A car travels 100 m downhill along a road that is inclined at 15° to the horizontal. Calculate the vertical distance through which the car travels.

The 100 m travelled by the car is the hypotenuse, H, of a right-angled triangle. The vertical distance to be calculated is the opposite side, O, using the angle of $\theta = 15°$.

$$\sin(15°) = \frac{O}{H} = \frac{O}{100}$$

giving:

$$O = 100 \times \sin(15°) = 100 \times 0.259 = 25.9\,\text{m}$$

Q2.24 A tree casts a shadow that is 15 m long, when the sun is at an angle of 30° above the horizon.

Calculate the height of the tree.

2.3.5 Pythagoras's equation

$$H^2 = O^2 + A^2$$ [2.13]

(The square on the hypotenuse of a right-angled triangle is equal to the sum of the squares on the other two sides.)

Q2.25 One side of a rectangular field is 100 m long, and the diagonal distance from one corner to the opposite corner is 180 m. Calculate the length of the other side of the field.

2.3.6 Small angles

When the angle, θ, is small (i.e. less than about 10°, or less than about 0.2 radians) it is possible to make some approximations.

Figure 2.4. Small angles.

In Figure 2.4:

- the length of the arc, s, in (a) will be approximately equal to the length of the opposite side, O, in the right-angled triangle in (b): $O \approx s$

- the lengths of the adjacent side, A, and the hypotenuse, H, in (b) will be approximately equal to the radius, r, in (a): $H \approx r$, and $A \approx r$.

If the angle, θ, is measured in *radians* and the *angle is small*, then:

$$\sin(\theta) = O/H \approx s/r = \theta \qquad \text{hence } \sin(\theta) \approx \theta$$
$$\tan(\theta) = O/A \approx s/r = \theta \qquad \text{hence } \tan(\theta) \approx \theta \qquad [2.14]$$
$$\cos(\theta) = A/H \approx r/r = 1 \qquad \text{hence } \cos(\theta) \approx 1.0$$

Q2.26 In the following table, use a calculator to calculate values for $\sin(\theta)$, $\cos(\theta)$ and $\tan(\theta)$ for each of the angles listed.

Use [2.11] to calculate the angle θ in radians.

Check whether the values of $\sin(\theta)$, $\cos(\theta)$ and $\tan(\theta)$ and θ in radians agree with [2.14].

The calculations for $\theta = 20°$ have already been performed:

$\theta°$ (degrees)	$\sin(\theta)$	$\cos(\theta)$	$\tan(\theta)$	θ (radians)
20	0.3420	0.9397	0.3640	0.3491
10				
5				
1				
0				

E2.27

A right-angled triangle has an angle, $\theta = 5°$, and an hypotenuse of length 2.0.

(i)   Calculate the length of the opposite side using a trigonometric function.

We know that $\theta = 5°$ and $H = 2.0$. Using $O = H \times \sin(\theta)$:

$$O = 2.0 \times \sin(\theta) = 2.0 \times \sin(5°) = 2.0 \times 0.08716 = 0.174$$

(ii)   Assume that the triangle is approximately the same as a thin segment of a circle with a radius equal to the hypotenuse, and estimate the length of the arc using a 'radian' calculation.

Converting $\theta = 5°$ into radians: $5° = 5 \times \pi/180$ radians $= 0.087\,27$ radians.
The arc length of a circle segment with radius $r = 2.0$ is given by $s = r \times \theta$.
$s = 2.0 \times 0.087\,27 = 0.175$.

The calculations for a triangle with a very small angle can often be made more easily using radians than using a trigonometric function.

Q2.27   Estimate the diameter of the Moon using the following information:
The Moon is known to be 384 000 km away from the Earth, and the apparent disc of the Moon subtends an angle of about 0.57° for an observer on the Earth – as illustrated in Figure 2.5.

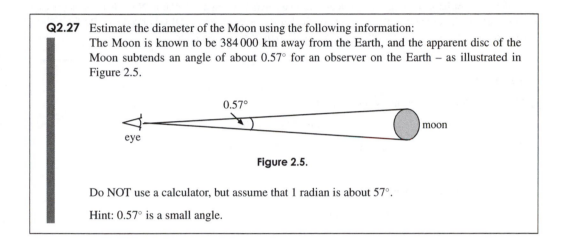

Figure 2.5.

Do NOT use a calculator, but assume that 1 radian is about 57°.

Hint: 0.57° is a small angle.

2.3.7 Inverse trigonometric functions

The angle can be calculated from the ratios of the sides of a triangle by using the 'inverse' functions:

$$\theta = \sin^{-1}(O/H)$$
$$\theta = \cos^{-1}(A/H) \qquad [2.15]$$
$$\theta = \tan^{-1}(O/A)$$

Note that the above are not the reciprocals of the various functions, for example, $\sin^{-1}(O/H)$ does NOT equal $1/\{\sin(O/H)\}$.
    The 'inverse' function can also be written with the 'arc' prefix.

$$\theta = \arcsin(O/H)$$
$$\theta = \arccos(A/H) \qquad [2.16]$$
$$\theta = \arctan(O/A)$$

Q2.28 The three sides of a right-angled triangle have lengths, 3, 4 and 5, respectively.
Calculate the value of the *smallest* angle in the triangle using:

(i) sine function

(ii) cosine function

(iii) tangent function

2.3.8 Computing angular measurements

The calculation of basic angular measurements can be carried out on a calculator. Note that it is necessary
to set up the 'mode' of the calculator to define whether it is using degrees (DEG) or radians (RAD).

Example E2.28 gives some examples of angle calculations on a calculator.

E2.28 🖥

Converting $36°$ to radians using $36 \times \pi/180$:

$$36° = 0.6283\ldots \text{ radians}$$

Converting 1.3 radians to degrees using $1.3 \times 180/\pi$:

$$1.3 \text{ radians} = 74.48\ldots°$$

Setting the calculator to DEG mode:

$$\sin(1.4) = 0.024 \ldots \qquad \cos^{-1}(0.21) = 77.88 \ldots°$$

Setting the calculator to RAD mode:

$$\sin(1.4) = 0.985 \ldots \qquad \cos^{-1}(0.21) = 1.359 \ldots \text{ radians}$$

When using Excel for angle calculations, it is important to note that **Excel uses *radians*** as its unit of angle,
NOT *degrees*.

🖥 Computer Techniques > Excel Tutorials: Skills

To convert an angle, θ, in radians *to* degrees, use the function **DEGREES**, and to convert from degrees *to* radians use the function **RADIANS**. Alternatively it is possible to use formulae derived from [2.10] and [2.11].

Excel uses the functions **SIN**, **COS** and **TAN** to calculate the basic trigonometric ratios. The inverse
trigonometric functions are **ASIN**, **ACOS** and **ATAN**.

The value of π in Excel is obtained by entering the expression '= PI()'.

E2.29 🖥

The following functions and formulae in Excel:

• '=DEGREES(B4)' converts the angle held in cell B4 from a value given in radians to a value
given in degrees.

- '=B4*180/PI()' also converts the angle held in cell B4 from a value given in radians to a value given in degrees.

- '=SIN(RADIANS(B4))' gives the sine of the angle (in degrees) held in cell B4.

- '=DEGREES(ACOS(B10/B11))' gives the angle (in degrees) for a triangle where the length of the adjacent side is held in B10 and the length of the hypotenuse is held in B11.

Q2.29 Refer to Q2.24, where a tree casts a shadow that is 15 m long, when the sun is at an angle of 30° above the horizon.

Write out the formula that would be used in Excel to calculate the height of the tree, assuming that the length of the shadow was entered into cell D1 and the angle of the sun in degrees was entered into cell D3.

3
Equations in Science

Overview

An equation in mathematics is a model for a *relationship* in science, and within the model it may be necessary to use letters, or other symbols, to represent the different variables of a scientific system. If we then manipulate the equation using the rules of algebra, we can generate new equations and relationships that might tell us something new about the science itself!

It is common practice to use letters from the Greek alphabet to supplement the 26 letters available from the Roman alphabet. Letters and pronunciations are given on the web site:

💻 Study Resources > Greek Symbols

For example, we can relate density, ρ (Greek letter, *rho*), mass, m, and volume, V, using an equation:

$$\rho = \frac{m}{V}$$

We can then use the same equation and substitute different sets of values to calculate the densities of different samples of materials.

This chapter reviews the basic rules of algebra, and then applies them specifically to the common task of rearranging an equation. The real power of algebra lies, not only in being able to model a relationship in science using an equation, but in being able to use the power of the mathematics to derive a solution to a problem. Using mathematics to logically find an answer that satisfies the problem is called an **analytical solution**.

The Units on 'Solving equations' and 'Simultaneous equations' develop ways of solving some of the most common analytical problems encountered in science.

3.1 Basic equation handling

3.1.1 Introduction

Handling and rearranging equations in science requires the development of some understanding of algebra. It is the purpose of this unit to introduce some basic algebraic techniques in the context of their use in handling simple equations. The following Units, 'Introduction to rearranging equations' and 'Rearranging complicated equations', develop the general approach to rearranging equations.

Working with scientific equations can involve a variety of mathematical approaches, the most common of which are given in these notes. The real skill is to know which approach to use at any given stage of a problem, and *this skill can only be gained by practice*. Some general principles are given, but the student should work through the different examples and questions and build up his/her own experience.

3.1.2 Rules of algebra

The basic rules for algebra are the same as for arithmetic. Revision of basic arithmetic is available on the web site.

⌨ Study Resources > Revision Mathematics

E3.1

Comparing arithmetic and algebraic expressions:

<table>
<tr><td align="center"><u>Arithmetic</u></td><td align="center"><u>Algebra</u></td></tr>
<tr><td align="center">$3 \times 2 = 2 \times 3 = 6$</td><td align="center">$cd = d \times c = c \times d = dc$</td></tr>
<tr><td align="center">$3^2 = 3 \times 3 = 9$</td><td align="center">$a^2 = a \times a$</td></tr>
<tr><td align="center">$4 \times 3^2 = 4 \times 3 \times 3 = 36$</td><td align="center">$cd^2 = c \times d \times d$</td></tr>
<tr><td align="center">$2 \times (4 + 3^2) = 2 \times 4 + 2 \times 3^2 = 8 + 18 = 26$</td><td align="center">$q(r + f^2) = (q \times r) + (q \times f \times f) = qr + qf^2$</td></tr>
<tr><td align="center">$\dfrac{4+2}{1} = 4 + 2 = 6$</td><td align="center">$\dfrac{a+b}{1} = a + b$</td></tr>
<tr><td align="center">$\dfrac{3}{4} \times \dfrac{5}{7} = \dfrac{3 \times 5}{4 \times 7} = \dfrac{3 \times 5}{7 \times 4} = \dfrac{3}{7} \times \dfrac{5}{4}$</td><td align="center">$\dfrac{a}{b} \times \dfrac{c}{d} = \dfrac{a \times c}{b \times d} = \dfrac{a \times c}{d \times b} = \dfrac{a}{d} \times \dfrac{c}{b}$</td></tr>
<tr><td align="center">$3/4 = \dfrac{3}{4} = \dfrac{1}{4/3}$</td><td align="center">$a/b = \dfrac{a}{b} = \dfrac{1}{b/a}$</td></tr>
</table>

Note that the '×' sign is omitted between letters: $a \times b = ab$

Where there is a mixed product of a number and one or more symbols, the number *always* goes first when the multiply sign is dropped:

$$p \times 3 = 3 \times p = 3p \quad \text{(do not write '}p3\text{' for } p \times 3)$$

Q3.1 Replace x with the number 2 and y with the number 3 in the following expressions and calculate the values:

(i) $3x^2$

(ii) $(3x)^2$

(iii) $4(y - x)$

(iv) $\dfrac{1}{x/y}$

(v) $4(y - x^2)$

(vi) $2xy^2$

(vii) $xy - yx$

(viii) $\dfrac{x+y}{1}$

A **term** in algebra is a number, symbol, or combination of numbers and symbols, which can be treated in the same way as a number in arithmetic.

E3.2

In the following equation:

$$z = \frac{3y\,(a+b)}{(2+b)} + 6p$$

$3y$, $6p$ the brackets $(a+b)$ and $(2+b)$ are all algebraic *terms*.

The **reciprocal** of x is the value of '1 over x' $= \dfrac{1}{x} = 1/x = x^{-1}$ [3.1]

E3.3

Reciprocal of 50 $= \dfrac{1}{50} = 0.02$

Reciprocal of 0.04 $= \dfrac{1}{0.04} = 25$

Reciprocal of a $= \dfrac{1}{a}$

Reciprocal of $(a+b)$ $= \dfrac{1}{(a+b)}$ but NOT $= \dfrac{1}{a} + \dfrac{1}{b}$

The **fraction** 'a over b' equals the **ratio** of 'a to b', and can be written in a number of equivalent ways:

$$\frac{a}{b} = {}^{a}\!/_{b} = a/b = a \div b = a \times \left(\frac{1}{b}\right) \qquad [3.2]$$

The upper term is the **numerator**, and the term underneath is the **denominator**.

⌨   Study Resources > Revision Mathematics

If the top (*numerator*) and bottom (*denominator*) are multiplied (or divided) by the *same* amount, the value of the fraction (or ratio) *stays the same*.

E3.4

Multiplying both top and bottom by '2' gives the same ratio (fraction):

$$\frac{3}{4} = \frac{2 \times 3}{2 \times 4} = \frac{6}{8} \qquad \text{and similarly} \qquad \frac{a}{b} = \frac{2a}{2b}$$

Dividing both top and bottom by '2' gives the same ratio (fraction):

$$\frac{6}{8} = \frac{6/2}{8/2} = \frac{3}{4} \qquad \text{and similarly} \qquad \frac{2a}{2b} = \frac{2a/2}{2b/2} = \frac{a}{b}$$

Cancelling occurs by dividing the top and bottom by the same term:

$$\frac{6}{8} = \frac{2 \times 3}{4 \times 2} = \frac{\not{2} \times 3}{4 \times \not{2}} = \frac{3}{4} \qquad \text{and similarly} \qquad \frac{a \times b}{3 \times a} = \frac{\not{a} \times b}{3 \times \not{a}} = \frac{b}{3}$$

See also cancelling in section 3.1.7.

Q3.2 Cancel where possible in the following fractions:

(i) $\dfrac{3 \times 3}{4 \times 3}$ 　　　　　　　　　　　(v) $\dfrac{a \times a}{b \times a}$

(ii) $\dfrac{9}{12}$ 　　　　　　　　　　　　　　(vi) $\dfrac{a^2}{b \times a}$

(iii) $\dfrac{8}{2}$ 　　　　　　　　　　　　　　(vii) $\dfrac{c \times d}{d}$

(iv) $\dfrac{5}{15}$ 　　　　　　　　　　　　　(viii) $\dfrac{g}{h \times g}$

3.1.3 Reading equations

When 'reading' a scientific equation, it is important to remember that it is a mathematical *representation* of relationships that may exist in a real world system.

Example E3.5 shows how an equation can represent a *general* series of relationships, and how it is then possible to calculate a *specific* outcome, given a *particular* condition (i.e. the value of *n* in this example).

E3.5 🖳

The *statement*,

'I think of a number, *n*, double it, add 6, divide by 2 to obtain a final number, *N*', can be *represented* by the equation:

$$N = \frac{2n + 6}{2}$$

If the original number, *n*, were 4, the answer, *N*, will be 7, and if the original number, *n*, were 6, the answer, *N*, will be 9.

Q3.3 Write equations to represent the following processes, and then use them to calculate a result for the data supplied:

    (i)   I think of a number, n, and add 3.
           Use the equation to calculate the result for $n = 6$.

    (ii)   I think of a number, n, and double it.
           What is the result if $n = 4$?

    (iii) I think of a number, n, double it, and then add 3.
           What is the result if $n = 2$?

    (iv) I think of a number, n, add 3, and then double the sum.
           What is the result if $n = 2$?

    (v)   I think of a number, n, add 3, double the sum, take away 6, and then divide by 2.
           What are the results if $n = 2, 6, 11$?

    (vi) I think of a number, n, double it, add 6, divide by 2, and then take away the number I first thought of.
           What are the results if $n = 2, 6, 11$?

    (vii) I think of a number, n, square it and add 2.
           What is the result if $n = 3$?

    (viii) I think of a number, n, subtract 3 and then take the square root.
           What is the result if $n = 28$?

Q3.4 Write equations to represent the following relationships.
For example, the statement,

> 'The area, A, of a circle can be calculated by multiplying the constant, π (Greek letter, *pi*), with the square of the radius, r',

can be represented by the equation: $A = \pi r^2$.

    (i)   The area, A, of a triangle is half base, b, times height, h.

    (ii)   The distance, d, covered by a runner is equal to the product of the average speed, v, and the time, t.

    (iii) The body mass index, BMI, of a person is equal to his/her weight (mass), m, divided by the square of his/her height, h.

    (iv) The volume, V, of a spherical cell is four-thirds the constant, π, multiplied by the cube of the radius, r.

    (v)   The velocity, v, of an object that has fallen through a distance, h, is given by the square root of twice the product of h and g, where g is the acceleration due to gravity.

3.1.4 Substitution of values

We can replace the letters in an equation by the *known* values of the equivalent variables, and then 'solve' the equation to find out the value of a remaining *unknown* variable.

E3.6 🖳

The quadratic equation:

$$3x^2 - 4x - 2 = 0$$

can be derived from the standard form:

$$ax^2 + bx + c = 0$$

by *substituting* the values, 3, −4 and −2 in place of a, b and c respectively.

The 'solutions' to the original quadratic equation are then given by *substituting* these values into the formulae (see 3.4.4):

$$x = \frac{-b + \sqrt{b^2 - 4ac}}{2a} \quad \text{or} \quad x = \frac{-b - \sqrt{b^2 - 4ac}}{2a}$$

which give (using '±' to condense the two formulae into one equation):

$$x = \frac{-(-4) \pm \sqrt{(-4)^2 - 4 \times 3 \times (-2)}}{2 \times 3} = \frac{+4 \pm \sqrt{16 + 24}}{6} = \frac{4 \pm \sqrt{40}}{6} = \frac{4 + 6.32}{6} \text{ or } \frac{4 - 6.32}{6}$$

with the final solutions: $x = 1.72$ or -0.387.

Q3.5   Substitute the values, $x = 2$, $y = 3$, and $z = 4$ into the following expressions, and evaluate the results:

     (i) $x + 2y - 0.5z$ (iv) $\dfrac{z}{x} + \dfrac{y^2}{z}$

     (ii) $z(x + y)$ (v) $x^3 + 3y - z^2$

     (iii) $\dfrac{x^2}{yz}$ (vi) $\dfrac{x + z}{2y - z}$

3.1.5 Using brackets (see also 3.3.4)

Brackets (also called *parentheses*) are used to *group together* the terms that are multiplied by the same *factor* (in the example below the factor is 'y'), e.g.

$$2y + 3y = (2 + 3)y = 5y$$

i.e. 2 lots of 'y' plus 3 lots of 'y' give a total of 5 lots of 'y'.

Similarly, *counting* 'a's in the expression $ba + ca$ gives us the equation:

$$ba + ca = (b + c)a = a(b + c)$$

i.e. b lots of 'a' plus c lots of 'a' give a total of $(b + c)$ lots of 'a'.

When *multiplying out brackets* the factor outside the bracket multiplies *every term* inside the bracket:

$$a(b + c) = ab + ac$$

If the *multiplying factor* is '−', then *every term* inside the bracket is multiplied by '−1':

$$-(3a - 2b) = -3a - (-2b) = -3a + 2b$$

E3.7

Multiply out the brackets in the following expressions:

$$3(2x + 3) = 6x + 9$$
$$a(a + b) = a^2 + ab$$
$$a(2x + 3) = 2ax + 3a$$
$$-3(a - b) = -3a + (-3) \times (-b) = -3a + 3b = 3b - 3a$$
$$-(4p - 3q) = (-1) \times (4p - 3q) = -4p - (-1) \times 3q = -4p + 3q = 3q - 4p$$
$$-(x - 2b) = -x + (-1) \times (-2b) = -x + 2b = 2b - x$$

Q3.6 Multiply out the following brackets:

(i) $3(2 + x)$ (iv) $a(x + 2)$

(ii) $3(2 + 4x)$ (v) $-3a(2 - x)$

(iii) $-2(4x - 3)$ (vi) $a(x + a)$

3.1.6 Factors

A **factor** of an expression is a term that divides *exactly* into the expression.

E3.8

Check understanding of each of the following:

'3' and '10' are both factors of '30':	$30 = 3 \times 10$
'2', '5', '6' and '15' are other factors of '30':	$30 = 2 \times 15$ and $30 = 5 \times 6$
'2' and 'y' are both factors of '2y':	$2y = 2 \times y$
'2', '5' and 'y' are factors of '3y + 7y':	$3y + 7y = 10y = 2 \times 5 \times y$
$(2 + p)$ is a factor of '$(2 + p)y$':	$(2 + p)y = (2 + p) \times y$
'y' is a factor of '$(2 + p)y$':	$(2 + p)y = (2 + p) \times y$
'p' is NOT a factor of '$(2 + p)y$':	The expression cannot be rearranged to give $p \times$ (other factors)

3.1.7 Cancelling (simplifying)

It is possible to cancel a term from both the numerator (top) and the denominator (bottom) of a fraction – as long as that term is a *factor of both the numerator and the denominator*:

E3.9

Check understanding of each of the following:

(i) $\dfrac{30}{25} = \dfrac{\cancel{30}^{6}}{\cancel{25}_{5}} = \dfrac{6}{5}$

'5' is a factor of both '30' and '25' and can be cancelled from top and bottom.

(ii) $\dfrac{8}{4} = \dfrac{\cancel{8}^{2}}{\cancel{4}_{1}} = \dfrac{2}{1} = 2$

'4' is a factor of both '8' and '4'.
'2' divided by '1' is just written as '2'.

(iii) $\dfrac{6x}{2y} = \dfrac{\cancel{6}^{3}x}{\cancel{2}_{1}y} = \dfrac{3x}{y}$

'2' is a factor of both '6x' and '2y' and can be cancelled from top and bottom.

(iv) $\dfrac{3x}{2x} = \dfrac{3\cancel{x}}{2\cancel{x}} = \dfrac{3}{2}$

'x' is a factor of both '3x' and '2x' and can be cancelled from top and bottom.

(v) $\dfrac{x(4+p)}{2x} = \dfrac{\cancel{x}(4+p)}{2\cancel{x}} = \dfrac{(4+p)}{2}$

Note: the '4' is NOT a factor of the top and cannot be cancelled with the '2' below.

(vi) $\dfrac{(4x+p)}{2x}$ cannot be simplified

The 'x' cannot be cancelled because it is not a factor of the numerator (top).

(vii) $\dfrac{3(x+p)}{x(x+p)} = \dfrac{3\cancel{(x+p)}}{x\cancel{(x+p)}} = \dfrac{3}{x}$

The bracket $(x+p)$ can be cancelled top and bottom – it is a factor of both.

Q3.7 Simplify the following equations by cancelling *where possible*.

(i) $x = \dfrac{4a}{2b}$

(ii) $x = \dfrac{4ap}{2pb}$

(iii) $x = \dfrac{y(4a+1)}{2by}$

(iv) $x = \dfrac{2y(2a+3)}{4by}$

Q3.8 Decide whether the answer is Yes or No for each of the following questions, whatever the values of x, y, a and b. None of x, y, a and b are zero.

(i) Does $\dfrac{x \times a}{a \times y}$ equal $\dfrac{x}{y}$?

(ii) Does $\dfrac{2x}{2y}$ equal $\dfrac{x}{y}$?

(iii) Does $\dfrac{x+2}{y+2}$ equal $\dfrac{x}{y}$?

(iv) Does $\dfrac{x/a}{y}$ equal $\dfrac{x}{y \times a}$?

(v) Does $\dfrac{x}{y}$ equal $\dfrac{1}{y/x}$?

(vi) Does $\dfrac{x/a}{y/b}$ equal $\dfrac{bx}{ay}$?

3.2 Introduction to rearranging equations

3.2.1 Introduction

When rearranging equations, the objective is usually to make one variable the **subject** of the equation. For example, if we wish to make x the subject of the equation, we would want to get 'x' on its own on the LHS of the equation and everything else on the RHS, i.e. '$x = \ldots\ldots\ldots$'.

(N.B. LHS stands for left-hand side and RHS for right-hand side.)

In these notes, we will usually be making 'x' *the subject of the equation*, but it is important to remember that any variable could be made the subject.

The 'equals' sign in an equation states that each side of the equation will have the same value as the other side – the equation is *balanced*.

Provided that the same action is carried out simultaneously on both sides of an equation, then the equation will remain balanced and will continue to be *true*.

In this Unit we will consider rearrangements that involve

- moving numbers or simple algebraic terms from one side to the other,

- cancelling in basic fractions, and

- handling squares and square roots.

The following Unit, 'Rearranging complicated equations', includes equations with fractions and more complicated algebraic expressions. The Unit, 'Solving equations', also introduces the use of inverse functions.

3.2.2 Moving terms by addition or subtraction

Algebraic terms (see E3.2) can be 'moved' from left to right (and vice versa) by *adding* or *subtracting* the *same value* from both sides.

E3.10

Make x the subject of the equation: $x + a = 3$

Subtract a from both sides: $x + a - a = 3 - a$

The 'a' has 'moved' from LHS to RHS: $x = 3 - a$

Another way of looking at the above process is to say that:

'If we move a term from one side of the equation to the other, then we *change its sign*, for example, $+a$ will become $-a$ (as in E3.10), $+4$ will become -4, $-3x$ will become $+3x$, etc.'

Q3.9 Rearrange the following equations to make x the subject.

(i) $x + 4 = 5$ (iv) $3 - a = 4 - x$

(ii) $6 + x = 4$ (v) $a + x - 3 = 4$

(iii) $a + x = 6$ (vi) $2.5 - b - c = 3.1 - x$

It is often useful just to switch an equation around, and *swap* **the LHS and RHS**:

E3.11

Make x the subject of the equation on the LHS: $p + 4 = x$

Swap over the LHS and RHS: $x = p + 4$

It is often also useful to multiply both sides of an equation by '-1' (provided that *all* terms are multiplied). This will change '$-x$' on the LHS to '$+x$': $(-1) \times (-x) = +x$

E3.12

Make x the subject of the equation: $4 - x = p$

Subtract 4 from both sides: $-x = p - 4$

Multiply both sides by '-1': $(-1) \times (-x) = (-1) \times (p - 4)$

x is now a positive value: $x = -p + 4 = 4 - p$

The order of terms can be changed to avoid a negative sign at the beginning of an expression – see the last line in E3.12.

Q3.10 Rearrange the following equations to make x the subject on the LHS.

    (i) $a - x = 6$ (iii) $3 - x = 0$

    (ii) $3 - a = x - 4$ (iv) $3a - 4 = 4 + x + 2a$

3.2.3 Moving terms by multiplying or dividing

It is possible to **multiply (or divide)** *both* sides of an equation by *any* number or variable, provided that this action is applied to *all* terms and any divisor is *not* zero.

After multiplying or dividing in an equation, it is often necessary to **cancel factors** in the top and bottom of fractions. See 3.1.7 for cancelling.

E3.13

Make x the subject of the equation: $3x = 6 + p$

Divide both sides by 3: $\dfrac{3x}{3} = \dfrac{(6 + p)}{3}$

Note use of brackets to group terms on RHS

Cancel the '3' on top and bottom of LHS: $\dfrac{\cancel{3}x}{\cancel{3}} = x = \dfrac{(6+p)}{3}$

The '3' can be divided into the bracket on the RHS if required:

$$x = \frac{(6+p)}{3} = 2 + \frac{p}{3}$$

E3.14

Make x the subject of the equation: $\dfrac{x}{2} = 3 + p$

Multiply both sides by 2: $\dfrac{x}{2} \times 2 = (3+p) \times 2 = 2(3+p)$

Cancel the '2' on top and bottom of LHS: $\dfrac{x}{\cancel{2}} \times \cancel{2} = x = 2(3+p)$

The bracket on the RHS can be multiplied out if required: $x = 2(3+p) = 6 + 2p$

It is possible to divide (or multiply) both sides of the equation by a '**bracket**' term.

E3.15

Make x the subject of the equation: $(3+p)x = 6 + p$

Group RHS into a bracket first, then:

Divide both sides by the bracket term $(3+p)$: $\dfrac{(3+p)x}{(3+p)} = \dfrac{(6+p)}{(3+p)}$

Cancel the '$(3+p)$' on top and bottom of LHS: $\dfrac{\cancel{(3+p)}x}{\cancel{(3+p)}} = \dfrac{(6+p)}{(3+p)}$

Giving: $x = \dfrac{(6+p)}{(3+p)}$

Q3.11   Rearrange the following equations to make x the subject.

(i) $5x = a - 6$

(ii) $mx = y - c$

(iii) $\dfrac{x}{3} = p - 4$

(iv) $(2-p)x = p + 4$

(v) $\dfrac{x}{3-p} = 5$

(vi) $\dfrac{(2-p)x}{3-p} = 2$

3.2.4 Collecting terms

If there is more than one term involving x, then it is necessary to **collect all the x terms** on one side of the equation:

E3.16

Make x the subject of the equation: $6 - 4p - 2x = 1 + 2a - 3x$

Move '6' and '$-4p$' from LHS to RHS: $-2x = 1 + 2a - 3x - 6 + 4p$

Move '$-3x$' from RHS to LHS: $-2x + 3x = 1 + 2a - 6 + 4p$

Simplify the LHS: $-2x + 3x = x$

Giving: $x = -5 + 2a + 4p = 2a + 4p - 5$

Q3.12   Rearrange the following equations to make x the subject.

       (i) $2x + 4 = x - 2$ (iv) $x = 2(p - x)$

      (ii) $3 - a + x = 6 - 2x$ (v) $x(2 - a) = a(3 - x)$

     (iii) $x - 3 - 2p = 5 + p - 3x$ (vi) $2(x - 1) = 3(3 - x)$

It is often necessary to **group terms together** using a bracket:

E3.17

Using a bracket to group x-terms in the expression: $3x + px - 2x - qx + x$, we have:

$$3x\text{'s} + \text{'}p\text{'}x\text{'s} + (-2)x\text{'s} + (-q)x\text{'s} + \text{one } x$$

giving:

$$3x + px - 2x - qx + x = (3 + p - 2 - q + 1)x = (2 + p - q)x$$

Similarly:

$$3x + ax \qquad\qquad = (3 + a)x$$

$$3x - bx + x \qquad = (3 - b + 1)x = (4 - b)x$$

$$3x - ax - 2x \quad = (1 - a)x$$

Q3.13   Simplify each of the following expressions, by collecting the x terms together. In each expression, find the value, or expression, to write in place of the question mark, '?'. For example if:

$$2x + px = (?) \times x$$

we would write that $? = 2 + p$ so that: $2x + px = (2 + p) \times x$.

(i) $5x - 2x = (?) \times x$ (iv) $5x - px - 2x = (?) \times x$

(ii) $3px + 2px = (?) \times x$ (v) $5x - px - 7x = (?) \times x$

(iii) $mx - 3x = (?) \times x$ (vi) $3x + px - 4x + 2kx = (?) \times x$

E3.18

Make x the subject of the equation: $x + 4 = a - px$

Collect all x terms on the LHS: $x + px = a - 4$

'Count' the x's (see E3.17): $(1 + p)x = a - 4$

It is now possible to *divide both sides* by $(1 + p)$: $\dfrac{\cancel{(1+p)}x}{\cancel{(1+p)}} = \dfrac{a - 4}{(1 + p)}$

Cancelling $(1 + p)$ on the LHS (see 3.1.7): $x = \dfrac{a - 4}{(1 + p)}$

It is important, when rearranging equations, to separate x terms from non-x terms, and to do this it is often necessary to first **expand (or open up) brackets** which contain both types of terms.

E3.19

Make x the subject of the equation: $3(3 + x) = 2(x + 4)$

Multiply out both brackets: $9 + 3x = 2x + 8$

Collect all x terms on the LHS: $3x - 2x = 8 - 9$

Giving: $x = -1$

Q3.14   Simplify each of the following expressions, by collecting the x terms together. It will be necessary to *multiply out the brackets first* before simplifying the whole expression.

(i) $3(2 + x) - 2x + 4$ (iv) $a(b - 2x) + b(2x - a)$

(ii) $3(2 + x) + 2(3x + 1)$ (v) $x(2 + a) - 3(a - x)$

(iii) $2(3 - 3x) - 4(1 - 4x)$ (vi) $2(a - 4) - 4(a - 2 - x) + 2a - 3x$

Q3.15 Rearrange the following equations to make x the subject.

(i) $2 + x = 5 - x$ (vi) $4 - qx + p = 2x - px + 9$

(ii) $2x - a = b + 4x$ (vii) $x(1 - 2a) = 4a(x - 3)$

(iii) $x = 3(2 - x)$ (viii) $(2 - x)p = 6$

(iv) $3x + 3 = px + 8$ (ix) $(2 - x)p = 6 - x$

(v) $3x + 3 = px + 8 - p$

3.2.5 Squares and square roots

Scientific equations often include square terms and/or square roots. It is possible to clear these out of the equations by either taking the square or square root of both sides of the equation, and using the fact that:

$$\sqrt{x^2} = \pm x \qquad\qquad [3.3]$$

Note that the square root of a positive number may be either positive or negative, and this can be represented by the symbol, '$\pm$'. Excel uses the function SQRT to calculate a square root.

E3.20

Make x the subject of the equation: $x^2 = a + b$

Take the square root of both sides: $\sqrt{x^2} = \pm\sqrt{(a + b)}$

Giving: $x = \pm\sqrt{(a + b)}$

E3.21

Make x the subject of the equation: $a + b = \sqrt{(x^2 + b^2)}$

Square both sides: $(a + b)^2 = (x^2 + b^2)$

Multiply out brackets on LHS: $(a + b)^2 = a^2 + 2ab + b^2$

Hence: $a^2 + 2ab + b^2 = x^2 + b^2$

Rearrange: $x^2 = a^2 + 2ab$

Take the square root of both sides: $x = \pm\sqrt{(a^2 + 2ab)}$

Q3.16 Rearrange the following equations to make 'x' the subject.

    (i) $x^2 = 9$ (iv) $a = \sqrt{(x + 3)}$

    (ii) $x^2 = 2 + a$ (v) $a = \sqrt{(x^2 + 25)}$

    (iii) $2x^2 = 6a - x^2$

3.3 Rearranging complicated equations

3.3.1 Introduction

In this Unit, we attempt to give some *general* approaches to rearranging equations in science. The previous Units, 'Basic equation handling' and 'Introduction to rearranging equations', introduced some basic algebraic skills; students who do not have a good understanding of such skills should study those Units before trying to rearrange more complicated equations.

After students have completed this Unit, they are also recommended to work through section 3.4.3 which introduces the use of *inverse functions* to rearrange equations.

3.3.2 General principles

The following are some very general principles for the rearrangement of equations:

1   If there are *any fractions* in the equation, try first to rearrange the equation to get all terms on a single equation line with no fractions – see section 3.3.3, 'Removing fractions'.

2   Any *brackets* that contain a mixture of x and non-x terms, should be opened up to separate the x and non-x terms – see 3.3.4, 'Multiplying brackets'.

3   *Collect all the x terms on the LHS side* of the equation, and non-x terms on the RHS – see 3.2.2, 3.2.3 and 3.2.4. on moving and collecting terms.

4   *Group* the x terms, e.g. $3x$, $(4 + 2p)x$ – see section 3.1.5, 'Using brackets'.

5   *Divide both sides of the equation by the multiplier of x* to get 'x' on its own on the LHS – see 3.2.3, 'Moving terms by multiplying or dividing'.

E3.22

Make x the subject of the equation: $3 = \dfrac{x}{x + 2} + p$

1 *Remove fractions* by multiplying both sides of the equation by $(x + 2)$. Note use of brackets. $3 \times (x + 2) = \dfrac{x \times \cancel{(x+2)}}{\cancel{(x+2)}} + p \times (x + 2)$

Cancel the $(x+2)$ in the fraction: $3(x + 2) = x + p(x + 2)$

2 *Multiply out the brackets*: $3x + 6 = x + px + 2p$

3 *Collect 'x' terms* on LHS: $3x - x - px = 2p - 6$

4 *Group 'x' terms* using brackets: $(2 - p)x = 2p - 6$

5 *Divide* by $(2 - p)$ to get 'x' on its own: $x = \dfrac{2p - 6}{2 - p}$

Example E3.22 illustrates the use of the general principles in rearranging a complicated equation. However, competence in knowing which approach to use in any given situation can only be achieved by experience and through exposure to different examples.

This Unit gives a variety of different examples in the rearrangement of equations, together with some further algebraic techniques that are often encountered. The student should work through the examples and questions as a way of gaining experience.

3.3.3 Removing fractions

A fractional term in an equation can be 'cleared' by multiplying *both sides* of the equation by the **denominator** of the fraction.

N.B. Provided that a term (including a bracket term) is a factor of both the top and bottom of a fraction, then it can be cancelled from the fraction – see 3.1.7.

E3.23

Make x the subject of the equation: $6 = \dfrac{2x}{x - 1} + \dfrac{p}{2}$

We must first remove the $(x - 1)$ term from the denominator of the first fraction:

Multiply *every term* on both sides of the $6 \times (x - 1) = \dfrac{2x \times \cancel{(x - 1)}}{\cancel{(x - 1)}} + \dfrac{p \times (x - 1)}{2}$
equation by $(x - 1)$. Note use of brackets.

Cancelling the $(x - 1)$ term in the first $6 \times (x - 1) = 2x + \dfrac{p \times (x - 1)}{2}$
fraction:

Multiply both sides of the equation by '2' to $12(x - 1) = 4x + p(x - 1)$
remove the second fraction:

Multiply out brackets: $12x - 12 = 4x + px - p$

Collect x terms on LHS: $12x - 4x - px = 12 - p$

Group the x terms with a bracket: $(8 - p)x = 12 - p$

Divide both sides by $(8 - p)$: $x = \dfrac{(12 - p)}{(8 - p)}$

Q3.17 Rearrange the following equations to make x the subject.

(i) $\quad 2 = \dfrac{4}{x}$

(ii) $\quad 2 = \dfrac{4-a}{x}$

(iii) $\quad 2 = \dfrac{4}{x} - a$

(iv) $\quad \dfrac{1}{x} + \dfrac{1}{a} = 3$

(v) $\quad \dfrac{1}{(x+1)} + \dfrac{1}{a} = 3$

(vi) $\quad \dfrac{x}{x+1} + \dfrac{1}{a} = 3$

(vii) $\quad \dfrac{3}{x+1} = \dfrac{4a}{x-2}$

(viii) $\quad \dfrac{x+1}{x-1} = y$

3.3.4 Multiplying brackets

For a basic introduction to brackets (or *parentheses*), see 3.1.5.

When multiplying out the product of *two* brackets, first multiply the second bracket *separately* by *every* term in the first bracket, and then multiply out *all* the different terms as in E3.24.

E3.24

To multiply out $(a - b + c)(p - b)$, multiply the second bracket by every term in the first bracket:

$$(a - b + c)(p - b) = a(p - b) - b(p - b) + c(p - b)$$

and then separately multiply out each of the three new bracket terms:

$$= ap - ab - bp + b^2 + cp - cb$$

Q3.18 Multiply out the following brackets:

(i) $\quad (a - 2)(a - 2)$ (same as $(a - 2)^2$)

(ii) $\quad (a - b)(a + b)$

(iii) $\quad (a + 4)(3 - ab)$

(iv) $\quad (a + b)(3c + d)$

(v) $\quad (x - 2y)(p - q)$

(vi) $\quad (a - b + 2)(3 + a + b)$

Q3.19 Rearrange the equations to make x the subject. It will be necessary to *multiply out the brackets first* before simplifying the whole expression.

(i) $\quad 3(2 + x) = 2x + 4$

(ii) $\quad 3(2 + x) = 2(3x + 1)$

(iii) $\quad 2(3 - 3x) = 4(1 - 4x)$

(iv) $\quad a(b - 2x) = b(2x - a)$

(v) $\quad x(2 + a) = 3(a - x)$

(vi) $\quad 2(a - 4) - 4(a - 2 - x) = 2a - 3x$

(vii) $\quad (2 - a)(4 - x) = 4a + 2x$

3.3.5 Adding or subtracting fractions

A basic introduction to fractions is given in the Revision Notes on the web site:
🖥 Study Resources > Revision Mathematics

Before adding (or subtracting) fractions, all fractions must have a **common denominator**.

It is always possible to find a *common denominator* (not necessarily the simplest) by multiplying all the *different* denominators together. Each fraction can then be multiplied top and bottom (as necessary) by an appropriate factor so that they all end up with the same denominators:

E3.25

Add $\dfrac{a}{x} + \dfrac{3}{y}$

A common denominator will be: $\qquad x \times y = xy$

Multiplying both fractions top and bottom, with 'y' and 'x' respectively,

gives: $\qquad\qquad\qquad \dfrac{a}{x} = \dfrac{ay}{xy} \quad \text{and} \quad \dfrac{3}{y} = \dfrac{3x}{yx} = \dfrac{3x}{xy}$

Then: $\qquad\qquad\qquad \dfrac{a}{x} + \dfrac{3}{y} = \dfrac{ay}{xy} + \dfrac{3x}{yx} = \dfrac{ay + 3x}{xy}$

E3.26

Simplify to a single fraction: $\dfrac{3}{(x+1)} - \dfrac{2}{x} + \dfrac{x}{(x+1)}$

- There are two *different* denominators 'x' and '$x + 1$'.

- A *common denominator* will be $x \times (x + 1)$.

- Multiplying each fraction top and bottom with 'x' or '$x + 1$' as appropriate, gives:

$$\text{First term: } \frac{3}{(x+1)} = \frac{3 \times x}{(x+1) \times x} = \frac{3x}{x(x+1)}$$

$$\text{Second term: } -\frac{2}{x} = -\frac{2 \times (x+1)}{x \times (x+1)} = -\frac{2(x+1)}{x(x+1)}$$

$$\text{Third term: } \frac{x}{(x+1)} = \frac{x \times x}{(x+1) \times x} = \frac{x^2}{x(x+1)}$$

- Add them all together: $\dfrac{3x}{x(x+1)} - \dfrac{2(x+1)}{x(x+1)} + \dfrac{x^2}{x(x+1)} = \dfrac{3x - 2(x+1) + x^2}{x(x+1)}$

- Simplify the numerator: $= \dfrac{x^2 + x - 2}{x(x+1)}$

Q3.20 Simplify each of the following to a single fraction:

    (i) $\dfrac{1}{2} + \dfrac{1}{3}$ (iv) $\dfrac{1}{p} + \dfrac{1}{q} - \dfrac{1}{pq}$

    (ii) $\dfrac{1}{2} + \dfrac{2}{3} - \dfrac{5}{6}$ (v) $\dfrac{1}{p} - \dfrac{1}{p+2}$

    (iii) $\dfrac{1}{p} + \dfrac{1}{q}$ (vi) $\dfrac{1}{p} + \dfrac{1}{p(p+2)}$

3.3.6 Factorizing expressions

A **factor** is a term that divides *exactly* into an expression – see 3.1.6.
Factorization is the process of splitting up an expression into its various factors.

E3.27

What are the factors of: $9p + 3pz$

                      '3' is a factor: $9p + 3pz = 3(3p + pz)$
             'p' is also a factor: $3(3p + pz) = 3p(3 + z)$

Check that the expressions are equivalent by multiplying out the brackets again,
i.e. $3p(3 + z) = 9p + 3pz$

Q3.21 What are the factors of each of the following expressions.

For example the expression $6px + 3py$ can be rearranged to give $3p(2x + y)$ giving the
factors: 3, p and $(2x + y)$
Hint: multiply the factors together to check that the factors that give the original expression:
$3p(2x + y) = 6px + 3py$.

    (i) $4x + 4y$ (iii) $4x + 6x^2$

    (ii) $4x + 6y$ (iv) $pqx + pbx$

3.4 Solving equations

3.4.1 Introduction

An algebraic equation is solved when a value for the variable is found that will make the equation true. For
example, substituting the value $x = 1$ will make the following equation true:

$$(3 - x) = 2 \text{ becomes } (3 - 1) = 2 \text{ which is TRUE}$$

The value of the variable, $x = 1$, is the *solution* of the equation.

There is no single method that is used to solve *all* equations – the choice of method depends on the type of equation involved.

Some of the equations encountered in this Unit include logarithmic and exponential functions. An introduction to these functions is given in Chapter 5.

3.4.2 Rearranging the equation

Provided that the equation has no power terms or special functions, then it is usually possible to rearrange the equation directly to make the variable the subject of the equation (see Units 3.2 and 3.3) and hence calculate its value.

E3.28

Solve the equation to find the value of p: $2p - 5 = 7 - p$

Rearranging the equation to make p the subject gives: $2p + p = 7 + 5$

$$3p = 12$$

$$p = 12/3 (= 12 \div 3) = 4$$

The solution to the equation is: $p = 4$

Once the *solution* to an equation has been derived, it is always advisable to substitute the value back into the original equation – just to make sure that the value *does indeed* make the equation true, and that there have been no mistakes.

For example, in E3.28, substitute the value for $p = 4$ back into the equation:

$$2 \times 4 - 5 = 7 - 4 \text{ becomes } 3 = 3 \text{ which is TRUE} - \text{the solution is correct!}$$

Q3.22   Solve the following equations, finding the value of x that will make each equation true:

     (i) $2x - 8 = 22$ (iv) $x - 32 = 9 \times 50 \div 5$

     (ii) $x + 4 = 12 - 3x$ (v) $x \div 3 = 2 - x$

     (iii) $3(2 - x) = 12$

3.4.3 Inverse functions

There are many examples where we need to take an *inverse function* to solve an equation.

To solve the equation: $\dfrac{1}{x} = 4$ we take the *inverse* of both sides:

$$\frac{1}{1/x} = \frac{1}{4} \quad \text{which becomes} \quad x = \frac{1}{4} = 4^{-1} = 0.25$$

where x and $1/4$ are the *inverse* (reciprocal) functions of $1/x$ and 4 respectively.

The inverse function for a 'square' is a 'square root', so that the solution to the equation $x^2 = 16$ would be $x = \sqrt{(16)} = \pm 4$, i.e. x equals $+4$ or -4.
    See 3.2.5 for rearranging equations with squares and square roots.

The inverse function for a 'log' is a 'power of 10'. If $\log(x) = 3.4$, then $x = 10^{3.4} = 2511.9$.
    See Unit 5.1 for the mathematics of logarithms.

On a calculator the *inverse* of a function is often found on the same key as the original function by previously pressing a 'shift' or '2nd function' key. For example, $\log(x)$ and 10^x usually use the same key.
    Some examples of common inverse functions are given in Table 3.1.

Table 3.1. Inverse functions.

Function	Inverse function
$1/x = 4$	$x = 4^{-1} = 1/4 = 0.25$ (reciprocal 4)
$x^2 = 18.34$	$x = \sqrt{(18.34)} = \pm 4.283$
$\sqrt{(x)} = 3.66$	$x = 3.66^2 = 13.40$
$\cos(x) = 0.79$	$x = \cos^{-1}(0.79) = 37.8°$
Similarly for *sin* and *tan*	(or $x = \arccos(0.79) = 37.8°$)

The following functions are introduced in Chapter 5.

$\ln(x) = -0.13$	$x = e^{-0.13} = 0.878$
$e^x = 45.66$	$x = \ln(45.66) = 3.82$
$\log(x) = 3.4$	$x = 10^{3.4} = 2511.9$
$10^x = 67.78$	$x = \log(67.78) = 1.831$

Q3.23 Solve the following equations, finding the value(s) of x that will make each equation true:

(i) $x^2 = 25$ (v) $e^x = 26.2$

(ii) $x^2 = 0.025$ (vi) $\sin(x) = 0.34$ (give x in degrees)

(iii) $\log(x) = 1.3$ (vii) $10^x = 569$

(iv) $\sqrt{(x)} = 9$ (viii) $10^x = 0.01$

Sometimes it is necessary to carry out a mixed process of rearrangement as well as taking an inverse function. In these cases, it is necessary to get the x terms separated from non-x terms *as soon as possible*. Sometimes the mixed terms are within the function itself, e.g. $\log(x + 200)$: and it is necessary to take the inverse of the function first. Alternatively it may be necessary to carry out some rearrangement first so that the *function is on its own* on the LHS – see E3.29.

E3.29

(i)　Solve:　　　　　　　　　　　　$\log(x + 200) = 2.6$

　　　Take inverse function of log *first*:　$(x + 200) = 10^{2.6} = 398.1$ (4sf)

　　　Rearrange:　　　　　　　　　　$x = 398.1 - 200 = 198$ (3sf)

(ii)　Solve:　　　　　　　　　　　　$\log(2x) + 0.3 = 2.5$

　　　Rearrange equation *first*:　　　$\log(2x) = 2.5 - 0.3 = 2.2$

　　　Take inverse function of log:　　$2x = 10^{2.2} = 158.49$ (5sf)

　　　Rearrange equation *again*:　　　$x = 158.5/2 = 79.24$ (4sf)

Q3.24　Solve the following equations, finding the value(s) of x that will make each equation true:

(i)　$x^2 + 9 = 25$　　　　　　　　　(iv)　$e^{3x} = 0.87$

(ii)　$(x + 9)^2 = 25$　　　　　　　(v)　$\log(2x + 1.2) = 0.45$

(iii)　$3 = \sqrt{(3 + 2x)}$　　　　　(vi)　$10^{x+2} = 246$

3.4.4　Quadratic equations

A common form of equation where a power is involved is a 'quadratic equation'. The term 'quadratic' means that the highest 'power' in the equation is 2. For example:

$$2p^2 - p - 6 = 0$$

is a *quadratic equation* in p.

In general, there will be TWO possible values of p that will make a quadratic equation true. There are generally two *possible solutions* to a quadratic equation.

E3.30

Check, by substituting values, that both $p = 2$ and $p = -1.5$ will make the equation: $2p^2 - p - 6 = 0$ true, i.e. $2p^2 - p - 6$ will equal 0.

Putting $p = 2$　　　$2 \times 2^2 - 2 - 6 = 8 - 2 - 6 = 0$　　　　　　　　　　　　　True

Putting $p = -1.5$　　$2 \times (-1.5)^2 - (-1.5) - 6 = 2 \times 2.25 + 1.5 - 6 = 4.5 + 1.5 - 6 = 0$　　True

The **general** method for finding the solutions starts by rearranging the quadratic equation into a standard form:

$$ax^2 + bx + c = 0 \qquad [3.4]$$

where a, b and c are constants in the equation, and x is the unknown variable.

With the equation in the above form, the solutions can be found by substituting the values of a, b and c into the formula:

$$x = \frac{-b \pm \sqrt{b^2 - 4ac}}{2a} \qquad [3.5]$$

The $\pm$ sign in the numerator of the equation, tells us to take *two solutions* – one with the '+' sign and one with the '−' sign.

It is a straightforward process to use the above formula to solve any quadratic equation:

1 Rearrange the equation into the form $ax^2 + bx + c = 0$.

2 Work out the values in the equation equivalent to a, b and c.

3 Substitute a, b and c into the formula for the solutions.

🖳 Computer Techniques > Excel Tutorial: Solver

E3.31 🖳

Solve the equation: $\qquad\qquad 2.4x^2 = 3.2x + 6.6$

Rearrange as in [3.4]: $\qquad\qquad 2.4x^2 - 3.2x - 6.6 = 0$

Compare directly with [3.4] $\qquad ax^2 + bx + c = 0$

giving the equivalences: $\qquad\quad a = 2.4, b = -3.2, c = -6.6$

Substitute these values in [3.5]: $\quad x = \dfrac{-(-3.2) \pm \sqrt{(-3.2)^2 - 4 \times 2.4 \times (-6.6)}}{2 \times 2.4}$

$$x = \frac{+3.2 \pm \sqrt{10.24 + 63.36}}{4.8} = \frac{+3.2 \pm \sqrt{73.6}}{4.8} = \frac{+3.2 \pm 8.58}{4.8}$$

Giving two solutions for x: $\qquad\quad x = 2.45 \; or \; x = -1.12$

Q3.25 Solve the following equations:

    (i) $\; 2x^2 - 3x + 1 = 0$ (iii) $\; 4x = 5 - 2x^2$

    (ii) $\; 3.2x^2 - 2.5x - 0.8 = 0$ (iv) $\; (x - 2)^2 = 2x$

When the quadratic equation represents a scientific relationship, then the correct solution in the *scientific context* may be just one, or both, of the possible mathematical solutions. It is necessary to use other information about the scientific problem to decide which solution (or both) is correct.

E3.32

A cricket ball is hit directly upwards with an initial velocity of 16.0 m s^{-1}. The equation for the time, t (seconds), at which it passes a height of 5.0 m is given by

$$5.0 = 16.0t - 4.9t^2$$

Solving the quadratic equation gives *two values*, $t = 0.35$ and 2.92 s.

In the *scientific context*, the two values are the times (in seconds) at which the ball passes the height 5.0 m, first going up and then coming down.

E3.33 🖳

The refractive index, n, of a piece of glass is given by solving the equation $n(n - 1) = 0.75$.

Find the value of n.

Expanding the equation gives $n^2 - n = 0.75$ or $n^2 - n - 0.75 = 0$.

This quadratic equation can be solved giving $n = 1.5$ *or* $n = -0.5$.

However the refractive index of a material is always positive so in this case $n = 1.5$ is the only solution.

In some equations, the value, $b^2 - 4ac$, inside the square root may give a *negative value*. It is not normally possible to take the square root of a negative number. Hence:

$$\text{If } b^2 < 4ac \text{ the equation has } \textit{no real solutions} \qquad [3.6]$$

Using *complex numbers* it is possible to derive a 'square root' from a negative number, but these techniques are beyond the mathematical scope of this book.
When the value, $b^2 - 4ac$, is *zero*, the square root term also becomes zero:

$$\text{If } b^2 = 4ac \textit{ both } \text{solutions to the equation have the } \textit{same value} \qquad [3.7]$$

Q3.26 Solve the following equations:

(i) $x^2 - 4x + 4 = 0$ (iii) $2x^2 - 3x + 4 = 0$

(ii) $x^2 + 0.5x - 1.5 = 0$ (iv) $4x^2 + 12x + 9 = 0$

3.4.5   Using logarithms to solve power equations

An important property of logarithms (see Unit 5.1) is that they can be used to drop a 'power' down onto the equation line:

$$\log(P^x) = x \times \log(P) \qquad\qquad [3.8]$$

In the following equation, note what happens when logs are taken of both sides of a 'power' equation:

$$P^x = Q$$
$$\log(P^x) = \log(Q)$$
$$x \times \log(P) = \log(Q)$$
$$x = \frac{\log(Q)}{\log(P)} \qquad\qquad [3.9]$$

We can use the above properties of logarithms to solve equations where x appears as a power.

E3.34 💻

Solve the equation $3^x = 14$ in *two* ways – using log and ln separately:

Take logarithms of both sides of the equation:

(using **log**)	(using **ln**)
$\log(3^x) = \log(14)$	$\ln(3^x) = \ln(14)$

which gives, using [3.8]:

$x \times \log(3) = \log(14)$	$x \times \ln(3) = \ln(14)$

Rearranging using [3.9] gives:

$x = \log(14)/\log(3)$	$x = \ln(14)/\ln(3)$
$x = 1.146/0.477$	$x = 2.639/1.099$
$x = 2.402$	$x = 2.402$

Notice that, in E3.34, it does not matter which type of logarithm is used, provided that the same logarithm is used on both sides of the equation. Both types of logarithms give the *same answer* to the problem.

See 5.1.6 for further examples of the use of logarithms in solving equations.

Q3.27   Solve the following equations, finding the value of x that will make each equation true:

(Use both 'log' (log to base 10) and 'ln' (natural log) to solve each equation, and check that both methods give the same results.)

(i) $5^x = 32$ (ii) $3.2^{(5-2x)} = 5.6$

3.5 Simultaneous equations

3.5.1 Introduction

The simplest example of a *'simultaneous equations'* problem has:

- two equations, which contain
- the same two variables, e.g. x and y, and
- with the requirement that the two equations must both be true (simultaneously) for the same values for x and y.

For example, we may have:

$$3x + 2y = 7 \quad \text{and} \quad 4x + y = 6$$

Both of the above equations are true (simultaneously) if $x = 1$ and $y = 2$. Check by substituting the values into the equations and finding that they both 'balance'.

$$3 \times 1 + 2 \times 2 = 7 \qquad \text{True}$$
$$4 \times 1 + 2 = 6 \qquad \text{True}$$

The solution for this pair of simultaneous equations is $x = 1$ and $y = 2$.

If the equations have three variables, e.g. x, y and z, then we will need three equations to find a solution. In general, if we have n different variables then we will need n different equations to find solutions for all n variables.

There are several different ways of finding the solution such that all the equations become true simultaneously. Analytical methods will give an exact solution, but may be mathematically difficult to perform. A graphical method can be used to find an (approximate) solution for more complicated equations.

3.5.2 Analytical solution

The most reliable analytical method for solving simultaneous equations aims to rearrange the equations so that one of the variables can be 'eliminated' from the equation.

The process follows FOUR basic steps, which are illustrated by solving the simultaneous equations in E3.35:

E3.35

Find values of p and q that make both to the following equations true:

$$4.2p - 1.68q = 9.66 \qquad \text{(A)}$$
$$2.6p + 6.5q = 2.34 \qquad \text{(B)}$$

See text for calculations.

Step 1. Rearrange the equations to make **one variable** the subject for both equations:
Equation (A) becomes:

$$4.2p = (9.66 + 1.68q) \qquad \text{which gives}$$
$$p = (9.66 + 1.68q)/4.2$$
$$p = 2.3 + 0.4q \qquad \text{(C)}$$

Equation (B) becomes:

$$2.6p = (2.34 - 6.5q) \quad \text{which gives}$$
$$p = (2.34 - 6.5q)/2.6$$
$$p = 0.9 - 2.5q \tag{D}$$

Step 2. Equations (C) and (D) both give expressions for p. Write these expressions as an equation:

$$2.3 + 0.4q = 0.9 - 2.5q \tag{E}$$

Step 3. Rearrange equation (E) to make q the subject of the equation:

$$0.4q + 2.5q = 0.9 - 2.3$$
$$2.9q = -1.4$$
$$q = -0.482\,76$$

Note. We have 'rounded off' the value of q to five significant figures. It is important to be careful about 'rounding off' too much in the middle of a problem in case a subsequent step calculates a *small difference* between *two large numbers*.

Step 4. Substitute this value of q back into either equation (C) or equation (D) to calculate the value for p. Using equation (C):

$$p = 2.3 + 0.4 \times (-0.48276) \quad \text{giving} \quad p = 2.107$$

Using the above FOUR steps, the solutions to the equations are:

$$p = 2.107 \quad \text{and} \quad q = -0.4828$$

(to four significant figures).

Checking the results by substituting back into the original equations to make sure that they are both *true*:

$$4.2 \times 2.107 - 1.68 \times (-0.4828) = 9.6605 \quad \text{and} \quad 2.6 \times 2.107 + 6.5 \times (-0.4828) = 2.3400$$

The calculated values on the right agree with the values in the original equations, (A) and (B), allowing for the 'rounding off' of values during the calculation. We can be sure that the calculated values for p and q are the solutions to the original equations.

Q3.28 Two walkers, Alex and Ben, start walking towards each other along a path. Alex starts 0.5 km from a village on the path and Ben starts 3.2 km from the village. Alex walks at 1.2 $\mathrm{m\,s^{-1}}$ and Ben walks at 0.9 $\mathrm{m\,s^{-1}}$.

We can describe the position of each man by giving the distance, d (in metres), of each man from the village as a function of time, t:

Alex: $d = 500 + 1.2 \times t$

Ben: $d = 3200 - 0.9 \times t$

Solve the simultaneous equations using an analytical method, and calculate:

(i) The time taken before they meet.

(ii) Their distance from the village when they meet.

Q3.29 We can sometimes calculate the individual concentrations, C_1, and C_2, of two compounds in a chemical mixture by measuring the spectrophotometric absorbance, A_λ, of the mixture at *different* wavelengths, λ.

The results of a particular experiment produce the following simultaneous equations:

$$0.773 = 2.25 \times C_1 + 2.0 \times C_2$$
$$0.953 = 0.25 \times C_1 + 6.0 \times C_2$$

Use the analytical method to solve the equations and derive the values for C_1 and C_2.

It is possible to use 'Tools > Solver ...' in Excel to solve simultaneous equations.

⌨ Computer Techniques > Excel Tutorial: Solver

3.5.3 Graphical method

In the graphical method, the equations are plotted as lines on the same graph. The solution is given by the co-ordinates of the point where the lines cross.

E3.36 ⌨

Solve the simultaneous equations:

$$y = x^2$$
$$y = 1.5 - 0.5x$$

Plot these equations as lines on a graph by choosing specific values of x and calculating the equivalent values of y for the two equations. See table below.

x	-2	-1	0	1	2
$y = x^2$	4	1	0	1	4
$y = 1.5 - 0.5x$	2.5	2	1.5	1	0.5

Plotting the above data gives the graph in Figure 3.1.

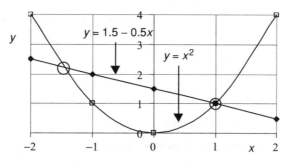

Figure 3.1. Simultaneous solutions.

The solutions to the simultaneous equations are given by the two points where the lines cross. The intersection points are identified by two circles that have approximate values: ($x = -1.45$, $y = 2.2$) and ($x = 1$, $y = 1$). These give two *approximate* solutions to the equations:

$$\text{Solution 1:} \quad x = -1.45 \text{ and } y = 2.2$$

$$\text{Solution 2:} \quad x = 1.0 \text{ and } y = 1.0$$

There will always be uncertainties in reading the co-ordinates of the crossing point from the graph. However, it is possible, once an approximate value is obtained, to redraw the graph to a much larger scale just around the crossing point.

The particular problems in E3.36 can also be solved analytically by first eliminating y from the two equations, giving

$$x^2 = 1.5 - 0.5x$$

This is now a quadratic equation that can be solved using [3.5] to give two solutions for x – see Q3.30.

Q3.30 Solve the simultaneous equations (given in E3.36)

$$y = x^2$$
$$y = 1.5 - 0.5x$$

using an analytical method.

Q3.31 Use a graphical method to solve the problem in Q3.29.

4

Linear Relationships

Overview

The linear relationship is one of the most common mathematical relationships used in science, and is often described on an x–y graph by the equation:

$$y = mx + c$$

where m is the *slope* of the line, and c is the *intercept* of the line where it crosses the y-axis. Note that, in various contexts, the equation of the straight line is also often written as '$y = a + bx$' where a is the intercept and b is the slope. However, for consistency in this book, we will always use the form '$y = mx + c$'.

As a simple example of a straight-line relationship, the values for temperature in the Celsius, C, and Fahrenheit, F, scales are related by the linear equation:

$$F = 1.8 \times C + 32 \qquad\qquad [4.1]$$

When plotted on an x–y graph, with F on the y-axis and C on the x-axis, the above relationship appears as a straight line, as in Figure 4.1.

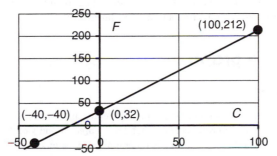

Figure 4.1. Relationship between Fahrenheit, F, and Celsius, C, temperature scales.

The slope, m, and intercept, c, of the straight line in Figure 4.1 are given by: $m = 1.8$ and $c = 32$. The plotted data points show the temperatures for:

- Melting ice, (0,32): $C = 0°C$ and $F = 32°F$

- Boiling water, (100,212): $C = 100°C$ and $F = 212°F$

- Point of 'equal values', $(-40, -40)$: $C = -40°C$ and $F = -40°F$

In Figure 4.1, the line does not pass through the origin (0, 0) of the graph.

In other scientific situations, we often find that the line does pass through the origin of the graph, i.e. the intercept, c, will be zero. In such a case, the value of one variable is *directly proportional* to the value of the other variable. For example, Beer–Lambert's law (for dilute solutions) states that the absorbance, A, of light by a solution in a spectrophotometer is directly proportional to the concentration, C, of the solution:

$$A = k \times C \qquad [4.2]$$

where $k = \varepsilon b$, is a constant for a particular measurement, ε is the absorptivity of the solute and b is the path length of the light through the solution.

There are very many situations in science where it is believed that a linear (straight-line) relationship exists between two variables, but when the experimental data is plotted, the data points on the graph show random scatter away from any straight line. If the aim of the experiment is to measure the slope and/or intercept of the straight line, then the scientist has the problem of trying to interpret which straight line is the *'best-fit'* for the given data. We see in Unit 4.2 that the process of linear regression is a powerful mathematical procedure, which allows *information from every data point* to contribute to the identification of the ('best-fit') line of regression.

Many scientific systems do not yield simple linear relationships, but it is possible, in a number of specific situations, to manipulate the non-linear data so that it can be *represented* by a linear equation – this is the process of *linearization* introduced in Unit 4.3. The transformed data can then be analysed using the familiar process of linear regression.

As an example of linearization, a spectrophotometer measures the *transmittance*, T, of light, which is related to concentration, C, by the non-linear equation of the form:

$$T = 10^{-kC} \qquad [4.3]$$

Equation [4.3] can be 'linearized' to [4.2] by taking the logs of both sides of the equation and defining absorbance by the equation:

$$A = -\log(T) \qquad [4.4]$$

However, is important to note that the process of transforming non-linear data to 'linear' data can distort the significance of errors in the original data. Further advice should be sought in interpreting the possible errors that might arise in the regression results from a process of linearization.

4.1 Straight-line graph

4.1.1 Introduction

The straight line is one of the most common mathematical representations used in science.

A 'straight-line', or linear, relationship occurs when the *change* in one variable (e.g. y) is *proportional* to the *change* in another variable (e.g. x), and is commonly represented by the 'straight-line' equation: $y = mx + c$. The *slope*, m, of the line, gives the *rate of change* of y with respect to x. The point at which the line crosses the y-axis is called the *intercept*, c – see Figure 4.2.

4.1.2 Plotting the graph

In an experiment, the **independent** variable is the one whose values are *chosen*, and the **dependent** variable is the one whose values are *measured*. For example, the pH (dependent variable) of the soil might be *measured* for *known* amounts (independent variable) of added lime.

An x–y graph is normally plotted with x as the *independent* variable on the horizontal axis (*abscissa*) and y as the *dependent* variable on the vertical axis (*ordinate*). We say that y *is plotted against* x, or the graph is of the form y *versus* x.

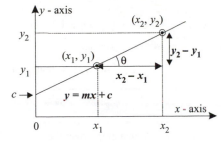

Figure 4.2. Straight-line graph.

Q4.1 In an experiment to measure plant growth as a function of light exposure, different plants are exposed to different levels of light exposure, L, and the resultant growth, G, is recorded. When the results are plotted, which of the following statements are true?

   (i) G should be plotted against L (i.e. L on the x-axis)   *True / False*

   (ii) L should be plotted against G (i.e. G on the x-axis)   *True / False*

   (iii) It does not matter which way round the results are plotted.   *True / False*

⌨ Computer Techniques > Excel Tutorial: XY-Graph

4.1.3 Straight-line equation

A point is located on an x–y graph by using its **co-ordinates:** (x,y).

Note that the x value is placed first.

Only ONE straight line can be drawn through any TWO particular points, as illustrated in Figure 4.2. In this book we will describe the straight line by using the common equation:

$$y = mx + c \qquad [4.5]$$

where m and c are constant values that define the particular line:

- **Slope of the line** is given by the coefficient of x: m.

                                                                           [4.6]

- **Intercept on the y-axis when $x = 0$** is given by the constant: c.

E4.1

Calculate the slope and intercept of the straight line with the equation: $2y = 4x + 3$

The first step is to *rearrange* the equation into the form, $y = mx + c$:

Divide both sides of the equation by '2': $y = (4/2)x + (3/2)$

Giving: $y = 2x + 1.5$

Comparing with the standard equation, [4.5] $y = mx + c$

we can see that: Slope, $m = 2$, and Intercept, $c = 1.5$

Q4.2 Start by plotting the line $y = 2x + 3$ on a graph over the range $-2 \leq x \leq +2$:

   (i)   For each of the values of x in the table below, calculate the value of y given by the equation, $y = 2x + 3$ – this gives the co-ordinates (x, y) of points that will be on this straight line.

x	-2	-1	0	$+1$	$+2$
y					

   (ii)   Plot the co-ordinates from (i) on an x–y graph, and connect with a line.

   (iii)   Where does the graph intercept the y-axis?

Do the following points lie on the line?

   (iv) $(1.5, 5.5)$ *Yes / No*

   (v) $(0.5, 4)$ *Yes / No*

   (vi) $(-0.5, -1.0)$ *Yes / No*

Without carrying out any new calculations, sketch the following lines on the same graph as (ii)

   (vii) $y = -2x + 3$

   (viii) $y = 2x + 1$

   (ix) $y = x + 3$

Q4.3 The length, L (m), of a simple metal pendulum as a function of the ambient temperature, T (°C) is given by the equation:

$$L = \{1 + \alpha \times T\} \times 1.2100$$

where α is the coefficient of linear expansion of the pendulum material, and 1.210 is the length of the pendulum when $T = 0\,°\text{C}$.

Assume $\alpha = 0.000\,019\,°\text{C}^{-1}$

   (i)   Multiply out the bracket to obtain a straight line equation of the form:

$$y = mx + c$$

   (ii)   Calculate the slope, m, and intercept, c, in the equation in (i).

Q4.4 A straight line is given by the equation:

$$3x + 4y + 2 = 0$$

Rearrange the equation to make y the subject of the equation.

Hence, calculate the slope and intercept of the line.

If a particular point with co-ordinates (x, y) lies on the line, then the values of x and y will make the equation *balance*, i.e. the value of 'y' will equal the value of '$mx + c$' and the equation will be TRUE.

If the point (x, y) does *not* lie on the line, then the equation will *not balance*, and the value of 'y' will *not* equal the value of '$mx + c$'

E4.2

The point (3,10) *lies on* the straight line $y = 2x + 4$:

Check by replacing x with '3': $2x + 4 = 2 \times 3 + 4 = 6 + 4 = 10$ which *balances* with $y = 10$ – the equation is TRUE.

The point (4,14) does NOT lie on the straight line: $y = 2x + 4$:

Check by replacing x with '4': $2x + 4 = 2 \times 4 + 4 = 8 + 4 = 12$ which does NOT *balance* with $y = 14$.

It is often necessary to calculate a value for x, corresponding to a value of y on a known straight line. We need then to rearrange [4.5] to make x the subject of the equation:

$$x = \frac{(y - c)}{m} \qquad [4.7]$$

Q4.5 A straight line ($y = mx + c$) has a slope of $+4$ and an intercept of -3, calculate the value of x when $y = 4$

Q4.6 The cooking time for a joint of meat is written as 40 minutes per kilogram plus 20 minutes.

(i)   Express this as an equation relating time, T (in minutes), and the mass, W (kg).

(ii) If there is only 2 hours available, what is the heaviest joint that could be cooked?

E4.3 💻

A car is travelling at a constant speed of 30 m s^{-1}, along a straight road *away* from a junction.

I start a stopwatch with, $t = 0$, when the car is 60 m away from the junction.

(i)   Write down an equation that will then relate the distance, s, of the car from the junction and the time, t, in seconds on my stopwatch.

(ii) Calculate the distance of the car from the junction when $t = 3$ s.

(iii) What will be the time when the car is 210 m from the junction?

Answer:

(i)   The rate of change of distance, s, with time, t, is 30, and this will be a *slope*, $m = 30$ in the equation – see Figure 4.3.

The *intercept*, c, is given by the value of s when t is zero. This is given in the question as 60 m. Hence $c = 60$.

The equation is therefore:

$$s = 30 \times t + 60$$

See Figure 4.3 for the above equation plotted on a graph of s against t

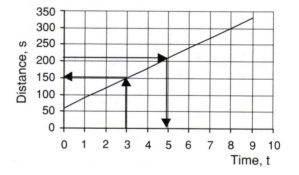

Figure 4.3. Graph for Example E4.3.

(ii) When $t = 3$, we can use the above equation to calculate:

$$s = 30 \times 3 + 60 = 90 + 60 = 150\,\text{m}$$

(iii) To find a value for t it is necessary to rearrange the above equation into the form given by [4.7]:

$$t = \frac{(s - 60)}{30}$$

Substituting values gives: $t = (210 - 60)/30 = 150/30 = 5$ s.

Q4.7 A car is travelling at a constant speed of 20 m s^{-1}, along a straight road *towards* a junction It is at a distance, $s = 140$ m, away from the junction when I start my stopwatch at time, $t = 0$ s.

(i) Which of the following equations will now describe the motion of the car?
(a) $s = 20t + 140$
(b) $s = 20t - 140$
(c) $s = -20t + 140$
(d) $s = -20t - 140$

(ii) What will be the time when the car passes the junction?

(iii) What will be the position of the car, when $t = 20$ s?

4.1.4 Calculating slope and intercept

The **slope, m**, of the straight line (see Figure 4.2) that passes through points (x_1, y_1) and (x_2, y_2) is given by the equation:

$$m = \frac{(y_2 - y_1)}{(x_2 - x_1)} \quad \text{or} \quad m = \frac{(y_1 - y_2)}{(x_1 - x_2)} \qquad [4.8]$$

The **intercept, c**, of the straight line is the value of y at the point where the line passes through the 'y-axis' (see Figure 4.2), i.e. the value of y when $x = 0$.

To calculate the *equation* of a straight line that passes through the two points (x_1, y_1) and (x_2, y_2):

1 Calculate the slope, m, using [4.8]

2 Substitute the value of m and the co-ordinates of one point (x_1, y_1) or (x_2, y_2) into [4.5] – this gives an equation where the intercept, c, is the only unknown value. Rearrange the equation to make c the subject,

$$c = y_1 - m \times x_1 \quad \text{or} \quad c = y_2 - m \times x_2$$

and calculate the value of the intercept, c.

E4.4 🖥

Calculate the equation of the line that passes through the points $(-2, 3)$ and $(2, 1)$.
Calculating the slope:

$$m = \frac{(y_2 - y_1)}{(x_2 - x_1)} = \frac{(1 - 3)}{(2 - (-2))} = \frac{-2}{(2 + 2)} = \frac{-2}{4} = -0.5$$

Substitute m into the equation, $y_2 = mx_2 + c$, with the co-ordinates $x_2 = 2$, and $y_2 = 1$:

$$1 = (-0.5) \times 2 + c = -1 + c$$

Rearranging gives the intercept: $c = 1 + 1 = 2$

The equation of the straight line is therefore: $y = -0.5x + 2$

Q4.8 Calculate the *slopes* of the straight lines which pass through each of the following pairs of points:

    (i) (1, 1) and (2, 3)   (iv) (2.0, 3.0) and (1.0, 3.5)

    (ii) (−1, 1) and (2, 3)   (v) (0, 1) and (1, 0)

    (iii) (1, 1) and (2, −3)   (vi) (2.0, 3.0) and (1.0, −3.5)

In some cases where it is necessary to calculate the coefficients of a straight line, the slope, m, may already be known. It is then possible to go directly to step 2 in the calculation process outlined above.

Q4.9 Derive the *equations* of the straight line that has:

    (i)   a slope of 2.0 and passes through the point (0, 3)

    (ii)   a slope of 0.5 and passes through the point (2, 3)

    (iii) a slope of −0.5 and passes through the point (2, 3)

Q4.10 If C represents the temperature in degrees Celsius and F represents the temperature in degrees Fahrenheit, water boils at a temperature given by $C = 100$ and $F = 212$ and water freezes at a temperature given by $C = 0$ and $F = 32$. Derive a straight-line equation which will give F as a function of C.

Hence calculate the value of C when $F = 0$.

(Hint : F and C become the y- and x-axes of a graph.)

4.1.5 Intersecting lines

If two straight lines, $y = m_A x + c_A$ and $y = m_B x + c_B$ meet at a point (x_O, y_O), then the values of x_O and y_O will make both equations true *simultaneously*:

$$y_O = m_A x_O + c_A$$
$$y_O = m_B x_O + c_B$$

Any unknown variables can (usually) be found by the methods of simultaneous equations.

E4.5 🖳

Calculate the *co-ordinates* of the crossing point for the two straight lines:

$$y = -0.5x + 2$$
$$y = x - 1$$

Using simultaneous equations (Unit 3.5) to 'solve' the above equations, we find that the values $x = 2$ and $y = 1$ will make both equations true simultaneously. Substituting these values back into the above equations make both equations *true*.

The point (2, 1) therefore lies on both lines. This is the point where the lines *cross*.

4.1.6 Parallel and perpendicular lines

Parallel lines have the *same* slope: $m_A = m_B$ [4.9]

(but *different* intercepts, $c_A \neq c_B$).

Perpendicular lines have slopes, m_A and m_B, given by the relationships:

$$m_A = -\frac{1}{m_B} \quad \text{and} \quad m_B = -\frac{1}{m_A} \qquad [4.10]$$

E4.6 🖳

Calculate the equation of the line which is perpendicular to the line $y = -0.5x + 2$ and passes through the point $(-2, 3)$.

The slope of the first line, $m_A = -0.5$.

Thus the slope of the *perpendicular* line, $m_B = -\left(\dfrac{1}{m_A}\right) = -\left(\dfrac{1}{-0.5}\right) = 2$.

Substitute this value of m into the equation, $y = m_B x + c_B$, with the co-ordinates $(-2, 3)$, and calculate the value of the intercept, c_B:

$$3 = 2 \times (-2) + c_B = -4 + c_B$$

Hence $c_B = 7$, and the equation of the line becomes: $y = 2x + 7$.

Q4.11   A second line is *parallel* to the line $y = 3x + 2$, and passes through the point (1,1). What is the equation of this second line?

Q4.12   A second line is *perpendicular* to the line $y = 3x + 2$, and passes through the point (1,1). What is the equation of this second line?

E4.7

Write down the equation of the line that is parallel to the x-axis (horizontal) and passes through the y-axis at $y = 2.5$.

A line that is parallel to the x-axis has slope, $m = 0$.

If the line passes through the y-axis at $y = 2.5$, then $c = 2.5$.

Hence the equation of line is

$$y = 2.5$$

Whatever, the value of x the horizontal line gives the same value, $y = 2.5$.

Q4.13   Write down the equation of the line that is parallel to the y-axis (vertical) and passes through the x-axis at $x = -1.5$.

4.1.7 Interpolation and extrapolation

Interpolation – finding the co-ordinates of a point on the line *between* existing points.
Extrapolation – finding the co-ordinates of a point on the line *outside* existing points.

In both interpolation and extrapolation, it is necessary to first calculate the equation of the line that passes through the two points, and then find the value of x (or y) equivalent to a new value of y (or x).

Q4.14   A straight line passes through the points (2, 3) and (4, 4). Calculate the values of:

   (i) y when $x = 3.8$ (iii) y where the line crosses the y-axis

   (ii) x when $y = 5.6$ (iv) x where the line crosses the x-axis

---

Q4.15 Initially, at time, $t = 0$, the height, h, of a plant is 15 cm, and it then grows linearly over a 20-day period, reaching a height of 20 cm. Derive a straight-line equation which will give the height, h, as a function of the time, t, in days.

(i) What is the height when $t = 5$ days?

(ii) Estimate the height of the tree 8 days after the end of the 20-day period, assuming that it continues to grow at the same rate.

4.1.8 Angle of slope

The angle, θ, between the line and the x-axis (see Figure 4.2) is given by the equation:

$$\tan(\theta) = \frac{(y_2 - y_1)}{(x_2 - x_1)} = m \qquad [4.11]$$

The angle, θ, can then be calculated using the inverse tan function (2.3.7):

$$\theta = \tan^{-1}(m) \text{ or } \arctan(m) \qquad [4.12]$$

Q4.16 A map shows the height contours on the side of a hill of fairly uniform slope. A point on the 250-m contour line is seen to be 400 m horizontally from a point on the 200-m contour line.

(i) Calculate the average slope, m, between the two points.

(ii) Calculate the average slope angle of the ground with respect to the horizontal.

4.2 Linear regression

4.2.1 Introduction

There are very many situations in science where it is believed that a linear (straight-line) relationship exists between two variables. However, when experimental data is plotted, the data points on the graph show some random scatter away from any straight line. If the aim of the experiment is to measure the slope and/or intercept of the straight line, then the scientist has the problem of trying to interpret which straight line is the *best-fit* for the given data.

The process of *linear regression* is a powerful mathematical procedure that calculates the position of the 'best-fit' straight line. The procedure finds the position of the line that will minimize the sum of the squares of the residual differences between each point and the optimum straight line. For this reason, the procedure is often called the 'method of least squares'. The mathematics of the technique are covered more fully in 13.1.5.

It is fortunate that many software packages (including Excel) will perform the calculations necessary for linear regression, and report directly both the slope, m, and the intercept, c, of the best-fit line. Calculations by hand are not required.

This Unit is mainly concerned with the *practicalities*, using appropriate software, of performing a linear regression on scientific data.

From the point of view of experiment design, it is important to understand that the process of linear regression uses *information from every data point*. This means that the additional data points (beyond the minimum requirement of two) become *replicate* data points in the calculation of the parameters of the best-fit straight line and improve the accuracy of the overall fit.

4.2.2 Linear regression

Linear regression is a mathematical process for finding the *slope*, m, and *intercept*, c, of a 'best-fit' straight line. Example E4.8 illustrates the use of Excel to perform this process.

E4.8 💻

The following x–y data is entered into an Excel spreadsheet:

      x 0.4   0.8   1.2   1.6   2

      y 0.48   0.72   0.8   1.07   1.16

Using the Chart Wizard, it is easy to plot the graph as in Figure 4.4.

The best-fit straight line is drawn using the Trendline Option under the Chart Menu, which also produces a print-out of the equation of this line.

💻 Computer Techniques > Excel Tutorial: XY-Graph

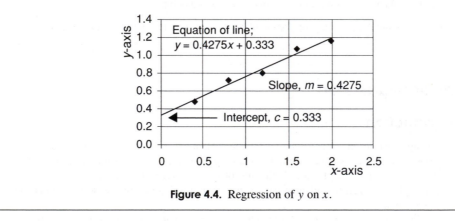

Figure 4.4. Regression of y on x.

The 'best-fit' straight line in Figure 4.4 is called the **line of regression of y on x**.

The process of linear regression assumes that the only *significant uncertainties are in the y direction*. The x-values for each data point are assumed to be *accurate*.

The line of regression of x on y (note reversal of x and y) will usually give a slightly different line.

Q4.17   The heart rates of athletes were measured 3 minutes after completing a series of 'step-ups' at different rates. If a process of linear regression is to be used to analyse the relationship between step-up rate and heart rate, decide which variable should be plotted on the y-axis and which variable on the x-axis.

4.2.3 Regression and correlation

It is important to know the difference between *regression* and *correlation*.

Correlation is a term that is used to indicate that the change in one variable is *related to* the change in a second variable. For example, the probability of developing lung cancer increases with the number of cigarettes smoked – the probability and number are known to be *correlated*.

Linear correlation is a specific measure of the extent to which the values of two variables are known to change in a way that could be described by a *straight-line equation*. The use of linear correlation as a statistical test is introduced in Unit 10.3.

Linear regression is a process that can be used to calculate the *slope* and *intercept* of a best-fit linear equation *after* it has been shown that two variables are indeed linearly correlated.

Q4.18 In each of the following situations, one variable is plotted against another, but it is necessary to decide whether to *test for correlation* between the variables or carry out a *regression analysis*. (The process of carrying out a test for correlation is given in Unit 10.3)

  (i)   Aluminium levels are recorded against sizes of fish pop-   Correlation/Regression
        ulations to investigate whether the level of aluminium in
        river water affects the size of the fish population.

  (ii) The increase in oxygen consumption of a species of small   Correlation/Regression
        mammal is measured as a function of body weight, in
        order to derive an equation that relates the two factors.

  (iii) A researcher records the abundance of heather plants as   Correlation/Regression
        a function of soil pH in different areas to see if the pH
        has an effect on the survival of the plant.

  (iv) Student marks are recorded for both coursework and ex-   Correlation/Regression
        amination to see if good students generally perform well
        in both and weaker students perform badly in both.

4.2.4 Software calculations

Excel and other statistical software can be used to calculate both the slope and intercept of the 'best-fit' line of regression through a set of data points.

Excel Functions, *fx* :

- SLOPE – calculates the best-fit slope (with no restriction on the intercept)

- INTERCEPT – calculates the best-fit intercept.

(Functions in Excel give *dynamic* results – they change if the input values change.)

Excel Data Analysis Tools: Tools > Data Analysis > Regression:

- Calculates slope and intercept (plus other data), and can force the intercept to be zero ($c = 0$).

(Data analysis tools in Excel are *non-dynamic*, 'once off', calculations – see Appendix I.)

Excel Trendline: Chart Menu > Add Trendline

- Calculates slope and intercept and can 'force' the intercept to be any chosen value.

The slope and intercept can be displayed on the chart but not entered into cells for use in calculations. It may be necessary to format the Trendline in order to show an appropriate number of significant figures for these values.

💻 Computer Techniques > Excel Tutorial: XY-Graph

Other statistics software packages will produce similar results. For example, MiniTab produces results for slope, m, and intercept, c, in the 'equation' form:

$$C2 = c + mC1$$

where C1 and C2 are the columns in MiniTab that hold the x and y data respectively.

💻 Computer Techniques > Statistics Software

E4.9 💻

In an experiment to measure the relationship between the variables P and Q, the values of P were measured for specific values of Q, and the data entered into rows 1 and 2 of an Excel spreadsheet as below:

	A	B	C	D	E	F
1	Q	1	3	5	7	9
2	P	3.9	7.2	7.9	12.5	13.6
3	Slope =	1.235	Intercept =	2.845		

The entry '= SLOPE(B2:F2,B1:F1)' into cell B3 of the Excel spreadsheet gives the slope of the best-fit line, $m = 1.235$.

The entry '= INTERCEPT(B2:F2,B1:F1)' into cell E3 of the Excel spreadsheet gives the intercept of the best-fit line, $c = 2.845$.

The two expressions for SLOPE and INTERCEPT assume that cells B2 to F2 hold the 'y-data' and cells B1 to F1 hold the 'x-data'.

Q4.19 Using the data from E4.9

(i)   Write down the equation of the line of regression, P on Q.

(ii) Plot the original data, plus the line of regression, on a graph.

(iii) Using the results from (i), calculate the value of P on the line of regression equivalent to a value of $Q = 3.4$.

Q4.20 The data in the table below gives the result of a spectrophotometric measurement, where the y-variable is the absorbance, A, and the x-variable is the concentration, C (mmol L^{-1}).

C	(x)	10	15	20	25
A	(y)	0.37	0.48	0.7	0.81

Use Excel and/or other statistics software, to perform a linear regression to obtain values for:

(i) Slope of 'best-fit' straight line.

(ii) Intercept of 'best-fit' straight line.

4.2.5 Forcing zero intercept

Sometimes it is known that the 'best fit' straight line should *pass through the origin* of the graph – i.e. have zero intercept, $c = 0$.

Excel can be used to perform a linear regression while forcing a zero intercept using any of the following methods:

- Function LINEST

- Excel Tools > Data Analysis > Regression

- Excel Trendline (can force intercept to be any chosen value)

In MiniTab it is possible to force a zero intercept by clicking 'Options. ...' in the Regression dialogue box and checking the 'Fit Intercept' option.

Q4.21 Using the same data as in Q4.20, make the assumption that the 'best-fit' straight line should pass through the origin of the A versus C graph. Calculate the slope of the line of regression with zero intercept using:

(i) Excel function LINEST

(ii) Excel Tools > Data Analysis > Regression

(iii) Excel Trendline

(iv) Other statistics software

4.2.6 Calibration line

The process of linear regression is frequently used to produce a best-fit **calibration line**. For example, spectrophotometric measurements often measure the absorbances, A, of solutions of known concentrations, C, and plot the 'best-fit' straight line of A against C. The linear relationship can then be used to calculate the concentration, C_O, of the unknown solution from a measured value of its absorbance, A_O.

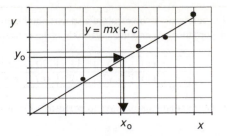

Figure 4.5. Regression of y against x.

A line of linear regression (Figure 4.5) is drawn using n data points of known values of y, corresponding to known values of x (calibration data). The coefficients of the line of regression are calculated:

$$\text{slope} = m$$

$$\text{intercept} = c$$

The aim of the experiment is then to estimate the unknown value, x_O, for a given value of y_O.

The 'best-estimate' x-value, x_O, of the unknown sample is then calculated by using the equation (see also [4.7]):

$$x_O = \frac{(y_O - c)}{m} \qquad\qquad [4.13]$$

E4.10

Using the data in E4.9 calculate the value of Q on the line of regression equivalent to a value of $P = 8.5$.

The equation of the line is: $P = 1.235 \times Q + 2.845$.

Rearranging this equation as in [4.13]:

$$Q = \frac{(P - c)}{m} = \frac{8.5 - 2.845}{1.235} = 4.579$$

For many calibration lines, the science of the problem is such that it is known that the calibration line must pass through the origin of the graph.

For example, in a spectrophotometric measurement it may be known that

- the absorbance of the solution obeys Beer's law ($A = k \times C$), and that

- the spectrophotometer was accurately set up so that a solution of zero concentration records an absorbance of zero,

In this case we can assume that the calibration line *must* pass through the origin of the graph of A against C. In this case, we can force the line of regression to pass through the origin and ensure that the intercept, $c = 0$.

Q4.22 Different masses w (in grams) of a sugar were dissolved in 100 mL of water and the angle of rotation θ (in degrees) of polarized light in the solution was measured in a polarimeter. The following results were obtained.

w (g)	1.5	3	4.5	6	7.5
θ (degrees)	1.3	2.7	4	5.2	6.6

(i) Assuming that pure water gives no rotation, find the best-fit equation for θ in terms of w.

(ii) Calculate the best-estimate of w for a sugar solution that gives a rotation of 4.9°.

4.2.7 Interpolation and extrapolation errors

The processes of interpolation and extrapolation were introduced in 4.1.7.

There is inherent uncertainty in the values for the slope, m, and intercept, c, calculated using linear regression. Hence there will also be uncertainty in the calculation of the value, x_O, when using [4.13].

The calculation for the experimental *uncertainty* in the final value for x_O is given in 13.1.7. From this calculation it is possible to draw some general conclusions.

The calibration data should be chosen such that the unknown test value falls within the central *half* of the calibration range. For example, if the calibration data covers absorbances from 0.2 to 0.6 (not forcing through origin), then, ideally, the absorbance of the test solution should fall within the range from 0.3 to 0.5. Provided that the test data falls within the central calibration range, then [13.31] is a good estimation of the overall uncertainty in the final value of x_O.

As the interpolated test values move towards the ends of the calibration range, then the uncertainties will increase – see Figure 13.3(a).

If the test values fall *outside* the range of calibration data (extrapolation) then the uncertainties will begin to increase very rapidly, and great care must be taken in using extrapolated values.

Good experimental design should take the above considerations into account when planning the choice of calibration values.

4.3 Linearization

4.3.1 Introduction

We have seen (Unit 4.2) how the mathematics of linear regression is a simple, yet powerful, procedure that can be used to identify a 'best-fit' straight line in a set of data that is expected to follow a linear relationship. However, many scientific systems do not yield simple linear relationships. What can be done about those?

There are statistical methods for fitting data to specific types of curves, i.e. *non-linear* regression. Some examples of the use in common statistics software are given on the web site:

🖥 Study Resources > Additional Material

However, these methods are not directly available in current versions of Excel and are beyond the scope of this book.

As an alternative to the use of *non-linear* regression, it is possible, in a number of specific situations, to manipulate the non-linear data so that it can be *represented* by a linear equation – this is the process of linearization. The transformed data can then be analysed using the familiar process of linear regression.

It is important to note (see 4.3.7) that the process of transforming the non-linear data can distort the significance of errors in the original data. Further advice should be sought in interpreting the possible errors that might arise in the regression results from a process of linearization.

Some of the equations and processes encountered in this Unit include logarithmic and exponential functions. An introduction to these functions is given in Chapter 5.

4.3.2 General principles

Linearization is a mathematical process whereby a non-linear equation (e.g., involving variables, P and Q) can be represented as a straight line on a suitable graph. The original equation must be manipulated in such a way that it appears in the form:

$$\mathrm{f}(P) = m \times \mathrm{f}(Q) + c \qquad\qquad\qquad [4.14]$$

where $\mathrm{f}(P)$ and $\mathrm{f}(Q)$ are functions of the original variables, P and Q, respectively.

The above equation is in the same form as the straight-line equation,

$$y = m \times x + c$$

and we can see that plotting $\mathrm{f}(P)$ against $\mathrm{f}(Q)$ (in place of y and x) will give a straight line with a slope, m, and an intercept, c.

There are three common ways of *linearizing* a non-linear function:

- Change of variable.

- Using natural logarithms for an exponential equation.

- Using logarithms (log or ln) to bring powers down onto the equation line.

⌨ Computer Techniques > Excel Tutorial: XY-Graph

4.3.3 Change of variable

Where possible, the simplest **linearization procedure** is to plot the functions of P and/or Q directly on the y- and x-axes of the graph.

E4.11 ⌨

The following data gives the area, A, of a circle as a function of its radius, r, and as a function of r^2:

A	0	0.79	3.14	7.07	12.57	19.63	28.27
r	0	0.5	1	1.5	2	2.5	3
r^2	0	0.25	1	2.25	4	6.25	9

Figure 4.6(a), shows the non-linear curve when we plot A directly against r. However, if we now plot A (as the y variable) against r^2 (as the x variable), we get a straight line that passes through the origin – Figure 4.6(b).

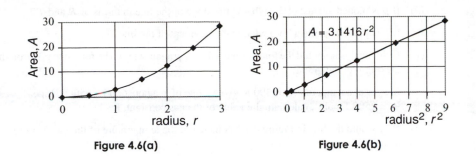

Figure 4.6(a)   **Figure 4.6(b)**

Compare the two equations,

$$A = \pi \times r^2$$

$$y = m \times x + c$$

If we plot A against r^2, we can see that the slope, m, will equal the constant π ($= 3.1416$), and that this equation has no intercept term, giving $c = 0$:

$$A = 3.1416 \times r^2$$

Q4.23 The relationship between the pressure, p (Pa), the volume, V (m³), and the absolute temperature, T (K), of one mole of an ideal gas is given by the equation:

$$p = RT \times \frac{1}{V}$$

where R is the gas constant.

The relationship between p and V, for constant T, is plotted on the two graphs below.

Figure 4.7(a) shows p against V and Figure 4.7(b) shows p against ($1/V$), both plotted for constant temperature, T.

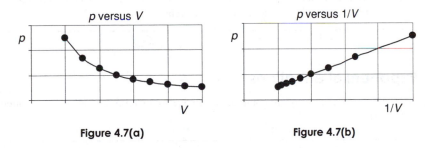

Figure 4.7(a)   **Figure 4.7(b)**

(i) Is p a linear function of V? – see Figure 4.7(a)

(ii) Is p a linear function of ($1/V$)? – see Figure 4.7(b)

(iii) If p is plotted against $(1/V)$ what is the slope of the line in terms of R and T?

(iv) If p is plotted against $(1/V)$ what is the intercept of the line?

(v) In this analysis, should the line of regression of p against $(1/V)$ be forced to go through the origin of the graph?

(vi) If a gas at a temperature of 300 K gives a line of p against $(1/V)$ with a slope, $m = 2500$ J mol^{-1}, calculate the value of the gas constant, R.

(vii) How would the line in Figure 4.7(b) change, if the temperature of the gas were to increase?

Q4.24 The Michaelis–Menten equation gives the initial velocity of an enzyme reaction, v, as a function of the substrate concentration, S,

$$v = \frac{v_{max} S}{K_M + S}$$

where K_M is the Michaelis–Menten constant and v_{max} would be the maximum reaction velocity for very large values of S.

Taking the reciprocal of both sides of the equation gives:

$$\frac{1}{v} = \frac{K_M + S}{v_{max} S} = \frac{K_M}{v_{max} S} + \frac{S}{v_{max} S}$$

which rearranges to:

$$\frac{1}{v} = \left(\frac{K_M}{v_{max}}\right) \times \frac{1}{S} + \frac{1}{v_{max}}$$

(i) The variables in the above equation are v and S. Compare the above equation to the straight-line equation: $y = mx + c$, and decide what function should be plotted on the y-axis and what function on the x-axis.

(ii) Having decided to plot the data as in (i), show how each of the constants K_M and v_{max} can be calculated from the slope and intercept.

4.3.4 Exponential equations

Equations involving growth or decay are common in science (see Unit 5.2), and they very often produce equations of the form:

$$P = A\, e^{kQ} \quad \text{or} \quad P = A\, \exp(kQ) \qquad [4.15]$$

where A and k are constants.

The **linearization procedure** here is to take the *natural logs* (log$_e$ or ln) of both sides of the equation, giving:

$$\ln(P) = \ln(A\ e^{kQ})$$

then (see 5.1.5)

$$\ln(Ae^{kQ}) = \ln(A) + \ln(e^{kQ}) = \ln(A) + kQ$$

giving

$$\ln(P) = k \times Q + \ln(A) \qquad\qquad [4.16]$$

which is of the same form as:

$$y = m \times x + c$$

with $y = \ln(P)$, $x = Q$, $m = k$, and $c = \ln(A)$.

Plotting $\ln(P)$ against Q will then give a straight line with slope, $m = k$, and intercept, $c = \ln(A)$.

$$k \text{ can be derived } directly \text{ from the slope: } k = m \qquad\qquad [4.17]$$
$$A \text{ can be } calculated \text{ from the intercept, c, using } A = e^{C} = \exp(c) \qquad\qquad [4.18]$$

If the data is plotted using *logs to base ten*, i.e. $\log(P)$ against Q, then the data will still give a straight line, but with different coefficients:

$$\text{slope, } m = k/2.303 \text{ and intercept, } c = \log(A)$$

giving:

$$k = 2.303 \times m \text{ and } A = 10^{C}$$

Q4.25 Most pharmokinetic processes are described by first-order kinetics. The drug concentration, C, in the body falls with time, t, according to the equation:

$$C = C_O \exp(-Kt)$$

where K is the elimination constant, and C_O the concentration at zero time.

In an experimental measurement, values of concentration, C, are measured as a function of time, t (hours), and a graph of $\ln(C)$ versus t is plotted. A regression analysis of the data gives:

$$\text{slope} = -0.61$$
$$\text{intercept} = 4.1$$

(i) Calculate the value of the elimination constant, K.

(ii) Calculate the value of the initial concentration, C_O.

> **Q4.26**   In the following equation, identify the functions that would be used for each axis to plot the
> equation as a straight line, and identify how the unknown constants can be determined from
> the slope and intercept of the regression line. Hint: treat $1/T$ as a variable.
>
> $$k = A \exp\left(-\frac{E}{RT}\right)$$
>
> k and T are the variables, R is the gas constant, A and E are unknown constants.

4.3.5   Equations with a variable as a power

One, or both, of the variables P and/or Q may appear as a (non-exponential) power, e.g.

$$P = A \times B^Q \tag{4.19}$$

(compare [4.19] with [4.23] to understand the difference between the equations).

The **linearization procedure** here is to take logarithms (log *or* ln) of both sides.
  Using natural logs we get

$$\ln(P) = \ln(A) + Q \times \ln(B)$$

giving

$$\ln(P) = \ln(B) \times Q + \ln(A) \tag{4.20}$$

which is of the same form as:

$$y = \mathrm{m} \times x + c$$

Plotting $\ln(P)$ against Q will then give a straight line with slope, $m = \ln(B)$, and intercept, $c = \ln(A)$:

$$B \text{ can be } \textit{calculated} \text{ from the slope: } B = e^m = \exp(m)$$
$$\tag{4.21}$$
$$A \text{ can be } \textit{calculated} \text{ from the intercept: } A = e^C = \exp(c)$$

Plotting $\log(P)$ against Q will also give a straight line with slope, $m = \log(B)$, and intercept, $c = \log(A)$:

$$B \text{ can be } \textit{calculated} \text{ from the slope: } B = 10^m$$
$$\tag{4.22}$$
$$A \text{ can be } \textit{calculated} \text{ from the intercept: } A = 10^C$$

> **Q4.27** For each of the following equations, identify the functions that would be used for each axis to plot the equation as a straight line. In each case, identify how the unknown constants can be determined from the slope and intercept of the regression line.
>
> (i) $W = K \times T^h$ W and h are the variables and T and K are unknown constants
>
> (ii) $N = N_O \, e^{kt}$ N and t are the variables and N_O and k are unknown constants

4.3.6 Equations with an unknown numeric power

In some experimental situations it may be believed that the system is described by an equation of the form:

$$P = A \times Q^B \qquad\qquad [4.23]$$

where the value of the power, B, is not known and needs to be *calculated* from the experimental results. (Compare [4.23] and [4.19] to understand the difference between the equations.)

The **linearization procedure** here is to take logarithms (log *or* ln) of both sides. Using natural logs we get

$$\ln(P) = \ln(A) + B \times \ln(Q)$$

giving

$$\ln(P) = B \times \ln(Q) + \ln(A) \qquad\qquad [4.24]$$

which is of the same form as:

$$y = m \times x + c$$

Plotting $\ln(P)$ against $\ln(Q)$ will then give a straight line with slope, $m = B$, and intercept, $c = \ln(A)$.

B can be derived *directly* from the slope: $B = m$

$[4.25]$

A can be *calculated* from the intercept: $A = e^C = \exp(c)$

Plotting $\log(P)$ against $\log(Q)$ will also give a straight line with slope, $m = B$, but with intercept, $c = \log(A)$.

B can be derived *directly* from the slope: $B = m$

$[4.26]$

A can be *calculated* from the intercept: $A = 10^C$

Q4.28   For each of the following equations, identify the functions that would be used for each axis to plot the equation as a straight line. In each case, identify how the unknown power, 'x', can be determined from the regression line.

   (i) $V = kA^x$ V is the volume of an animal of surface area, A, and k is an unknown constant.

   (ii) $E = k(T + 273)^x$ E is the intensity of radiation emitted from an object at a temperature, T °C, and k is an unknown constant.

4.3.7 Error warning

When using a linearization technique for non-linear data, the transformation process will also act on the errors in the data points, and this can have the effect of distorting the importance of some of the points in the final regression process.

   It is always important to be careful about interpreting the accuracy of results obtained by a linearization process, and possibly seek further guidance, particularly when relying on the accuracy of an *extrapolated* value.

5

Logarithmic and Exponential Functions

Overview

Logarithmic and exponential functions in mathematics are particularly useful (and common) in science because they can be used as accurate models for many real-world processes.

The mathematics of these functions can be handled very easily, provided that a few basic rules are learnt. The first Unit, 'Mathematics of e, ln and log' concentrates on developing an understanding that will directly underpin the use of the functions in a scientific context.

The exponential function is particularly important in modelling growth or decay systems in all branches of science, e.g. elimination of a drug from the bloodstream, radioactive decay, bacterial growth, etc. Although these processes are all based on the same mathematics, the scientists in different areas have developed different ways of quantifying the processes, for example,

- '1 log' decay (or '2 log', '3 log', etc.) in drug elimination,

- 'generation time' in bacterial growth,

- 'time constant' in electronic circuits.

The second Unit, 'Exponential Growth and Decay', demonstrates that it is easy to compare numerical calculations based on these different historical methods.

We see in Unit 5.1 that a logarithm is essentially the inverse function of an exponential function. However, logarithms also have an important modelling role in their own right, particularly in systems where a *multiplication* of an input to a system produces an *additive* effect on the output, e.g. in decibel and pH scales.

The student will also often meet the use of logarithmic scales on graphs. These may be used to linearize an exponential function (Unit 4.3) or may simply be used to condense data which covers a wide data range.

5.1 Mathematics of e, ln and log

5.1.1 Introduction

The exponential constant, 'e', and logarithms are used extensively in science, but many students find the purpose of the exponential function and the mathematics of logarithms very confusing. However, the underlying 'rules' are quite simple and it is worthwhile spending some time getting a good understanding of the basic concepts.

5.1.2 Powers and bases

$$M^X \text{ is } pronounced \text{ '}M \text{ to the power of } X\text{'} \tag{5.1}$$

where M is the **base**, and X is the **exponent**, **power** or **index**.

In this Unit we are mainly concerned with the two most commonly used 'bases':

- **Base '10'** – this is useful because we use a decimal system of counting.

- **Base 'e'** (= 2.71828 ...) – this is useful because the number, e, has properties which simplify the mathematics for many problems of growth and decay in science.

The specific value, 2.71828..., is called Euler's number and denoted by 'e'.

Note that, in electronic format (calculators and computers), the 'to the power of' is often expressed using '^', and sometimes by using '**', for example, $8.0^{3.1}$ may be written as 8.0^3.1 or 8.0**3.1.

5.1.3 Properties of powers (exponents)

The properties of powers are the same for any base:

$$e^A \times e^B = e^{A+B} \qquad\qquad 10^A \times 10^B = 10^{A+B} \tag{5.2}$$

$$\frac{e^A}{e^B} = e^A \div e^B = e^{A-B} \qquad \frac{10^A}{10^B} = 10^A \div 10^B = 10^{A-B} \tag{5.3}$$

$$(e^A)^B = e^{A \times B} \qquad\qquad (10^A)^B = 10^{A \times B} \tag{5.4}$$

$$e^1 = e \qquad\qquad 10^1 = 10 \tag{5.5}$$

$$e^0 = 1 \qquad\qquad 10^0 = 1 \tag{5.6}$$

Q5.1 Simplify each of the following:

(e.g. $e^2 \times e^3 = e^5$)

  (i) $10^2 \times 10^3$ (v) $e^2 \times e^0$

  (ii) $(10^2)^3$ (vi) $e^3 \times e^2$

  (iii) $10^3 \div 10^2$ (vii) $e^5 \div e^3$

  (iv) $10^3 \div 10^{-2}$ (viii) $(e^2)^3$

5.1.4 Exponential functions, e^x, 10^x

In these notes we will sometimes write 'e^x' as '$\exp(x)$', for clarity; for example, '$\exp(kT)$' is clearer to read than 'e^{kT}'. Note also that the Excel function EXP(x) is used to calculate exponential value, e^x.

Using a calculator:

- e^x is calculated by using the 'e^x' key

- 10^x is calculated by using the '10^x' key

It is important to note that the 'EXP' key on a *calculator* is used to enter the *power of ten* in scientific notation (2.1.2), for example, pressing the keys 1.6 'EXP' 3 on a *calculator* will enter the value 1.6×10^3.

Using Excel:

- e^x is calculated by using the EXP function '=EXP(XX)'

- 10^x is calculated by typing in the formula '=10^XX'

where XX is either the value of 'x' entered directly, or the cell address where the data value of 'x' is held.

💻 Computer Techniques > Excel Tutorial: Skills

It is important to note the difference between the EXP button on a calculator and the EXP function in Excel.

Q5.2 Use a calculator and/or Excel to evaluate:

(i) 10^3 (iv) $e^{4.2}$

(ii) $10^{3.4}$ (v) e^0

(iii) 10^{-3} (vi) e^1

5.1.5 Logarithms

Logarithms are *defined* so that they have the following key properties:

$$
\begin{array}{ll}
\text{If } e^A = B & \text{If } 10^A = B \\
\text{then} & \text{then} \\
A = \log_e(B) = \ln(B) & A = \log_{10}(B) = \log(B)
\end{array}
\qquad [5.7]
$$

A logarithm performs the *inverse process* (see 3.4.3) to taking the power. On a calculator, 'e^x' and '10^x' usually appear as the *second functions* on the keys for 'ln' and 'log' respectively.

$\log_{10}(x)$ is the logarithm of x to base '10' – often written simply as $\log(x)$

$\log_e(x)$ is the logarithm of x to base 'e' – often written simply as $\ln(x)$

$\log_e(x)$, or $\ln(x)$, is also called the **natural**, or Naperian, logarithm.

The values for $\log(x)$ and $\ln(x)$ for any *positive* value of x can be found directly by using the 'log' and 'ln' keys on a calculator, or by using the functions LOG or LN in Excel. It is not possible to calculate the logarithm of a *negative* number.

Q5.3 Using a calculator and/or Excel, evaluate:

(i) log(348)   (vi) log(0.5)

(ii) log(34.8)   (vii) log(2)

(iii) log(3.48)   (viii) log(20)

(iv) log(100)   (ix) $\ln(e^1)$

(v) log(0.01)   (x) ln(10)

We can use the definition of logarithms given in [5.7] to solve simple equations.

Q5.4 Calculate x in the following by using [5.7]. For example, if $e^x = 18$, then $x = \ln(18) = 2.89$

(i) $e^x = 22$ (iv) $10^x = 18$

(ii) $e^{3x} = 22$ (v) $10^{2x} = 18$

(iii) $2e^{3x} = 22$ (vi) $1.2 \times 10^{2x} = 18$

Some important *values* for logarithms include:

$$\log_e(e) = \ln(e) = 1 \qquad \log_{10}(10) = \log(10) = 1 \qquad [5.8]$$
$$\log_e(1) = \ln(1) = 0 \qquad \log_{10}(1) = \log(1) = 0 \qquad [5.9]$$
$$\log_e(0) = \ln(0) = -\infty \qquad \log_{10}(0) = \log(0) = -\infty \qquad [5.10]$$

Other important relationships for logarithms include:

$$\mathbf{\ln(e^A) = A \qquad \log(10^A) = A} \qquad [5.11]$$

E5.1 🖳

$$\ln(e^{kT}) = kT \qquad \log(1000) = \log(10^3) = 3 \qquad \log(0.01) = \log(10^{-2}) = -2$$

$$\mathbf{\ln(p^q) = q \times \ln(p) \qquad \log(p^q) = q \times \log(p)} \qquad [5.12]$$

E5.2 🖳

$$\ln(8^3) = 3 \times \ln(8) = 6.238\ldots \qquad \log(8^3) = 3 \times \log(8) = 2.709\ldots$$

$$\mathbf{\ln(pq) = \ln(p) + \ln(q) \qquad \log(pq) = \log(p) + \log(q)} \qquad [5.13]$$

E5.3 🖳

$$\ln(3 \times 8) = \ln(3) + \ln(8) \qquad \log(200) = \log(2 \times 100)$$

$$= 1.099 + 2.079 \qquad = \log(2) + \log(100)$$

$$= 3.178 \qquad = 0.301 + 2 = 2.301$$

$$\ln(p/q) = \ln(p) - \ln(q) \qquad \log(p/q) = \log(p) - \log(q) \qquad \text{[5.14]}$$

E5.4 🖳

$$\ln(3/8) = \ln(3) - \ln(8) \qquad \log(0.02) = \log(2/100)$$

$$= 1.099 - 2.079 \qquad = \log(2) - \log(100)$$

$$= -0.98 \qquad = 0.301 - 2 = -1.699$$

$$\log(2) \approx 0.30 \qquad \text{[5.15]}$$

$$\ln(x) = 2.30 \times \log(x) \qquad \text{[5.16]}$$

This equation is true (to 3 sf) for any value of x.

Q5.5 Use equations [5.8] to [5.16] to evaluate the following expressions WITHOUT using a calculator.

(i) $\log(10^{-0.3})$ Hint: $\log(10^A) = A$

(ii) $\ln(e^{0.62})$ Hint: $\ln(e^A) = A$

(iii) $\log(2)$ Hint: use an approximate value

(iv) $\log(20)$ Hint: $\log(20) = \log(2 \times 10) = \log(2) + \log(10)$

(v) $\log(200)$ Hint: $\log(200) = \log(2 \times 100) = \log(2) + \log(100)$

(vi) $\log(0.5)$ Hint: $\log(0.5) = \log(1 \div 2) = \log(1) - \log(2)$

(vii) $\ln(1000)$ Hint: $\ln(x) = 2.30 \times \log(x)$

(viii) $\ln(2)$ Hint: use value of $\log(2)$ from (iii) and $\ln(x) = 2.30 \times \log(x)$

(ix) $\log(2^{3.1})$ Hint: use value of $\log(2)$ from (iii) and $\log(p^q) = q \times \log(p)$

(x) $\ln(0.407 \times 7.39)$ Hint: $\ln(0.407) = -0.90$ and $\ln(7.39) = 2.0$

5.1.6 Solving 'power' equations with logarithms

In solving equations, we need to make the unknown value (e.g. 'x') the subject of the equation. When the variable x is in a power term, we need to bring it down onto the equation line by using [5.11] or [5.12].

Which type of logarithm we use depends on the 'base' of the 'power'.

E5.5 🖥

To solve equations of the form: $e^{2x} = 3.7$.

The 'base' is 'e' so we take *natural logs* (to base 'e') of both sides to get:

$$\ln(e^{2x}) = \ln(3.7)$$

Using [5.11]

$$\ln(e^{2x}) = 2x$$

Hence

$$2x = \ln(3.7) = 1.308\ldots$$
$$x = 0.654\ldots$$

E5.6 🖥

To solve equations of the form: $10^{2x} = 3.7$

The 'base' is '10' so we take *logs* (to base '10') of both sides to get:

$$\log(10^{2x}) = \log(3.7)$$

Using [5.11]

$$\log(10^{2x}) = 2x$$

Hence

$$2x = \log(3.7) = 0.568\ldots$$
$$x = 0.284\ldots$$

E5.7 🖥

To solve equations of the form: $8^{6x} = 0.6$

The 'base' is neither 'e' nor '10' so we can take either ln (base 'e') or log (base '10') of both sides.

For example, taking logs (base 10) of both sides to get:

$$\log(8^{6x}) = \log(0.6)$$

which gives (using [5.12])

$$6x \times \log(8) = \log(0.6)$$

giving

$$x = \log(0.6)/\{6 \times \log(8)\} = -0.2218/\{6 \times 0.903\} = -0.0409$$

Q5.6 Solve each of the following equations (i.e. find the value of 'p'):

(i) $e^p = 0.006$ (iv) $33.0 = 10^{-4p}$

(ii) $10^{4p} = 3.0$ (v) $4^p = 33$

(iii) $e^{-2p} = 0.006$ (vi) $0.3 = 0.2^p$

5.1.7 Logarithmic scales

Some systems in science are measured as a **ratio** of values. For example, the *decibel scale* of loudness, L, appropriate to the human ear, compares the power density, P, of the sound with the power density, $P_O = 1.0 \times 10^{-12}$ W m^{-2}, which is the quietest sound that can just be heard in normal hearing.

Loudness, $L(\mathrm{dB}) = 10\log(P/P_O) = 10\{\log(P) - \log(P_O)\}$ [5.17]

When comparing two sounds the *difference in loudness*, $L_1 - L_2$, will be given by:

$$L_1 - L_2 = 10\{\log(P_1) - \log(P_O)\} - 10\{\log(P_2) - \log(P_O)\} = 10\{\log(P_1) - \log(P_2)\}$$

giving

Difference in loudness between two sounds: $L_1 - L_2 = 10\log(P_1/P_2)$ dB   [5.18]

The *difference* is 10 times the logarithm of the *ratio* of the power densities, P_1 and P_2. As P_1 and P_2 have the same units, the ratio has no units, and $10\log(P_1/P_2)$ is simply a number.

E5.8

If the power density of sound is doubled, $P_1/P_2 = 2$, calculate the increase in loudness.

Difference in loudness, $L_1 - L_2 = 10\log(2) = 10 \times 0.30 = 3.0$ dB

A *doubling* of power gives an addition of 3 dB.

Q5.7   A sound has an initial loudness of 70 dB. Calculate the new loudness using [5.18] if the power
density of the sound is

   (i) doubled (i.e. by a factor of 2)   (iv) halved

   (ii) increased by a factor of 8   (v) reduced by a factor of 8

   (iii) increased by a factor of 100   (vi) reduced by a factor of 100

Another important area where students will encounter the use of logarithms is in electrochemistry. The
acidity of a solution depends on the *concentration* of hydrogen ions [H^+], expressed in units of mol L^{-1}.
However, the acidity is normally measured using the logarithmic pH scale (to base 10), by taking the log of
the *numerical value* of [H^+]:

$$pH = -\log([H^+]) \hspace{3cm} [5.19]$$

Example E5.9 compares the logarithmic pH scale with the direct concentration scale.

E5.9

Values of [H^+] and pH for pure water and *example* values for a strong acid and a strong base:

	Strong acid (example)	Pure water	Strong base (example)
Hydrogen ion concentration: [H^+]/mol L^{-1}	0.1	1.0×10^{-7}	1.0×10^{-13}
pH value: $pH = -\log[H^+]$	$= -\log(0.1)$ $= 1.0$	$= -\log(1.0 \times 10^{-7})$ $= 7.0$	$= -\log(1.0 \times 10^{-13})$ $= 13.0$

The logarithmic pH scale can be considered to be a *linear scale* in 'acidity'; the *pH*-value increases for
decreasing acidity. The numbers, 13, 7, 1 are a far more convenient linear representation of 'acidity' than
the absolute scale of hydrogen ion concentration: $10^{-13}, 10^{-7}, 0.1$.

Q5.8   Using the values of [H^+] and *pH* given in E5.9 for pure water, and the examples for strong
acid and strong base, calculate the following:

   (i)   *Difference* in [H^+] between pure water and the strong base.

   (ii) *Difference* in [H^+] between the strong acid and pure water.

   (iii) Are the values in (i) and (ii) the same?

    (iv)   *Ratio* of $[H^+]$ values between the pure water and the strong base.

    (v)   *Ratio* of $[H^+]$ values between the strong acid and pure water.

    (vi)   Are the values in (iv) and (v) the same?

    (vii)   *Difference* in *pH* between the pure water and the strong base.

    (viii)   *Difference* in *pH* between the strong acid and pure water.

    (ix)   Are the values in (vii) and (viii) the same?

Question Q5.8 shows that, on the pH scale the *difference* between a strong acid and water $= 7 - 1 = 6$, which is the same as the *difference* between water and a strong base $= 13 - 7 = 6$. However, the concentration scale, $[H^+]$, is very *non*-linear. The *difference* in $[H^+]$ between a strong acid and water $= 0.1 - 1 \times 10^{-7} = 0.099\,999\,9$, which is *very different from* the *difference* between water and a strong base $= 1 \times 10^{-7} - 1 \times 10^{-13} = 0.000\,000\,099\,999\,9$.

Q5.9   (i)   The hydrogen ion concentration of a particular base is $[H^+] = 3.4 \times 10^{-9}$ mol L^{-1}. Calculate its pH-value.

      (ii)   The pH-value of a particular acid is 4.2. Calculate its hydrogen ion concentration in mol L^{-1}.

In spectrophotometric measurements, the absorbance, A, of a solution is related to its percentage transmittance, $T\%$, according to the equation:

$$A = -\log\left(\frac{T\%}{100}\right) \tag{5.20}$$

Q5.10   Using [5.20], calculate the missing values in the following table of equivalent values ($\infty =$ infinite absorbance):

Percentage transmittance, $T\%$	0		1	10	50	
Absorbance, A	∞	3		1		0

5.2 Exponential growth and decay

5.2.1 Introduction

There are many examples of exponential growth or decay systems in all branches of science, and different disciplines have devised different ways of quantifying very similar processes, e.g. half-life in radioactivity, the elimination constant for drug concentration, the amplification of a photomultiplier. However, in this Unit we see how many simple growth (or decay) systems can be represented, and analysed, by using one common exponential equation.

5.2.2 Exponential systems

There are many systems in science where the *future change* in the system depends on the *current* state of the system. For example:

- A population (e.g. bacterial cells in blood system) may increase by a constant *proportion* of the *current population*.

- In radioactive decay, the *rate of decrease* in radioactivity is proportional to the *current level* of radioactivity.

Q5.11　A poor (but consistent) gambler loses *exactly half* of his remaining money, M, every day. He starts with $M = £640$ at the beginning ($n = 0$) of the first day, and is down to £320 at the end of the first day ($n = 1$) and down to £160 at the end of the second day ($n = 2$). For *each value* of n in the table below calculate:

　(i)　how much money, M, is left by *halving the previous value*

　(ii)　the value of M, using the equation $M = 640 \times \exp(-0.693n)$

	End of day $n =$	(0)	1	2	3	4	5	6	7
(i)	Money, $M(£) =$	640	320	160					
(ii)	$M = 640 \times \exp(-0.693n) =$	640							

The answer to Q5.11 shows that a *proportionate* change can be mathematically represented by an exponential equation. The *exponential* decay for Q5.11 is reproduced in Figure 5.1.

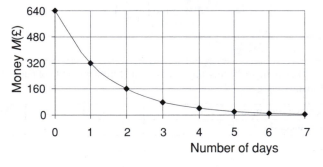

Figure 5.1.　Exponential decay.

5.2.3 Mathematics of growth and decay

Assume a population increases by a **factor**, g, within a given time period. For example,

- $g = 1.5$ gives a population *increase* of 50% in each period, and

- $g = 0.9$ gives a population *drop* of 10% in each period.

The population at the start ($n = 0$) is A_O.
In the nth period, the population will change by a *factor*, g, from A_{n-1} to A_n:

$A_1 = A_O \times g$ $A_1 = A_O \times g^1$
$A_2 = A_1 \times g$ substituting for A_1 $A_2 = (A_O \times g) \times g$ $A_2 = A_O \times g^2$
$A_3 = A_2 \times g$ substituting for A_2 $A_3 = (A_O \times g^2) \times g$ $A_3 = A_O \times g^3$
                       giving, after n steps $A_n = (A_O \times g^{n-1}) \times g$ $A_4 = A_O \times g^n$

After n periods: the population, A_n, is described by:

$$A_n = A_O \times g^n \qquad\qquad [5.21]$$

Figure 5.2 shows the **exponential growth** of a population with $A_O = 100$ and $g = 1.5$ when the time period is 1 day.

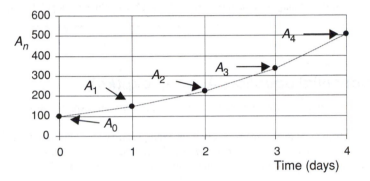

Figure 5.2. Exponential growth.

The growth is called 'exponential' because it is governed by the *exponent, n*.
   The equation, $A_n = A_O \times g^n$, can be used to describe the growth (decay) of *any exponential system* over a series of periods, n.
   For a *particular* system it would be necessary to choose appropriate values for each of the variables, A_O, g and n.

E5.10

The equation for radioactive decay (see 5.2.7) can be written as $A_t = A_O (1/2)^{t/T}$, where T is the half-life of the material and t is time. This equation is equivalent to the equation, $A_n = A_O \times g^n$, provided that $g = 0.5$ and $n = t/T =$ number of 'half-lives'.

Q5.12   In a 'chain letter', one person sends a letter to each of six people who are *each* 'encouraged' to forward the letter to six new people within one week. Assume that the chain is not broken and all six people actually forward the letter to *new* recipients.

(i) What is the 'period' of change in this example?

(ii) What is the value for g in this example?

(iii) Using an equation of the form: $A_n = A_O \times g^n$, estimate how many weeks, months or years it would take before at least 20 million people have become involved in the chain.

Q5.13 A bacterial culture in the 'death phase' has an initial population of 2.0×10^6 cells per mL, which decays exponentially by 50% in 0.4 hours.

(i) Calculate the population after 2 hours.

(ii) What is the value for g in this example?

(iii) Using an equation of the form: $A_n = A_O \times g^n$, calculate the population after 2.2 hours.

5.2.4 Exponential equation, $A_n = A_O \exp(kn)$

Growth and decay in different areas of science have produced different expressions of the basic equation, $A_n = A_O \times g^n$.

However, the problem with this form is that different examples may need different values for the *base*, g.

It is more *convenient* to use a *standard equation* that always uses the same 'base'. For reasons that may not be obvious at first, the best choice for a common 'base' is Euler's number, 'e':

$$A_n = A_O\, e^{(kn)} = A_O \exp(kn) \qquad [5.22]$$

The base, e, is used extensively because of its simplicity when used in the *differentiation* (Unit 6.2) of growth and decay equations.

If we write $g = e^k$ in [5.22] then:

$$A_n = A_O \times g^n = A_O \times (e^k)^n = A_O \times e^{kn} = A_O \times \exp(kn)$$

Hence equations [5.21] and [5.22] are equivalent, provided that:

$$g = e^k \qquad [5.23]$$

which is rearranged to give:

$$k = ln(g) \qquad [5.24]$$

E5.11 🖳

Equivalent values for g and k are calculated using [5.23] and [5.24], and are given in the table below.

Change in one 'period'	g	k	Applications
Increase by 10 times	10	2.303	
Increase by e = 2.718 times	2.718	1	
Increase by 2 times	2	0.693	Generation time: 5.2.6
No change	1	0	
Decrease to half of value	0.5	−0.693	Half-life decay: 5.2.7
Decrease to e^{-1} = 36.8% of value	0.368	−1	Time constant: 5.2.9
Decrease to tenth of value	0.1	−2.303	'log n' decay: 5.2.8

Note from E5.11 that, for an exponential *decay*, $g < 1$, and k will be *negative*.

Where the growth is measured as a *continuous* function of **time**, t, the equation would be rewritten (replacing n with t) as:

$$A_t = A_0 e^{(kt)} = A_0 \exp(kt) \qquad [5.25]$$

Note that k will have units of '1/time'.

E5.12 🖳

Describe the growth shown in Figure 5.2 using an equation of the form $A_n = A_0 \times e^{(kn)}$.

The growth factor $g = 1.5$ and $A_0 = 100$. We can calculate the equivalent value for k:

$$k = \ln(g) = \ln(1.5) = 0.405$$

Hence we can write

$$A_n = 100 \times \exp(0.405 \times n) = 100 \exp(0.405n) = 100\, e^{(0.405n)}$$

Q5.14 The exponential growth of a population, N, can be written as an equation:

$$N_t = N_0 e^{kt}$$

where, N_t is the population at time t days, and N_0 is the initial population at time $t = 0$ days. If $N_0 = 3500$ and $k = 0.02$ day^{-1}, calculate the population after

(i) 25 days   (iii) 75 days

(ii) 50 days

Q5.15 A disease is spreading exponentially, such that the number of cases is increasing by 10% every week. If there are 100 cases at the beginning of the first week (when $t = 0$), derive an equation of the form

$$N_t = N_O \exp(kt)$$

to give the number of cases, N_t, after a time, t, measured in weeks.

Calculate the values for N_O and k appropriate to this problem.

N.B. The units of k in this problem will be 'week^{-1}'

5.2.5 Substituting values to calculate 'k'

A common problem requires the calculation of the constant, k, when particular values of A_O, A_t and t are known. The solution is obtained by substituting the known values into [5.25] and then rearranging the equation to calculate the unknown, k.

E5.13 🖥

A bacterial population, initially with $N_O = 5.2 \times 10^5$ cells per mL, is decaying exponentially according to the equation $N_t = N_O \times e^{(kt)}$, and it is found that after $t = 2.0$ hours, the population has become $N_{2.0} = 3.5 \times 10^4$ cells per mL.

(i) Calculate the value of k in the equation.

(ii) Hence calculate the expected population after 3.0 hours.

(i) Substituting the known values for $t = 2.0$ into [5.25]

$$A_{2.0} = N_{2.0} = 3.5 \times 10^4 \quad \text{and} \quad A_O = N_O = 5.2 \times 10^5$$

giving:

$$3.5 \times 10^4 = 5.2 \times 10^5 \times \exp(k \times 2.0)$$

Dividing both sides by 5.2×10^5:

$$0.06731 = \exp(k \times 2.0)$$

Taking 'ln' of both sides:

$$-2.699 = k \times 2.0$$

which gives

$$k = -1.349$$

(ii) Substituting the value of N_O, k, and $t = 3.0$ we get:

$$N_{3.0} = 5.2 \times 10^5 \times \exp(-1.349 \times 3.0) = 9.09 \times 10^3$$

Q5.16 The population of a bacterial colony is initially measured as 450 and, after 10 hours, has grown to $N_{10} = 620$.

   (i) Assuming that the growth is exponential, find the values N_O and k that will be required to describe this growth when using the equation:

   $$N_t = N_O\, e^{kt}$$

   N.B. The units of k in this problem will be 'hour^{-1}'

   (ii) Estimate the population after 12 hours.

5.2.6 Generation time

The exponential growth of a bacterial population can be measured by the *generation time*, G, which is the time that the population takes to *double* in number.

The exponential growth of such systems is described by the equation:

$$N_t = N_O\,(2)^{t/G} \qquad\qquad [5.26]$$

After a time equal to the *generation time*, $t = G$, the number will increase to *double* the initial value:

$$N_G = N_O \times (2)^{G/G} = N_O \times (2)^1 = 2 \times N_O \qquad\qquad [5.27]$$

Comparing [5.26] with [5.21] we can see that g has the value:

$$g = 2$$

and from [5.24] and E5.11 we can see that the equivalent value of k is

$$k = 0.693$$

The equivalent 'exponential' equation for a 'generation' growth is:

$$N_t = N_O \exp\left(0.693 \times t/G\right) \qquad\qquad [5.28]$$

where t/G is also equal to the number of 'generation times' that have elapsed.
By rearranging [5.28], and taking natural logarithms of both sides:

$$\ln\left(\frac{N_t}{N_O}\right) = \ln\left\{\exp\left(0.693 \times t/G\right)\right\} = 0.693 \times \frac{t}{G}$$

which can be rearranged again to give:

$$G = 0.693 \times \frac{t}{\ln\left(N_t/N_O\right)} \quad \text{or} \quad G = 0.301 \times \frac{t}{\log\left(N_t/N_O\right)} \qquad\qquad [5.29]$$

Equation [5.29] allows the calculation of generation times, using either ln or log.

E5.14

(i) Calculate the generation time for a bacterial colony that increases from 2.0×10^4 cells to 1.7×10^6 in 180 minutes.

(ii) Express the growth of the colony in (i) using an 'exponential' equation.

Answer:

(i) Using [5.29] directly we can write:

$$G = 0.693 \times \frac{t}{\ln(N_t/N_O)} = 0.693 \times \frac{180}{\ln(1.7 \times 10^6/2.0 \times 10^4)} = 0.693 \times \frac{180}{4.443}$$

$G = 28$ minutes.

(ii) Substituting for G in [5.28] gives:

$$N_t = N_O \exp(0.693 \times t/G) = N_O \exp(0.0247 \times t)$$

Q5.17 Repeat the calculation in E5.14, but use logs instead of ln.

Q5.18 A bacterial colony has a generation time of 20 minutes. If the initial population is 1000 cells, calculate the population after

(i) 20 minutes   (iii) 10 minutes

(ii) 60 minutes   (iv) 45 minutes

5.2.7 Half-life decay

The exponential decay of a number of different systems (e.g. radioactivity, drug metabolism) are defined using the time (half-life), T, that it takes an activity, A_t, to decay to half ($= A_O/2$), of its initial value, A_O. The exponential decay of such systems is described by the equation:

$$A_t = A_O \left(\frac{1}{2}\right)^{t/T} \qquad\qquad [5.30]$$

After a time equal to the half-life, $t = T$ (one *half-life*), the activity will fall to *one half* of the initial value:

$$A_T = A_O(1/2)^{T/T} = A_O(1/2) = A_O/2 \qquad\qquad [5.31]$$

Comparing [5.30] with [5.21] we can see that g has the value:

$$g = (1/2) = 0.5$$

and from [5.24] and E5.11 we can see that the equivalent value of k is

$$k = -0.693$$

The equivalent 'exponential' equation for a 'half-life' decay is:

$$A_t = A_O \exp(-0.693 \times t/T) \qquad\qquad [5.32]$$

where the value (t/T) is also equal to the 'number of half-lives' that have elapsed.

E5.15 🖳

A radioactive isotope decays with a half-life of 30 days, calculate the proportion of activity remaining after 100 days.

Using [5.32] we can write directly:
$$A_t = A_O \exp(-0.693 \times t/30) = A_O \exp(-0.0231 \times t)$$
where t is the time in days. After 100 days:
$$A_{100} = A_O \exp(-0.0231 \times 100) = A_O \times 0.0993$$
Hence the *proportion* remaining:
$$A_{100}/A_O = 0.0993$$

Q5.19 The half-life of a particular radioactive isotope is 3 days.

Calculate the *proportion* of the original isotope left after

(i) 3 days   (iii) 9 days

(ii) 6 days

Q5.20 The activity of an unknown radioactive isotope, X, is found to decay to one tenth of its initial activity after a period of 26 hours.

Calculate the half-life of the isotope.

5.2.8 Decimal reduction time

The decay of a bacterial population can be described by the time taken for the population to drop by a *factor* of 10.

The exponential decay of such systems is described by the equation:

$$N_t = N_o \left(\frac{1}{10}\right)^{t/T} \qquad\qquad [5.33]$$

After a time, t, equal to the 'decimal reduction time', T, the activity will fall to *one tenth* of the initial value: Comparing [5.33] with [5.21] we can see that g has the value:

$$g = (1/10) = 0.1$$

and from [5.24] and E5.11 we can see that the equivalent value of k is

$$k = -2.303$$

The equivalent 'exponential' equation for decimal reduction time is:

$$N_t = N_O \exp(-2.303 \times t/T) \qquad\qquad [5.34]$$

E5.16 ⌨

A bacterial population has a decimal reduction time of $T = 20$ s.

If the initial population is 3.0×10^5, calculate the population after

(i)   20 s

(ii)   40 s

(iii)   30 s

Answer:

(i) $t = 20$ s is equal to one unit of 'decimal reduction time', thus the population will have fallen to $1/10$ of the initial value,

$$N_{t=20} = 3.0 \times 10^5/10 = 3.0 \times 10^4$$

(ii) $t = 40$ s is equal to two units of 'decimal reduction time', thus the population will have fallen to $(1/10) \times (1/10)$ of the initial value,

$$N_{t=40} = 3.0 \times 10^5/100 = 3.0 \times 10^3$$

(iii) $t = 30$ s is not an integer number of 'decimal reduction times', and it is necessary to use [5.34]:

$$N_{t=30} = 3.0 \times 10^5 \times \exp(-2.303 \times 30/20) = 9.48 \times 10^3$$

5.2.9 Decay time constant

If an exponential decay in time is written as

$$A_t = A_O e^{-t/\tau} \qquad\qquad [5.35]$$

then τ(tau) is called the *time constant*. When the time, t, equals the time constant, τ, the value, A, has dropped to:

$$A_\tau = A_O e^{-\tau/\tau} = A_O e^{-1} = 0.368 A_O$$

i.e. after one 'time constant', the value drops to 36.8% of the initial value.

Q5.21 The time constant, τ, of a capacitor of capacitance, C, which discharges through a resistor of resistance, R, is given by

$$\tau = CR$$

Calculate the time taken for a capacitor to discharge to 1% of its initial charge, given that $C = 0.01 \times 10^{-6}$ F and $R = 330 \times 10^3\ \Omega$.

A decay equation can 'decay' to a *non-zero* value. In this case the rate of change of the variable slows down as it approaches the final value. An example is illustrated by Q5.22 in which the value, V_t, approaches V_O with a time constant, T.

Q5.22 The following equation represents a form of growth that reaches a maximum saturation level, V_O.

$$V = V_O \times \left\{ 1 - \exp\left(-\frac{t}{T}\right) \right\}$$

(i) For $V_O = 10.0$ and $T = 1.5$ s, plot the behaviour of V_t on a graph against values of time, $t = 0, 1, 2, 3,$ and 4.

(ii) From the graph, *estimate* values for the ratio (V_t / V_O) when $t = T$ and when $t = 2T$.

(iii) Use the equation to *calculate* the values for the ratio (V_t / V_O) when $t = T$ and when $t = 2T$.

6

Rates of Change

Overview

Nature is not static, and much interesting science is concerned with how systems change.

We start the first Unit with familiar calculations of average speed, by recording change of position within a finite time interval. The calculation of speed is related to the slope of a line on a *distance* against *time* graph. We then develop the more general case for calculating speed at an instant, when the speed is represented by a continuously varying curve on a graph. This 'rate of change' mathematics is also applied to other systems.

Simple examples in Excel are used to demonstrate how continuously varying data can be *modelled* in a suitable spreadsheet format. The calculations for *approximate* models are relatively simple, but it is often possible to develop solutions to the required degree of accuracy.

The second Unit introduces the concept of 'differentiation' and gives a brief introduction to some ideas of calculus. The analytical mathematics (calculus) of differentiation is beyond the scope of this book, but the web site develops some examples in the use of Excel to model both differentiation and the associated topic of integration.

⌨ Study Resources > Additional Material

6.1 Rate of change

6.1.1 Introduction

The most familiar 'rate of change' is probably *speed*. We express speed, v, as a rate of change of distance, s, with time, t, e.g. 'miles per hour' or 'metres per second'. However, 'rates of change' are also found in all areas of science involving a variety of different variables.

We start with simple examples of average speeds within finite periods of time, and then progress to the more general case of a continuously varying $x-y$ relationship.

6.1.2 Rate of change with time

We can express a 'rate of change' with a simple equation:

$$\text{Average speed, } v = \frac{\text{Change of distance}}{\text{Change of time}} = \frac{\Delta s}{\Delta t} \qquad [6.1]$$

Note that in this context:

Change = Final value − Initial value

The Greek symbol, Δ (capital delta), is often used to denote a 'difference' in value.

E6.1 💻

A motorcyclist is initially at a distance of 5000 m (5 km) along a road from a petrol station.

(i)   She drives *away* from the petrol station reaching a distance of 9000 m (9 km) after 200 s. Calculate her *average* speed in this period.

(ii) She continues to drive *away* from the petrol station, but at a faster speed, reaching a distance of 12 000 m (12 km) after a further 100 s. Calculate her new *average* speed in this period.

(iii) She then turns round and drives *towards* the petrol station, reaching it after a further 400 s. Calculate her average speed in this last period.

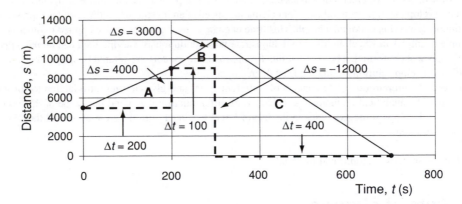

Figure 6.1. Distance–time graph for E6.1.

Figure 6.1 uses a *s–t* diagram to represent the distance along the road, *s*, of the motorcyclist in E6.1. The straight lines, which connect the points, assume that the motorcyclist travels at a constant speed in each section of the journey.

(i)   The first section of the journey is illustrated by triangle A in Figure 6.1, where:

Change in distance, $\Delta s = 9000 - 5000 = 4000$ m, and

Change in time, $\Delta t = 200$ s

The average speed in the first section is calculated using [6.1]:

$$v = \frac{\Delta s}{\Delta t} = \frac{4000}{200} = 20 \text{ m s}^{-1}$$

(ii) The second section of the journey is illustrated by triangle B in Figure 6.1, where the average speed is calculated using [6.1]:

$$v = \frac{\Delta s}{\Delta t} = \frac{12\,000 - 9000}{300 - 200} = \frac{3000}{100} = 30 \text{ m s}^{-1}$$

(iii)   In the third section of the journey, illustrated by the triangle C in Figure 6.1, the change in distance is *negative*, which gives a *negative* average speed

$$v = \frac{0 - 12\,000}{400} = \frac{-12\,000}{400} = -30 \text{ m s}^{-1}$$

The speed is *negative* in (iii) because the motorcyclist is travelling in a direction *opposite* to the direction in which the distance is being measured.

E6.2 🖳

Calculate the slopes of the straight line graph for each of the three triangular sections in the graph in Figure 6.1.

The slope, m, of the straight lines are calculated using [4.8], giving:

$$\text{Triangle A: } m = \frac{9000 - 5000}{200 - 0} = \frac{4000}{200} = 20 \text{ m s}^{-1}$$

$$\text{Triangle B: } m = \frac{12\,000 - 9000}{300 - 200} = \frac{3000}{100} = 30 \text{ m s}^{-1}$$

$$\text{Triangle C: } m = \frac{0 - 12\,000}{400} = \frac{-12\,000}{400} = -30 \text{ m s}^{-1}$$

Comparing the results in E6.1 and those in E6.2, it can be seen that:

Rate of change of s with t equals the **Slope of the graph** of s against t

---

Q6.1   Calculate the missing values in the table below:

Initial distance (m)	Final distance (m)	Initial time (s)	Final time (s)	Speed (m s^{-1})
2000	4000	100	150	40
2000		100	200	30
2000	6000		400	40
2000		100	200	−40
5000	3000	0	200	

6.1.3   Modelling a continuous curve

In E6.1 the data is given at *discrete intervals*. We calculate an average speed between the points by assuming a constant (straight line) speed between those points.

For a curve with a constantly changing slope we can still measure changes over finite intervals to produce an *approximate* model for the variation in slope.

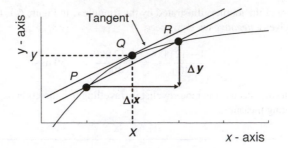

Figure 6.2. Approximation for rate of change.

For example, in Figure 6.2 the slope at the point Q can be found by taking the differences, Δy and Δx, between two points, P and R, on either side of Q.

In Figure 6.2, the slope at Q, m_Q, is given approximately by

$$m_Q \approx \frac{\Delta y}{\Delta x} \qquad\qquad [6.2]$$

If the curve is described by the co-ordinates of a series of points, then the slope of the curve can be calculated approximately by repeating the calculation in [6.2] between successive pairs of data points. This form of calculation can be *modelled* conveniently using Excel.

Example E6.3 illustrates the use of repeated calculations to *model* the change in surface gradient on a path that crosses a hill peak.

E6.3 💻

Figure 6.3 is a map of height contours around a hill peak. Adjacent contours differ in height by 20 m, with the outer contour at a height, $h = 60$ m. A straight path, given as the x-axis (in km), passes directly over the hill peak.

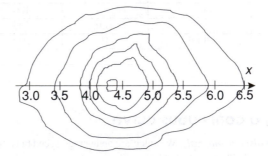

Figure 6.3. Height contours around a hill peak.

Calculate (approximately) the slope of the ground at the points where each contour (except the outer) crosses the x-axis.

See text for calculation.

The positions where each contour crosses the x-axis in E6.3 are given in Table 6.1, as an extract from an Excel worksheet.

Table 6.1. Heights along x-axis in E6.3.

	A	B	C	D	E	F	G	H	I	J	K	L	M
4	Height, h (m)	60	80	100	120	140	160	160	140	120	100	80	60
5	Distance, x (km)	2.9	3.4	3.8	4.0	4.1	4.3	4.4	4.9	5.2	5.4	5.9	6.5
6	Slope, m (m km^{-1})		44	67	133	133	67	−33	−50	−80	−57	−36	

The calculation of slope at the position where $x = 3.4$ km is achieved by using the data points on either side, $x = 2.9$ and $x = 3.8$, and entering appropriate differences into [6.2] (the results are formatted to integer values):

$$m_{x=3.4} \approx \frac{\Delta h}{\Delta x} = \frac{(100 - 60)}{(3.8 - 2.9)} = 44.4 \text{ m km}^{-1}$$

The entry in the Excel worksheet for cell C6 is '=(D4-B4)/(D5-B5)'.

The slopes at each subsequent point are calculated by repeating the above calculation, moving from point to point. This form of repetitive calculation is very easily performed in Excel.

Notice that the slope is positive leading up to the peak and then negative after the peak.

E6.4 🖳

The following data shows the results of a potentiometric titration, where the cell potential is recorded as more titrate is added.

Calculate the 'end-point' of the titration, where the *rate of change of cell potential is greatest.*

Volume added	V (cm^3)	8.0	8.5	9.0	9.2	9.4	9.6	9.8	10.0	10.2	10.4	10.6	10.8	11.0	11.5	12.0
Cell potential	E (mV)	154	161	173	178	188	197	211	228	237	248	255	260	265	269	273
Slope	$\dfrac{\Delta E}{\Delta V}$		19.0	24.3	37.5	47.5	57.5	77.5	65.0	50.0	45.0	30.0	25.0	12.9	8.0	

The rate of change (or slope) of the cell potential with volume is calculated using an equation derived from [6.2]. For example, the slope when volume added, $V = 8.5$ mL, is calculated:

$$\text{Slope (at } V = 8.5) \approx \frac{\Delta E}{\Delta V} = \frac{173 - 154}{9 - 8} = 19.0 \text{ mV cm}^{-3}$$

The graphs below show the cell potential and its slope plotted against the volume added:

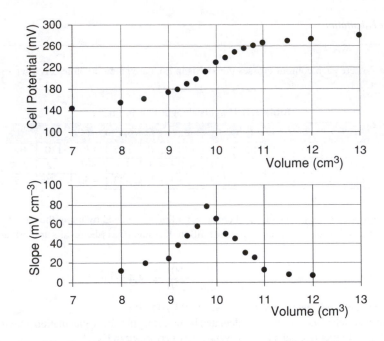

The volume, $V = 9.8$, where the slope is greatest, can be seen most easily from the graph of the **slope** of the curve, at the point where this graph reaches a maximum.

Q6.2 The data in the table below gives the speed of a car as a function of time, as it accelerates to overtake a slower vehicle.

Take **Acceleration = Rate of change of speed with time**, and use [6.2] to estimate the acceleration between the times 2 and 28 seconds inclusive, and plot your results on a suitable graph.

Time (s)	0	2	4	6	8	10	12	14	16	18	20	22	24	26	28	30
Speed (ms^{-1})	20	20	20	21	23	27	31	34	35	35	35	34	33	32	32	32
Acceleration (ms^{-2})																

6.1.4 Slope of a graph at a point

We can aim to measure the slope of the graph at *a particular point* by making the measurement interval *as small as possible*. By making the measurement interval 'infinitely' small, we can obtain a value for the slope of the curve at a point.

Figure 6.2 shows a continuous x–y curve with three points on the curve, P, Q and R. The points P and R are used to estimate the slope of the curve around Q, and are separated by small differences Δx and Δy in the x and y directions respectively

The **average** slope of the curve between P and R is given *approximately* by equation [6.2].

In Figure 6.2, a straight line drawn between the points P and R crosses the curve at the two points P and R. If we now bring P and R closer towards Q, we can imagine reaching a situation when the straight line *only just touches* the curve at the point Q.

The straight line that **just touches** a curve at a point Q, but *does not cross the curve* is called the **tangent to the curve at the point Q.**

When we bring P and R closer towards Q, and make Δx approach zero we rewrite the ratio, $\Delta y/\Delta x$, as dy/dx, and the slope of the curve at Q is given by:

$$\text{Limit}_{\Delta x \to 0} \frac{\Delta y}{\Delta x} \Rightarrow \frac{dy}{dx} \qquad\qquad [6.3]$$

dy/dx is called the **differential coefficient, derivative**, or **slope at the point**.

The value of the differential coefficient (or slope), dy/dx, will vary from point to point along a curve.

The rate of change with time is sometimes written with a 'dot' over the variable. For example the cardiac output in terms of volume, V, of millilitres of blood per minute can be written as

$$\dot{V} = \frac{dV}{dt} \text{ mL min}^{-1}.$$

Q6.3 The following graph shows the population, N_t, of a bacterial colony as a function of time, t.

The population at time, $t = 60$ min, is $N = 37.85 \times 10^4$, and the straight line is the tangent to the graph at that point.

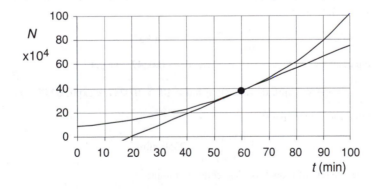

(i)   Using the drawn tangent, estimate the rate of growth of the colony at the time, $t = 60$ min.

(ii) Draw a tangent to the curve at time, $t = 90$ min, and hence estimate the rate of growth of the colony at the time, $t = 90$ min.

6.2 Differentiation

6.2.1 Introduction

The concept of the differential coefficient was introduced in 6.1.4 as being the slope of a curve at a single point. In this Unit we give some examples of equations that describe the slope of a function at any given point. These equations were derived using the mathematical processes of *calculus*. However, it is beyond the scope of this book to develop this advanced form of mathematics, but we include this section as a brief introduction into its power and uses.

We also see in 6.2.2 that an important property of Euler's number, e, is that the slope (or differential coefficient) of the curve $y = e^x$ is also equal to e^x. This is a unique property that only applies to the number $e = 2.718\,28\ldots$, and is the reason that 'e' is used so often in the mathematics of growth and decay.

6.2.2 Differentiation

Differentiation is the mathematical process (in calculus) for calculating the differential coefficient, dy/dx, *directly from the equation* that relates y to x.

The following Example E6.5 gives an illustration of the use of differentiation.

E6.5

The number, N_t, in a bacterial colony with a generation time of 28 minutes can be described [5.28] using the following equation, where t is given in minutes:

$$N_t = N_O \exp(0.0247 \times t)$$

If the initial colony size is $N_O = 8.6 \times 10^4$ (when $t = 0$), calculate the rate of growth (in cells per minute) when $t = 60$ minutes.

We will see from Table 6.2 that the differential coefficient of the general exponential equation, $y = A \exp(Bx) = Ae^{Bx}$, is given by

$$\frac{dy}{dx} = \frac{d}{dx}(Ae^{Bx}) = ABe^{Bx}$$

By direct comparison ($y \to N_t$, $x \to t$, $A \to N_O$, $B \to 0.0247$), we get

$$\frac{dN_t}{dt} = \frac{d}{dt}(N_O \exp(0.0247 \times t)) = 8.6 \times 10^4 \times 0.0247 \times \exp(0.0247 \times t)$$

Then substituting for $t = 60$ gives

$$\left(\frac{dN_t}{dt}\right)_{t=60} = 9350 \text{ min}^{-1}$$

i.e. when $t = 60$, the rate of growth is 9350 cells per minute.

In these notes, we only give a very brief introduction to the type of calculations in differentiation that can be performed using calculus. We give examples in Table 6.2 of some common forms of *equations*, together with the *differential coefficient* of those functions.

Note that the differential coefficient, $\dfrac{dy}{dx}$, can also be written as $\dfrac{d}{dx}(y)$.

Table 6.2. Differentials of common equations (c, m, A and B are constants)

Equation	Differential coefficient	Comment
$y = c$	$\dfrac{dy}{dx} = \dfrac{d}{dx}(c) = 0$	The rate of change of a constant is **zero**.
$y = mx$	$\dfrac{dy}{dx} = \dfrac{d}{dx}(mx) = m$	A straight line has a **constant slope**, m.
$y = mx + c$	$\dfrac{dy}{dx} = \dfrac{d}{dx}(mx) + \dfrac{d}{dx}(c) = m$	Each item in an 'addition' is differentiated separately. The slope of a straight line is independent of c.
$y = Ax^B$	$\dfrac{dy}{dx} = A \times B \times x^{B-1} = ABx^{B-1}$	This is a general case for the differentiation of a power.
$y = \exp(x) = e^x$	$\dfrac{dy}{dx} = \dfrac{d}{dx}(e^x) = e^x$	This is the special property of 'e' – it equals its own differential!
$y = A\,e^{Bx}$	$\dfrac{dy}{dx} = \dfrac{d}{dx}(A\,e^{Bx}) = AB\,e^{Bx}$	Note: power of 'e' does not change.

E6.6

The distance, z, travelled by a falling stone as a function of time, t (ignoring air resistance), is given by

$$z = 4.9 \times t^2$$

Given that Velocity = Rate of change of distance with time, calculate the velocity of the stone after 2 seconds.

We know that

$$\text{Velocity, } v = \text{rate of change of distance, } \frac{dz}{dt} = \frac{d}{dt}(4.9 \times t^2)$$

Comparing with the standard equation from Table 6.2

$$\frac{d}{dx}(Ax^B) = A \times B \times x^{B-1} = ABx^{B-1}$$

we can deduce, by putting $A = 4.9$, $B = 2$ and $x = t$, that

$$v = \frac{d}{dt}(4.9 \times t^2) = 4.9 \times 2 \times t^{2-1} = 9.8 \times t$$

When, $t = 2.0$, we can calculate that:

$$v = 9.8 \times 2.0 = 19.6 \text{ m s}^{-1}$$

Q6.4 The decay of a bacterial population is described by the equation:

$$N_t = 5.6 \times 10^8 \times \exp(-0.4 \times t)$$

where N_t is the population at a time t (s).

Calculate the number of bacteria dying per second at times:

(i) $t = 0$ s   (ii) $t = 5$ s   (iii) $t = 10$ s

(N.B. The differential coefficient will be negative as the numbers of bacteria are decreasing – see also E6.1(iii).)

6.2.3 Estimating changes

If the value of the differential coefficient can be calculated, it is then possible to estimate the effect of *small* changes, Δx and Δy, by using the approximation:

$$\frac{\Delta y}{\Delta x} \approx \frac{dy}{dx} \qquad\qquad [6.4]$$

which can be rearranged to give

$$\Delta y \approx \frac{dy}{dx} \times \Delta x \qquad\qquad [6.5]$$

E6.7

Using the results from E6.5, the rate of growth of a particular bacterial colony when $t = 60$ min is

$$\left(\frac{dN_t}{dt}\right)_{t=60} = 9350 \text{ min}^{-1}$$

Estimate the increase in colony size, ΔN, between $t = 58$ min and $t = 62$ min.
The time interval, $\Delta t = 62 - 58 = 4$ min.

Using [6.5] for this problem, we can write

$$\Delta N \approx \left(\frac{dN_t}{dt}\right)_{t=60} \times \Delta t$$

giving

$$\Delta N \approx 9350 \times 4 = 37\,400 \text{ cells}$$

Q6.5 The volumes, $V(\text{m}^3)$, of adult animals of a particular species are related approximately to their heights, h (m), by the equation:

$$V = 0.04 \times h^3$$

(i)   Derive an expression for the differential coefficient, $\dfrac{dV}{dh}$.

(ii) Hence estimate the increase in volume (ΔV) of an animal of height, $h = 1.6$ m, if it then grows by ($\Delta h =$) 0.02 m.

7
Statistics and Information

Overview

All scientific experiments suffer from subject variability and/or measurement variability (Unit 1.2). This variability can occur with the actual subjects being measured, e.g., different colonies of bacteria will grow at different rates even though the conditions appear to be the same. The variability can also appear in the actual measurement process itself, e.g., repeated measurements of fragments of the same glass may yield slightly different values for its refractive index.

However, using statistics gives us the ability to

- analyse variability and uncertainty,

- describe experimental variation in a meaningful way, and

- quantify our confidence in reaching conclusions.

This chapter gives an introduction to the basic concepts of statistics and probability, and starts by introducing methods that are used to describe and quantify the variability occurring in a simple set of experimental data. The first section highlights the fact that a set of experimental results only represents a *sample* of all the possible values that could be obtained if the experiment could be repeated over and over again. We then use statistics to *infer*, from a data sample, *best estimates* for the characteristics of the actual system being measured.

With more extensive data sets it is often more convenient to record the number (or *frequency*) of data values falling within given *data ranges*. The characteristics of the data are then described by the characteristics of the overall *frequency* distribution or probability distribution.

We introduce the basic rules for future probability and confidence based on the frequency of past events. We also relate this frequentist approach to the Bayesian approach to probability, which uses prior and posterior odds and likelihood ratios. Bayesian statistics are being used more extensively in managing probabilities in complex 'real-life' situations.

Finally, a section on factorials, permutations and combinations gives further basic statistics that underpin the later development of the binomial theorem and non-parametric tests.

Most of the calculations developed in this chapter can be carried out using Excel, MINITAB or other software. An overview of these techniques can be found on the web site.

7.1 Describing and inferring

7.1.1 Introduction

This Unit starts by addressing an important question that must be considered when applying statistics to any particular set of scientific data:

> Are the values in the set to be considered as a *self-contained set* of numbers (a *population*), or are they *representative* (a *sample*) of a bigger (*source population*) of other possible measurements?

The statistics that must be used for *samples* and *populations* are subtly different.

The results of an experiment are normally just a sample of the many different results that could be obtained if the experiment were to be repeated. Hence we should apply *sample statistics* to experimental data and not population statistics.

This Unit then introduces the basic parameters that are used to describe the general characteristics of statistical data. *Mean* and *median* are used to give the *location* (or position) of a set of data values along a value axis. *Standard deviation* and *interquartile range* are introduced as ways of describing the *dispersion* (or spread) of data around that location. *Variance* also becomes a useful parameter in science when assessing the combined effect due to variability and uncertainty from different experimental factors – see Unit 13.1.

Information about other common statistical descriptors (e.g. kurtosis, skewness) can be found on the web site:

⌨ Study Resources > Additional Materials

7.1.2 Populations and samples

We start by considering the data set given in E7.1.

E7.1 ⌨

Describe the *characteristics* of the data set A:

| **A** | 31 | 36 | 44 | 39 | 40 |

Calculations are given in the following text.

The way in which we treat the five numbers in E7.1 will depend on whether the data set A, is

- *a* population consisting of just the numbers 31, 36, 44, 39, 40, with no other numbers involved, or

- *a* sample of five replicate experimental measurements, which form part of a much larger *source population* of measurements that *could be created* if the experiment were repeated many times.

We can test whether a data set is a *sample* or a *population* by considering the effect of *repeating the process by which the data values were identified*.

If repeating the process of identification still gives the same values, then the data set is a *population*; but if different values could be produced, then the set is a *sample*.

A student collects **all** 200 frogs (a *population*) from a specific pond and, finding that 120 are female and 80 male, he concludes that the proportion of females in that pond is 0.60. A second student collects **all** the frogs again (the same *population*) and will record exactly the same proportion of females to males. By comparison, a student who randomly picks just 50 frogs only collects a *sample*, and may get a different proportion of females and males in every different sample of 50 frogs that he selects – see Q8.23.

A *sample* of values only gives us *best estimates* of the *population* values that we are trying to measure. However, it is possible, through good experiment design (Unit 13.3), to obtain results that are sufficiently accurate for the purpose, without the need to make an excessive number of measurements.

Q7.1 Identify, for each of the following data sets, whether the data set is a *sample* or a *population*.

  (i)   The four numbers: {5.1, 5.2, 5.1, 5.3}.   Sample/Population

  (ii) Four repeated measurements of the pH of a solution.   Sample/Population

  (iii) The numbers recorded in 37 spins of a roulette wheel.   Sample/Population

  (iv) The 37 numbers (including '0') on a roulette wheel.   Sample/Population

  (v)   The scores achieved by England in all their football matches   Sample/Population
        during the 2002 World Cup.

  (vi) The amount of pocket money received by a random selection   Sample/Population
        of children as part of a 'lifestyle' survey.

If data set A (E7.1) is a self-contained *population*, then we can use statistics to simply *describe* the characteristics of the data set.

If, however, the data set A is a *sample* of *experimental* values, then we will use statistics to *infer* the characteristics of the *source population* of *experimental* measurements from which the sample was derived. For example, we will use the average (mean) value of the sample data to *infer* what would be the average (mean) value if the experiment were repeated many more times.

A **parameter** is a *variable* that is used to describe some characteristic of a *population*. A parameter is usually given a Greek letter as a symbol, e.g. μ for population mean and σ for population standard deviation.

A **statistic** is a *variable* that is used to describe some characteristic of a *sample*, e.g. $\bar{x}$ for sample mean and s for sample standard deviation.

The value of the statistic of a sample (e.g. mean of the sample) is used to infer the value of the equivalent parameter of the population (i.e. mean of the population) from which the sample was drawn. Different samples from the *same* population will typically give different values for the same statistic (i.e. different *sample* means).

7.1.3 Statistics notation

Typically, a calculation in statistics may be dealing with very many data values, and it would be time-consuming to write out every single number. Instead, statistics often uses a form of shorthand to show what is happening in the calculations.

Subscripts are used to identify *particular values* in a set of data. For example the values from data set A in E7.1 could be written as: a_1, a_2, a_3, a_4, a_5, where $a_1 = 31$, $a_2 = 36$, etc.

The Greek symbol, Σ (capital sigma), is often used to indicate the *summation* (addition) of a range of numbers.

$\sum_1^5 (a_i)$ is shorthand in statistics for saying '*add together* a_1, a_2, a_3, a_4, a_5'.

E7.2 🖥

Using the values of data set A from E7.1, the summation is written as:

$$\sum_1^5 (a_i) = a_1 + a_2 + a_3 + a_4 + a_5 = 31 + 36 + 44 + 39 + 40 = 190$$

In general, for 'n' x-values to be added together, we would write:

$$\sum_1^n (x_i) = x_1 + x_2 + x_3 + \cdots + x_{n-1} + x_n \qquad [7.1]$$

The above equation says '*sum* all values of x_i for i ranging from $i = 1$ to $i = n$'.

Note that when the values of *all* the possible variables in a set are to be added together, the summation is often written in simplified forms:

$$\sum_1^n (x_i) = \sum_i (x_i) = \sum (x_i)$$

It is well worth taking a few minutes to understand each new symbol when it appears. The ability to 'read' statistics more fluently will be an enormous help in developing a good understand of statistics as a whole.

Q7.2 Calculate $\sum_1^5 (x_i)$ for the following set of x values:

$$x_1 = 4.9, x_2 = 4.7, x_3 = 5.1, x_4 = 4.9, x_5 = 4.4$$

7.1.4 Means and averages

The **mean** (or **average value**), for a *population* is usually written as μ (pronounced *mew*).

The **mean** (or **average value**), for a *sample* is usually written as $\bar{x}$ (pronounced *x-bar*),

The mean value, $\bar{x}$, of a *sample* of data is the **best estimate** for the unknown mean, μ, of the *source population* from which the sample was derived. Note that $\bar{x}$ is a *statistic*, which is used to *infer* a best estimate for the *parameter*, μ.

For example, opinion pollsters take samples of about 1000 people to get 'best estimates' for the way in which the population may choose to vote in a political election. The voting intentions of the population are *inferred* from the sample results.

The mean values, $\bar{x}$ and μ, calculated for sample data and population data have the same numerical values. Excel uses the function AVERAGE to calculate the mean value.

Inferring a mean value from sample data:

$$\text{Sample mean, } \bar{x} = \frac{\text{Sum of all data values}}{\text{Number of data values}} = \frac{\sum_1^n (x_i)}{n} = \frac{\sum (x_i)}{n} \qquad [7.2]$$

Describing a mean value from population data:

$$\text{Population mean, } \mu = \frac{\text{Sum of all data values}}{\text{Number of data values}} = \frac{\sum_1^n (x_i)}{n} = \frac{\sum (x_i)}{n} \qquad [7.3]$$

E7.3 🖥

The data set A in E7.1 could be treated as either a *sample* or a *population* of values:

$$\text{Sample mean, } \bar{x} = \frac{\sum_1^5 (a_i)}{5} = \frac{31 + 36 + 44 + 39 + 40}{5} = \frac{190}{5} = 38$$

$$\text{Population mean, } \mu = \frac{\sum_1^5 (a_i)}{5} = \frac{31 + 36 + 44 + 39 + 40}{5} = \frac{190}{5} = 38$$

The *mean value* for a data set is useful because it gives the *location of the data* set on the value axis. We now know, for example, that data set A is spread (dispersed) around a 'central' value of 38.

Q7.3 Calculate both mean values, $\bar{x}$ and μ, for each of the following sets of x values:

(i) $x_1 = 4.9$, $x_2 = 4.7$, $x_3 = 5.1$, $x_4 = 4.9$, $x_5 = 4.4$

(ii) $x_1 = 0.25$, $x_2 = 0.27$, $x_3 = 0.19$, $x_4 = 0.22$

7.1.5 Dispersion and deviations from the mean

The *mean* value gives the *location* of the data values. The next step is to get an idea of the *dispersion* of data values (or *how spread out* they are).

The **deviation**, d_i, of a specific data value, x_i, from the mean value, $\bar{x}$, of a set of sample data is given by:

$$d_i = x_i - \bar{x} \qquad [7.4]$$

For a set of *population* data the deviation is given by δ (pronounced *delta*): $\delta_i = x_i - \mu$

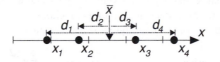

Figure 7.1. Deviation of data values from the mean.

The deviation, d_i, of the data value tells us how far that value, x_i, is from the mean value (or 'centre'), $\bar{x}$, of the data set – Figure 7.1.

We now need to develop a parameter/statistic that will give an overall impression of the *dispersion* (spread) of the *whole data set*.

We could try just adding the deviations of all the data values together. However, we find that:

The *sum of all deviations* in any data set is always zero.

$$\sum_{1}^{n} (d_i) = \sum_{1}^{n} (x_i - \bar{x}) = 0$$

$$\left\{ \text{similarly for a population: } \sum_{1}^{n} (\delta_i) = \sum_{1}^{n} (x_i - \mu) = 0 \right\}$$

Would the sum of the squares of the deviations be a useful measure of dispersion?

$$\textit{Sum of all (deviations)}^2 = \sum_{1}^{n} (d_i^2) = \sum_{1}^{n} (x_i - \bar{x})^2 \qquad [7.5]$$

$$\left\{ \text{similarly for a population: } \sum_{1}^{n} (\delta_i^2) = \sum_{1}^{n} (x_i - \mu)^2 \right\}$$

The sum of (deviations)^2 is also called the **sum of squares**, **SS** – see 13.1.2.

The sum of all (deviations)^2 is *not zero*, and gives a good basis for the measurement of dispersion. This application is developed in the following section, 'Variance and standard deviation'.

Q7.4 For the following set of x sample values:

$$x_1 = 4.9, x_2 = 4.7, x_3 = 5.1, x_4 = 4.9, x_5 = 4.4$$

(i)   Calculate the mean value, $\bar{x}$, of the sample.

(ii) Calculate each of the deviations,

$$d_1, d_2, d_3, d_4, d_5$$

(iii) Calculate the sum of all deviations,

$$\sum_{1}^{n} (d_i) = \sum_{1}^{n} (x_i - \bar{x})$$

(iv) Will the answer for (iii) always be the same for any set of data?

(v)   Calculate the sum of all (deviations)^2:

$$\sum_{1}^{n} (d_i^2) = \sum_{1}^{n} (x_i - \bar{x})^2$$

7.1.6 Variance and standard deviation

We now introduce *variance* and *standard deviation*, which are very common parameters/statistics used to describe the dispersion of data values.

The exact calculation of variance and standard deviation depends on whether we are treating the data set as a *sample* (normally the case for experimental data) or a *population*.

Variance is defined as the 'mean' value of the square of the deviations:

$$\text{Sample variance:} \qquad s^2 \quad \text{or} \quad \sigma_{n-1}^2 = \frac{\sum (deviation)^2}{(n-1)} \qquad [7.6]$$

$$\text{Population variance:} \qquad \sigma_n^2 = \frac{\sum (deviation)^2}{n} \qquad [7.7]$$

where n is the number of values in the data set, and σ is pronounced *sigma*.

Note that the denominator is '$(n-1)$' for a sample and 'n' for a population.
Standard deviation is defined as the *square root of the variance*:

$$\text{Standard deviation} = \sqrt{\text{Variance}} \qquad [7.8]$$

$$\text{Standard deviation of a sample} = \sqrt{\text{Sample variance}} = \sqrt{\frac{\sum(\text{deviation})^2}{(n-1)}}$$

$$\text{Standard deviation of a population} = \sqrt{\text{Population variance}} = \sqrt{\frac{\sum(\text{deviation})^2}{n}}$$

Variance is an important parameter when dealing with experimental uncertainty (Unit 1.2, Section 8.3.3 and Unit 13.1). However, variance cannot be used as a *direct* measure of *dispersion*, because it has units which are the *square* of the units for the data value. For example the variance of a measurement of time in seconds would have units of $(\text{seconds})^2$.

Standard deviation does have the same units as the data values, and has become the most commonly used indicator to describe the *dispersion* (*spread*) of data values.

Sample statistics	Population parameters
Variance, s^2 or σ^2_{n-1}	Variance, σ^2_n
$$s^2 = \sigma^2_{n-1} = \frac{\sum_1^n (x_i - \bar{x})^2}{(n-1)} \qquad [7.9]$$	$$\sigma^2_n = \frac{\sum_1^n (x_i - \mu)^2}{n} \qquad [7.10]$$
Standard deviation, s or σ_{n-1}	Standard deviation, σ_n
$$s = \sigma_{n-1} = \sqrt{\frac{\sum_1^n (x_i - \bar{x})^2}{(n-1)}} \qquad [7.11]$$	$$\sigma_n = \sqrt{\frac{\sum_1^n (x_i - \mu)^2}{n}} \qquad [7.12]$$

Using the above equations to calculate standard deviation requires the calculation of deviations for each data point. This can be time-consuming. However, it is also possible to use the following equation [7.13] to calculate the sample standard deviation, *without the need to calculate the mean value and all the deviations*:

$$s = \sigma_{n-1} = \sqrt{\frac{\sum_1^n (x_i)^2 - \dfrac{\left(\sum_1^n x_i\right)^2}{n}}{(n-1)}} \qquad [7.13]$$

It is only necessary to take the sum of the x values, the sum of the squares of the x values, and then substitute into [7.13]. In practice, equation [7.13] can be an easier method of calculating the standard deviation than equation [7.11].

The *population standard deviation*, σ or σ_n, is the actual standard deviation of the set of numbers in the given population. Excel uses the function STDEVP for a population standard deviation.

The *sample standard deviation*, s or σ_{n-1}, is the *best estimate* of the standard deviation, σ_n, of the *source population* of all experimental values from which the sample has been taken. Excel uses STDEV for a sample standard deviation.

Q7.5 For the following set of x values:

$$x_1 = 4.9, x_2 = 4.7, x_3 = 5.1, x_4 = 4.9, x_5 = 4.4$$

(i)   Calculate the sample standard deviation using [7.11].

(ii) Calculate the sample standard deviation using [7.13].

Q7.6 For each of the three (sample) sets of experimental data calculate the values in the table below and answer (iv):

	Mean	Sample variance	Sample standard deviation
(i) 8, 6, 7, 4, 7			
(ii) 68, 66, 67, 64, 67			
(iii) 418, 416, 417, 414, 417			

(iv) What similarities are there about the answers for (i), (ii) and (iii)?

The **coefficient of dispersion**, *CD*, is a statistic that is used to compare the 'spread' of a distribution to its mean value. *CD* is defined as the variance of the data divided by the sample mean value:

$$CD = \frac{s^2}{\bar{x}} \qquad [7.14]$$

Do not confuse coefficient of dispersion with *coefficient of variation* (see 8.3.2), which is the ratio of the *standard deviation* to the mean value of an experimental result.

Q7.7 The following set of x values represent experimental data values of the heights of five plant shoots

$$x_1 = 4.9\,\text{cm}, x_2 = 4.7\,\text{cm}, x_3 = 5.1\,\text{cm}, x_4 = 4.9\,\text{cm}, x_5 = 4.4\,\text{cm}$$

(i)   Calculate the sample standard deviation.

(ii)   What are the units of standard deviation in this case?

(iii) Do the units in (ii) have any physical significance in respect of height?

(iv)   Calculate the sample variance.

> (v) What are the units of variance in this case?
>
> (vi) Do the units in (v) have any physical significance in respect of height?
>
> (vii) Calculate the coefficient of dispersion.
>
> (viii) What are the units of the coefficient of dispersion in this case?

7.1.7 Ranked data

Ranked data has the data values *sorted* into ascending (or descending) order and assigned a *rank position*. Ranking is used extensively in *non-parametric* statistics (Chapter 12), where only the 'order' of the data values is important and not the actual (parametric) values.

For example, the following nine data values:

Data:	2.3	15.3	3.7	4.5	4.2	11.1	6.3	3.7	3.3

can be sorted into ascending order and assigned rank positions from 1 to 9.

Data:	2.3	3.3	3.7	3.7	4.2	4.5	6.3	11.1	15.3
Rank:	1	2	3	4	5	6	7	8	9

The two values of '3.7' in the above data raise the question of what to do with equal (or tied) values. In the example, we allocated rank values in sequence, but we will see in Chapter 12 that the allocation would be slightly differently when used for non-parametric tests.

The *location* of a set of, n, data values is described by:

- **Median** is the *middle* value in a set of ranked data values and gives the *location* of the data. The median is the value with the rank $0.50 \times (n + 1)$.

- **Lower quartile**, Q_1, is the value *one* quarter of the way from the lowest to the highest value. The lower quartile is the value with the rank $0.25 \times (n + 1)$.

- **Upper quartile**, Q_3, is the value *three* quarters of the way from the lowest to the highest value. The upper quartile is the value with the rank $0.75 \times (n + 1)$.

In the above data set, the number of data values, $n = 9$.

- Median value has the rank $= 0.5 \times (9 + 1) = 5$

  Median value with rank '5' $= 4.2$

- Lower quartile value has the rank $= 0.25 \times (9 + 1) = 2.5$

  Lower quartile value with rank '2.5' is half way between '3.3' and '3.7' $= 3.5$

- Upper quartile value has the rank $= 0.75 \times (9 + 1) = 7.5$

  Upper quartile value with a rank '7.5' is half-way between '6.3' and '11.1' $= 8.7$

The *dispersion* (spread) of non-parametric data is described by:

- **Interquartile range**, IQR, is the difference in value between the upper quartile and lower quartile.

$$IQR = Q_3 - Q_1$$ [7.15]

- **Total range** is the difference in value between the *lowest value* and the *highest value*.

	Median	Lower quartile	Upper quartile	IQR
Q7.8 For each of the sets below, complete the table:				
(i) 5.0, 8.2, 7.9, 6.6, 7.6, 5.7, 3.2, 7.5, 5.9				
(ii) 11, 8, 13, 9, 21, 24, 12, 22, 29, 43				
(iii) 45, 67, 23, 78, 67, 56, 98, 23, 49				

7.1.8 Box and whisker plots

A box and whisker plot is very useful way of visualizing raw experimental data.

Figure 7.2 shows box and whisker plots for two data samples, S and T, drawn against the data value axis.

- The *'middle' line* of each plot shows the position of the *median* value.

- The *ends of the 'box'* give the positions of the *upper and lower quartiles*.

- The *ends of the 'whiskers'* give the *maximum and minimum* values in the data.

In Figure 7.2, S and T have similar median values (location), but T has a larger interquartile range (dispersion) than S. The fact that the median of T is not at the centre of its 'box' shows that the T data is not symmetrical.

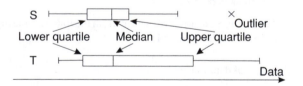

Figure 7.2. Box and whisker plots of data sets S and T.

The cross outside the range of the 'whiskers' for S shows a value that is an *outlier*. An outlier is a value that lies more than $1.5 \times$ IQR from the nearest quartile.

For *parametric* data that follow a *normal* distribution (see Unit 8.1) there is less than a 0.7% probability that a random data value will be an outlier. Hence outliers in experimental data should be examined carefully to identify whether they arise from errors in the experimental process or whether they may indicate a real experimental effect.

> **Q7.9** Represent all of the number sets in Question Q7.8, as box and whisker plots.

It is not possible to use standard Excel to draw box and whisker plots. However, statistics packages such as MINITAB and SPSS can produce these diagrams easily.

7.2 Frequency statistics

7.2.1 Introduction

In statistics, we often need to count the number of times that a particular outcome, x, occurs, e.g. the number of *large* eggs laid by a group of chickens, or the number of students who achieve an 'upper second' classification in their degree.

The number of times a particular outcome, x, occurs is called 'frequency'. The value of using the concept 'frequency' is that it provides a useful method of describing and handling large quantities of data. In this Unit we will write the frequency of x as $f(x)$ or f_x. The two expressions are equivalent, but it is sometimes easier to print one than the other.

We will also establish the link between the frequency, $f(x)$, with which particular events have occurred in the past, with the probability, $p(x)$, with which similar events may be expected to occur in the future.

7.2.2 Plotting data values

An important function of statistics is to handle large numbers of data values in such a way that the analyst can easily understand the essential characteristics of the data set.

A **stem and leaf diagram** is a quick way of displaying a spread of data values for initial analysis.

E7.4 ⌨

This stem and leaf diagram shows the examination marks, M, for 50 final-year students:

$$30 \mid 7\ 9\ 9$$
$$40 \mid 1\ 2\ 2\ 4\ 5\ 6\ 7\ 7\ 8\ 9$$
$$\textbf{50} \mid 1\ 1\ 1\ 1\ 2\ 2\ 3\ 3\ \textbf{4}\ 5\ 5\ 5\ 6\ 8\ 9\ 9\ 9$$
$$60 \mid 0\ 0\ 1\ 2\ 2\ 2\ 3\ 4\ 4\ 5\ 5\ 5\ 6\ 7$$
$$70 \mid 0\ 0\ 1\ 2\ 7$$

Figure 7.3. Stem and leaf diagram.

The first row shows that there were students with marks 37, 39 and 39 respectively, and, similarly, the final row shows that students had marks 70, 70, 71, 72 and 77.

The 'stem' is the column on the left, which shows a value that is common to all data values on that level. The 'leaves' are the individual marks that must be added to the stem level to get the actual data value. For example, the values *circled with bold italics* show that one student achieved a mark of $54 = 50$ (stem) plus 4 (leaf).

The value of a stem and leaf diagram is that it can be hand-drawn very quickly to review the spread of results as they are being recorded in an experiment.

Another quick way of displaying information is the **dot plot**, which just gives a 'dot' for each data value.

E7.5

The graph (Figure 7.4) shows a dot plot (produced using MINITAB) for the examination marks data for the 50 students given in Figure 7.3.

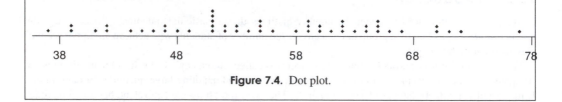

Figure 7.4. Dot plot.

Q7.10 The following data gives the weights, w (in grams), of a collection of 24 seeds.

48	46	46	66	55	55	47	44
49	40	33	57	42	25	58	32
49	51	39	67	61	42	62	27

(i)   Represent the data by drawing a stem and leaf diagram

(ii) Use the diagram to count the numbers of seeds that fall within the following ranges:

$$20 \leq w < 30; \quad 30 \leq w < 40; \quad 40 \leq w < 50; \quad 50 \leq w < 60; \quad 60 \leq w < 70$$

(N.B. '$20 \leq w < 30$' means that w is a number that is greater than *or equal to* 20, but also less than 30.)

7.2.3 Frequency data

When we have a large number of data values it can be unnecessarily cumbersome to treat them as *individual* data items. It is more convenient to divide the data range into a number of divisions, and then simply count how many data values fall into each division.

Each division, g, is called a **class** (or **bin**), and the number of data values in the class is called its **Class frequency**, f_g.

For example, Table 7.1, summarizes the examination marks, M, of the 50 final-year students given in E7.4. The class frequency is found by counting the number of values in each class.

The degree grade (class) is decided on the basis of the *numerical range* within which the student's aggregate mark, M, falls.

Each class may also be described as a *category* and given a **categorical name**:

'Fail (F)', 'Third (3rd)', 'Lower second (2.2)', 'Upper second (2.1)' and 'First (1st)'.

Table 7.1. Final-year performance of 50 students.

Mark range	Categorical name	Class frequency	Relative frequency
$M < 40$	Fail (F)	$f(F) = 3$	$3/50 = 0.06$
$40 \leq M < 50$	Third (3rd)	$f(3rd) = 10$	$10/50 = 0.2$
$50 \leq M < 60$	Lower second (2.2)	$f(2.2) = 18$	$18/50 = 0.36$
$60 \leq M < 70$	Upper second (2.1)	$f(2.1) = 14$	$14/50 = 0.28$
$M \geq 70$	First (1st)	$f(1st) = 5$	$5/50 = 0.1$
Total =		50	1.00

The results are expressed as the number, class frequency, $f(g)$ or f_g, of students achieving each grade (class, g) of degree: for example, three students fell into the 'Fail' category and five students were in the 'First' category.

The *proportion* of data values in each class is given by the **relative class frequency** (see Table 7.1).

$$Relative\ class\ frequency = \frac{Class\ frequency}{Total\ of\ all\ frequencies} \qquad [7.16]$$

Relative class frequency may often be expressed as a percentage.

Q7.11 A group of 24 graduates were surveyed to record their salaries, S, two years after graduation. The results, measured in £000s, were as follows:

20.1	17.7	22.2	15.0	22.0	23.7	15.4	16.7
14.7	18.7	20.4	24.5	26.2	24.9	17.7	21.3
16.2	21.7	17.1	17.0	19.0	12.0	18.8	18.5

Use the table below to record answers to the following questions:

(i)   Count the number (frequency) of graduates, $f(g)$, who achieve **salaries**, S, in the ranges (classes), g, that are given in the table below:

(ii) Calculate the relative frequency (rounding to three significant figures) for each of the classes, g.

(iii) Calculate the 'total' for all relative frequencies.

(iv) Is the answer in (iii) what was expected?

Salary range	Number (frequency)	Relative frequency
g	$f(g)$	
$12.0 \leq S < 16.0$		
$16.0 \leq S < 20.0$		
$20.0 \leq S < 24.0$		
$24.0 \leq S < 28.0$		
Totals:	24	

7.2.4 Plotting frequency data

If the data values are grouped into different *categories*, we can conveniently plot this **categorical** class data using a column graph or a pie chart:

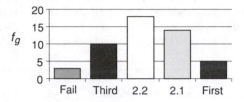

Figure 7.5. Column graph for the final-year performance of 50 students.

For example, Figure 7.5 is a **column graph** which records the performance (Table 7.1) of the cohort of 50 final-year students (from E7.4). Note that Excel draws a column graph with *vertical* 'columns' and a bar graph with *horizontal* 'bars'.

The overall shape of the column (or bar) graph conveys a lot of information, without the need to write down the mark of every individual student.

In a column graph, each class frequency is proportional to the **height** of the relevant column.

Figure 7.6 is a **pie chart** that records the same data as Figure 7.5, but the values for each class are recorded as *percentage relative frequencies*.

In a pie chart, each class frequency is proportional to the **angle** of the relevant slice.

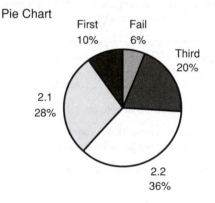

Figure 7.6. Pie chart for the final-year performance of 50 students.

Q7.12 The following data gives the areas of four regions of the United Kingdom, given in units of 1000 km^2.

	Area/1000 km^2	Fraction	Angle/degrees
England	130	0.533	192
Northern Ireland	14		
Scotland (incl. islands)	79		
Wales	21		
Total:	244		

(i)   Represent the area data by drawing a column graph.

(ii) Calculate the fraction of the total area covered by each region, by dividing the area of the region by the total, for example England covers a fraction $130/244 = 0.533$

(iii) Multiply each fraction by 360 to calculate the angles for a pie chart, for example, for England $0.533 \times 360 = 192°$

(iv) Plot the data as a pie chart.

7.2.5 Histograms

The data can be plotted as a histogram**,** provided that a *quantitative variable* can be plotted along the *x*-axis instead of *categorical data*, for example, plotting the actual degree marks instead of the degree classifications – see Figure 7.7.

It is very important to understand that, in a **histogram**:

- The **AREA of each block** (= height × width) is proportional to the *class frequency.*

- The data is *continuous* along the value axis – there are no *gaps between the blocks* (unless a block has a zero height).

- The *total area of a frequency histogram* is equal to the *total number, n*, of data values.

The *height* of each 'block' on the ordinate (*y*-axis) scale of the histogram is given by the **frequency density**, *fd*(*g*).

$$\text{Frequency density}, fd(g) \quad \text{or} \quad fd_g = \frac{\text{Class frequency}}{\text{Width of class}} = \frac{f_g}{\Delta x_g} \qquad [7.17]$$

The Greek symbol, Δ (pronounced *capital delta*), is often used to indicate a difference in a variable – see 6.1.2. In the above equation 'Δx' refers to a change in the *x*-variable.

The **area** of each 'block' is given by **class** *frequency*, *f*(*g*):

Class frequency, $f(g)$ or f_g = Average class frequency density, $fd(g)$ × Width of class, Δx_g

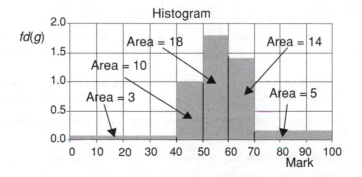

Figure 7.7. Histogram for the final-year performance of 50 students.

$$f(g) = f_g = fd(g) \times \Delta x_g \qquad\qquad [7.18]$$

E7.6 🖳

Figure 7.7 shows the **histogram** for the final-year performance of 50 students (from Figure 7.3): The **'heights'** (frequency density) of the histogram 'blocks' were calculated as follows:

Class g	Class frequency $f_g = f(g)$	Class range	Class width $\Delta x_g = \Delta x(g)$	Frequency density $fd_g = fd(g)$
Fail	3	$0 \leq x < 40$	40	$= 3/40 = 0.075$
Third	10	$40 \leq x < 50$	10	$= 10/10 = 1.0$
2.2	18	$50 \leq x < 60$	10	$= 18/10 = 1.8$
2.1	14	$60 \leq x < 70$	10	$= 14/10 = 1.4$
First	5	$70 \leq x \leq 100$	30	$= 5/30 = 0.167$

Compare the *shape* of the column graph (drawn using categories) in Figure 7.5 with the histogram (drawn using a continuous quantitative axis) in Figure 7.7. The differences occur when the *class widths are not the same*, i.e. for the 'Fail' and 'First' categories.

Q7.13 84 children in year 8 of school were given a project to record the distribution of their heights, h (cm), and they chose to count the number of pupils whose heights fell within specific ranges. However, they used different widths for the different ranges, with the result that the final data is recorded as follows:

Class range h (cm)	Class frequency $f_g = f(g)$	Class width $\Delta x_g = \Delta x(g)$	Frequency density $fd_g = fd(g)$
$144 < h \leq 154$	6	10	$= 6/10 = 0.6$
$154 < h \leq 158$	21	4	$= 21/4 = 5.25$
$158 < h \leq 160$	17		
$160 < h \leq 162$	15		
$162 < h \leq 166$	16		
$166 < h \leq 176$	9		

(i) Use the empty columns of the table to calculate the class width and class density $\{fd(g) = f(g)/\Delta x(g)\}$ for each class (two classes have already been completed).

(ii) Decide if it is possible to present the data graphically using a histogram (even though the class widths are not all the same).

(iii) If the answer to (ii) is 'yes', then draw the relevant histogram.

7.2.6 Mean and standard deviation

In 'frequency data' we often do not know the exact value of *each* data item. We only know that its value falls somewhere within a particular class range, for example a student with an upper second grade has a mark somewhere between 50 and 59.

We make the *approximation* that *every data item* in a particular class, g, has a value equal to the *mean class value*, $\bar{x}_g$, of that class.

The mean class value is obtained by calculating the average of all the data values that *could be* included in that class. For example, the mean class value of a class that has *any value* in the range $50 \leq x < 60$ would be $\bar{x}_g = 55$. Note, however, that the mean class value of a class that has *only integer values* 50 to 59 would be the mean value of 50, 51, 52, 53, 54, 55, 56, 57, 58 and 59, giving $\bar{x}_g = 54.5$.

$$\text{The } \textbf{approximate sum} \text{ of the } \boldsymbol{f_g} \text{ values in class } g = f_g \times \bar{x}_g \qquad [7.19]$$

For example, if the mean class value is $\bar{x}_g = 55$, and there are $f_g = 4$ items in that class (e.g. 51, 53, 55, 58), then the *approximate* sum of those four values $= f_g \times \bar{x}_g = 4 \times 55 = 220$. This is not the exact sum ($= 219$ for the example values) but will often give a sufficiently good approximation.

In the equations below, the symbol, $\sum_g$, means 'add up the following expression for all classes'. For example, $\sum_g (f_g)$ means 'add up the values of f_g for all classes'.

$$\textbf{Total number} \text{ of data values} : n = \sum_g (f_g) \qquad [7.20]$$

$$\textbf{Approximate sum} \text{ of ALL values in ALL classes} : \text{sum} = \sum_g (f_g \times \bar{x}_g) \qquad [7.21]$$

We can derive approximate equations for calculating the mean and standard deviation:

Mean values (population or sample):

$$\mu = \frac{\sum_g (f_g \times \bar{x}_g)}{n} \quad \text{or} \quad \bar{x} = \frac{\sum_g (f_g \times \bar{x}_g)}{n} \qquad [7.22]$$

Standard deviations (population or sample):

$$\sigma = \sqrt{\frac{\sum_g \{f_g \times (\bar{x}_g - \mu)^2\}}{n}} \quad \text{or} \quad s = \sqrt{\frac{\sum_g \{f_g \times (\bar{x}_g - \bar{x})^2\}}{n-1}} \qquad [7.23]$$

E7.7 💻

The number of students in a sample group obtaining (integer) marks in different ranges is given in the table below. Use Excel to calculate the mean and sample standard deviation for the distribution of frequency values.

Range	0–9	10–19	20–29	30–39	40–49	50–59	60–69	70–79	80–89	90+
Number	0	1	4	18	39	80	47	9	2	0

Note (for example) that the mean value, x_g, for the range 30–39 will be 34.5 when the possible scores have only integer values. A similar calculation applies for all ranges.

Using [7.22] and [7.23], the calculation in Excel gives

Mean value, $\bar{x} = 53.50$
Standard deviation, $s = 11.47$

See also E8.7 and E11.8.

Q7.14 In this question we compare *two ways* of calculating the *sum* and the *mean* of a set of data. We will use the same data for the salaries, S, of 24 graduates as used in Q7.11.

20.1	17.7	22.2	15.0	22.0	23.7	15.4	16.7
14.7	18.7	20.4	24.5	26.2	24.9	17.7	21.3
16.2	21.7	17.1	17.0	19.0	12.0	18.8	18.5

In the first part, we estimate the (iv) sum and (v) mean by first dividing the data values *into classes* and using the mean class values. We then compare these values to the (vi) sum and (vii) mean calculated by using each *individual* data value.

(i) Using classes, g, as in the table below, count the frequency, f_g, for each class, and the total, n, for all classes. (Check these with the answer to Q7.11.)

(ii) Calculate the mean class values, $\bar{S}_g$, for each class, g. (Take the mean of all possible values within each range – for example $\bar{S}_g$ for the first class is 14.0.)

(iii) For each class, g, multiply the number (f_g) in the class by the mean class value, $\bar{S}_g$, to give the class sum, $f_g \times \bar{S}_g$, and enter the value in the right-hand column. This is an *estimate* of the sum of salaries for all graduates in that class.

(iv) Find the sum, $\sum_g f_g \times \bar{S}_g$, for all classes. This gives an estimate of the *sum of all* 24 graduates' salaries.

(v) Estimate the *mean value*, $\bar{S}$, of *all* the salaries by dividing the value in (iv) by the number of graduates, n.

(vi) Find the true sum by adding all the *individual* salaries.

Is this value the same as the result in (iv)?

(vii) Find the true mean by dividing the value in (vi) by the number of graduates, n.

Is this value the same as the result in (iv)?

(viii) When might the method used in steps (i) to (v) be preferable to the more accurate method used in (vi) and (vii)?

Salary range Class, g	Frequency f_g	Mean class value $\bar{S}_g$	Class sum $f_g \times \bar{S}_g$
$12.0 \leq S < 16.0$		14.0	
$16.0 \leq S < 20.0$			
$20.0 \leq S < 24.0$			
$24.0 \leq S < 28.0$			
$n = \sum_g (f_g) =$		Sum $= \sum_g (f_g \times \bar{S}_g) =$	
		Mean value, $\bar{S} = \dfrac{\sum_g (f_g \times \bar{S}_g)}{n} =$	

7.2.7 Relative frequency and probability

The outcomes of random events (or 'trials') are often governed by particular probabilities, for example, the scores recorded by six-sided cubic dice are governed by a *1 in 6* chance.

Frequency data, *f(g)*, can be used to describe the number of times particular *outcomes*, *g*, *have occurred*.

Probability, ***p(g)***, is a *prediction* of the relative frequency with which *future outcomes*, *g*, can be *expected to occur*.

The **Relative frequency** with which particular events *have occurred* in the *past* can sometimes be used to predict the *probability* with which similar events *will occur* in the *future*.

$$p(g) = \frac{f(g)}{n} \hspace{3cm} [7.24]$$

A **frequency distribution** is easily transformed into a **probability distribution**.

The *area* under a *frequency* distribution equals *n*, the sum of the individual frequencies.

The *area* under a *probability* distribution equals 1.0, the sum of the individual probabilities.

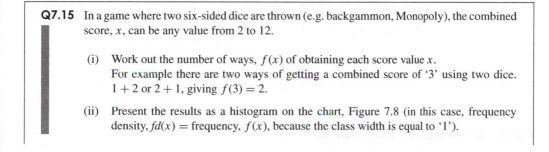

Q7.15 In a game where two six-sided dice are thrown (e.g. backgammon, Monopoly), the combined score, *x*, can be any value from 2 to 12.

(i) Work out the number of ways, $f(x)$ of obtaining each score value *x*.
For example there are two ways of getting a combined score of '3' using two dice. $1 + 2$ or $2 + 1$, giving $f(3) = 2$.

(ii) Present the results as a histogram on the chart, Figure 7.8 (in this case, frequency density, $fd(x) =$ frequency, $f(x)$, because the class width is equal to '1').

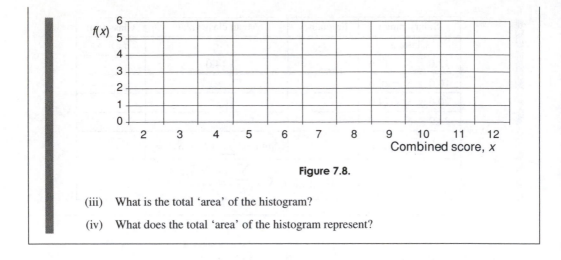

Figure 7.8.

(iii) What is the total 'area' of the histogram?

(iv) What does the total 'area' of the histogram represent?

Q7.16 Convert the 'frequency' histogram, $f(x)$, produced in Q7.15, into a 'probability' histogram, $p(x)$. Remember that, when rolling two dice, there are a total of 36 different possible outcomes.

(i) For each combined score, x, calculate the probability $p(x)$, that one roll of the two dice will give that particular result.

For example there are two ways of getting a combined score of '3' using two dice: $1 + 2$ or $2 + 1$. There are a total of 36 different possible outcomes. Hence the probability of getting a combined score of '3', $p(3) = 2/36 = 0.0556$

(ii) Present the results as a histogram on the chart, Figure 7.9.

(iii) What is the total 'area' of the histogram?

(iv) What does the total 'area' of the histogram represent?

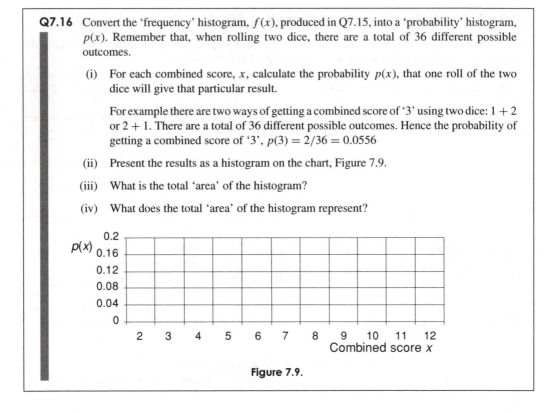

Figure 7.9.

7.2.8 Continuous distribution

A histogram is normally characterized by a number of clearly visible blocks, e.g. Figure 7.7.

If the *width* of each block is *reduced* and the *number* of blocks *increased* in proportion, then the envelope (outer curve) of the histogram will generally become 'smoother'; i.e. we change from a histogram to a '*continuous*' distribution.

For example, on the basis of past student performance, a particular university estimates the distribution of students who will gain different course marks. Figure 7.10 shows the **probability density**, *pd*(*x*), for the expected course marks.

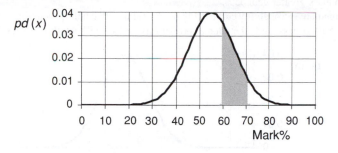

Figure 7.10. Probability density of course marks.

In a continuous probability distribution, the total area under the curve = 1.0, i.e. any student, who completes the course, has a mark that falls *somewhere* in the range covered by the curve. For a *continuous* distribution, we assume that the course mark can be calculated to any specific value and not just presented as an integer.

In a continuous probability distribution, the *probability* that a trial falls within a given range of *x*-values is given by the *area under the curve* between those values.

E7.8

In Figure 7.10, the probability that a student, selected at random, will have a mark between 60% and 70% is equal to the shaded area under the curve. We can make a rough estimate of this probability by noting that

- the width of the area = 70 − 60 = 10, and

- the average height of the area is about 0.025.

This gives a probability area of about $0.025 \times 10 = 0.25$ – i.e. there is about a 25% chance that any particular student will get a mark in this range (an upper second). An exact calculation gives a probability of 24.17%.

Q7.17   Use the graph in Figure 7.10 to make a **rough** estimate of the probability that a student, selected at random, will have a course mark between 30% and 40%.

7.2.9 Cumulative probability

Cumulative probability, *cp*(*x*), is the probability that the result of a trial may fall anywhere in the range from $-\infty$ to *x*.

The cumulative probability, *cp*(*x*), for a specific value of *x* is equal to the probability area under the *pd*(*x*) curve from $-\infty$ to *x*.

In situations where the value of x can only be *positive* (as for examination marks), the cumulative probability, $cp(x)$, is equal to the probability area from 0 to x.

The cumulative probability, $cp(x)$, will increase from 0 to 1.0 as the value of x increases. Using the example from Figure 7.11, the probability (= 0.84) of selecting a student with a mark between 0% and 65% is given by the shaded area in Figure 7.11. This probability is represented by the circled point on the cumulative probability curve in Figure 7.12: 0.84 of students will get a mark of 65% or less and so $1 - 0.84 = 0.16$ of students will get a mark greater than 65%.

Similarly, the cumulative probability curve in Figure 7.12 shows that half (probability = 0.5) of all students will get a mark of 55% or less, and all students (probability = 1.0) will get 100% or less!

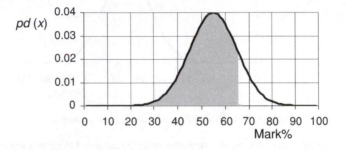

Figure 7.11. Probability area (shaded) for a mark between 0% and 65%.

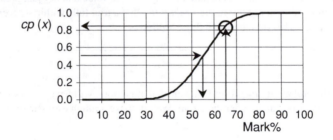

Figure 7.12. Cumulative probability of exam scores.

Q7.18   Use the data from the 'probability' histogram produced in Q7.16.

   (i)   For each combined score, x, calculate the cumulative probability that one roll of the two dice will give that score or any score below that value.

   (ii)   What is the cumulative probability that a roll of two dice will give a combined score of 6 or less?

   (iii)   What is the probability that a roll of two dice will give a combined score of at least 10?

7.3 Probability

7.3.1 Introduction

Probability is a measure of the *certainty* with which we may expect a particular outcome to occur. Probability is based on a scale of 0 to 1.

An *impossible* event, e.g., a cow jumping over the moon, has a probability of zero: p(cow will jump over moon) $= 0$.

A *certain* event, e.g., an iron bar sinking in water, has a probability of one: p(iron bar will sink in water) $= 1$.

The concept of probability is often used when the outcome of some event (or trial) *appears* to be governed solely by chance, for example, in rolling dice in a board game.

The fact that many outcomes *appear* to be governed by chance is often due to the complexity of the factors leading up to the particular outcome. The behaviour of playing dice, as they bounce and roll on the table, is actually governed by the laws of mechanics. However, the complexity of their motion is such that it is impossible to control their behaviour and predict the number that will be shown. In practice, the numbers shown by dice are *random* and unpredictable – each number has an equal chance of appearing (assuming that the dice are not biased!).

7.3.2 Probability and frequency

There are two main ways of estimating the probability with which a future event may occur:

- *Calculated on the basis of the science* of the situation, for example, as playing dice are symmetrical cubes, we would expect an *equal* probability for any number to appear.

- *Estimated from the frequency* with which particular events have *occurred in the past*, for example, past experience may show that one in three eggs laid by a particular chicken are small eggs and we may then expect that the future probability would also be 1 in 3.

Frequency, in statistics, is a measure of the number of times a particular outcome occurs.

$$f(X)\text{ represents the number of times the outcome, }X,\text{ occurs}$$

If we group the possible outcomes into different *classes*, then the *class frequency* is the number of times an outcome falls into a particular class (see also 7.2.3).

E7.9

For example, if we collect 180 eggs and divide them into 'small', 'medium' or 'large' categories (or classes), we may find that we have 60 'small' eggs, 80 'medium' eggs and 40 'large' eggs. The class frequencies in this case are described by the frequencies, f(small), f(medium), and f(large) respectively. This particular situation is described in Table 7.2.

Table 7.2. Example of a frequency (or probability) distribution of egg sizes

Class	Class frequency		Relative frequency		Probability		
Small	f(small) $=$	60	f(small)$/n =$	60/180	p(small)	$= 1/3$	≈ 0.333
Medium	f(medium) $=$	80	f(medium)$/n =$	80/180	p(medium)	$= 4/9$	≈ 0.445
Large	f(large) $=$	40	f(large)$/n =$	40/180	p(large)	$= 2/9$	≈ 0.222
Totals	$n =$	180				$= 1/3 + 4/9 + 2/9 = 1$	$= 1.000$

Relative frequency [7.16] is the observed frequency *divided* by the total, n, of all frequencies. The sum of all relative frequencies is 1.00.

In the 'egg' example in E7.9, there have been 180 'events' (sometimes called 'trials' in statistics), and, for each event or trial, there were three possible class 'outcomes': small, medium or large.

The **probability**, $p(X)$, of a future outcome, X, is a *prediction* of the *relative frequency* of that outcome *assuming the same situation could be repeated many times in the future*.

$$p(X) = \frac{f(X)}{n} \qquad\qquad [7.25]$$

Similarly, the future **frequency**, $f(X)$, of a particular outcome, X, can be calculated by multiplying the probability of that particular outcome, $p(X)$, by the number, n, of events or trials:

$$f(X) = p(X) \times n \qquad\qquad [7.26]$$

Taking the example of the eggs in E7.9, we know that the relative frequency, $f(\text{small})/n$, of getting a 'small' egg was 1/3, based on 180 previous eggs. We can then make a prediction for the probability that the *next* egg will be small, $p(\text{small}) = 1/3$. Similarly we can also predict the other probabilities, $p(\text{medium}) = 4/9$ and $p(\text{large}) = 2/9$ – see Table 7.2.

When we predict *future probabilities* based on *past frequencies*, we are assuming that there have been no changes that may affect the outcomes, for example a change in diet for the chickens may have reduced the proportion of small eggs.

A probability calculated on the basis of past experience is called an *experimental* probability.

Q7.19   Over a number of years it is found that the 140 graduates of the honours degree in horticulture include 34 with a third class degree, 50 with a lower second degree, 38 with an upper second degree and 18 with a first class degree.

   (i)   Calculate the *relative frequencies* for each class of degree.

   (ii)   Estimate the future *probability* that a particular student may gain a first class degree.

   (iii) What assumptions were made in order to answer (ii) above?

   (iv)   Estimate the number of students, in a cohort of 30 graduates, who would attain each of the four classes of degree.

In many other cases, we can use the science of the problem to *calculate* the relative frequencies without the need to carry out extensive trials. For example, we can use the *symmetry* of a six-sided die to calculate that the relative frequency (and hence probability) of getting any specific number (e.g. a '5') on a normal die. Each 'number' has an equal chance (1 in 6) of occurring:

$$p(1) = p(2) = p(3) = p(4) = p(5) = p(6) = 1/6$$

where $p(1)$, $p(2)$, etc. are the probabilities of recording a '1', '2', etc. when throwing the die.

A probability that is *calculated* using the science and mathematics of the problem is called a *theoretical* probability.

E7.10 🖳

The numbers observed of gene types AB, Ab, aB, ab are expected to be in the ratio 9:3:3:1 respectively, and each observation will give one of these four.

(i) Calculate the *probabilities* of observing each of the four types.

Starting with the ratios, we know that, in a total of $n = 9 + 3 + 3 + 1 = 16$ observations, we would expect to see 9, 3, 3, 1 of each type respectively. Using the equation $p(X) = f(X)/n$, we can calculate the future probabilities from the observed frequencies:

$$p(AB) = 9/16, \; p(Ab) = 3/16, \; p(aB) = 3/16, \; p(ab) = 1/16$$

(ii) Calculate the *number* of each type that would be expected out of a total of 200 observations.

Out of a new total of 200 observations, we can calculate the expected number for each type by using [7.26]: $f(X) = p(X) \times n$

$9/16 \times 200 = 112.5$

$3/16 \times 200 = 37.5$

$3/16 \times 200 = 37.5$

$1/16 \times 200 = 12.5$

See also E11.3.

Q7.20 A standard pack of 52 playing cards contains four suits (hearts, clubs, diamonds and spades), and each suit contains one each of cards numbered, 2 to 10, plus a jack, a queen, a king and an ace. The jack, queen, king are called picture cards.

Calculate the probabilities of selecting at random:

  (i) the king of clubs,   (iv) an ace,

  (ii) a king (of any suit),   (v) a card of the club suit.

  (iii) a picture card,

7.3.3 Combining independent probabilities with AND

If two *independent* trials occur, with the probability of getting A in the first trial being $p(A)$ and the probability of getting B in the second trial being $p(B)$, then the probability of getting A followed by B is:

$$p(A \text{ AND } B) = p(A) \times p(B) \qquad\qquad [7.27]$$

Note the use of '×' with the 'AND' combination.

E7.11

Calculate, using the egg example from Table 7.2, the probability, when picking two eggs, that the first egg is small and the second is large:

$p(\text{small AND large}) = p(\text{small}) \times p(\text{large}) = (1/3) \times (2/9) = 2/27$

Note that the multiplication of probabilities for the AND combination assumes that the two *probabilities are independent of one another*. The outcome for the first event or trial must have no influence over the outcome for the second event.

Q7.21 It is observed that in a particular group of plants: 75% are tall and 25% are short, and 75% have a round seed shape and 25% have a wrinkled seed shape.

Calculate the probability of finding each of the following four combinations (assume that the height and seed shape are not related):

(i)   tall with a round seed,

(ii)   short with a round seed,

(iii) tall with a wrinkled seed,

(iv)   short with a wrinkled seed.

7.3.4 Combining probabilities with OR

If A and B are two of the possible outcomes of a trial, which have probabilities of $p(A)$ and $p(B)$ respectively, then the probability of *either A or B* occurring is given by:

$$p(A \text{ OR } B) = p(A) + p(B) - p(A \text{ AND } B) \qquad [7.28]$$

Note the use of '+' for the 'OR' combination.

If the outcomes A and B are *mutually exclusive*, i.e. they cannot occur together and $p(A \text{ AND } B) = 0$, then we have the simple form of the equation:

$p(A \text{ OR } B) = p(A) + p(B)$

E7.12

(i) Calculate, using the data from Table 7.2, the probability of getting *either* a medium egg *or* a large egg:

$p(\text{medium OR large}) = p(\text{medium}) + p(\text{large}) = 4/9 + 2/9 = 6/9 = 2/3$

(ii)   Calculate the probability of getting a small egg *or* a medium egg *or* a large egg:

$$p(\text{small OR medium OR large}) = p(\text{small}) + p(\text{medium}) + p(\text{large}) = 1/3 + 4/9 + 2/9 = 1$$

The probability of '1' in (ii) shows that it is certain ($p = 1$) that the egg will definitely be one of the three possible sizes!

Q7.22   A trial has four possible outcomes, A, B, C and D, which have the following probabilities:

$$p(A) = 1/8, \; p(B) = 1/6, \; p(C) = 1/3, \; p(D) = 9/24$$

Calculate the probabilities that a single outcome will result in

(i) A or B, i.e. calculate the probability, $p(A \text{ or } B)$,

(ii) A or B or C,

(iii) A or B or C or D.

Q7.23   Using the standard pack of playing cards described in Q7.20, calculate the probability of selecting at random:

(i)   the king of clubs *or* the king of diamonds,

(ii)   the king of hearts *or* the king of clubs *or* the king of diamonds *or* the king of spades,

(iii) any picture card *or* any ace.

7.3.5   Combining probabilities with NOT

If A is one possible outcome of a trial, then the outcome 'NOT A' means that the outcome A has *not* occurred. 'NOT A' is normally written as $\bar{A}$, and sometimes (for convenience on web pages) as $\sim A$.

Do not confuse $\bar{A}$ (NOT A) with $\bar{x}$, which is the mean value of a set of x-data.

If an event or trial has possible outcomes A, B, C, then the 'NOT A' condition requires that either B or C must occur:

$$p(\bar{A}) = p(B \text{ OR } C) = p(B) + p(C)$$

In any event or trial where A is one possible outcome, then *either A or NOT A must* definitely occur, hence:

$$p(A \text{ OR NOT } A) = 1$$

which is the same as

$$p(A) + p(\bar{A}) = 1$$

giving:

$$p(\bar{A}) = 1 - p(A) \qquad\qquad [7.29]$$

E7.13

Calculate, using the data from Table 7.2, the probability of getting an egg that is NOT small. In the case of three categories of eggs, the outcome of 'NOT small' means that the egg produced must be either 'medium' or 'large'.

$$p(\text{NOT small}) = p(\text{medium OR large}) = p(\text{medium}) + p(\text{large})$$

Using the data from Table 7.2:

$$p(\text{NOT small}) = 1 - p(\text{small}) = 1 - 1/3 = 2/3$$

which gives the same result as using

$$p(\text{NOT small}) = p(\text{medium OR large}) = 4/9 + 2/9 = 6/9 = 2/3$$

Q7.24 Using the standard pack of playing cards described in Q7.20, calculate the probability of selecting at random:

   (i)   a king (of any suit),

   (ii)   a card that is not a king,

   (iii)   either a king or a card that is not a king.

7.3.6 Multiple outcomes

For more complicated probability outcomes, it is often possible that one *overall outcome* may be achieved by any one of several different separate outcomes. In this case, the overall probability is given by the *addition of the probabilities of the separate outcomes*.

E7.14

Calculate the probability, when two dice are thrown together, of getting a total score of '4'.

When two dice are thrown together, the important *overall outcome* is the *sum* of the two dice values. The same overall *sum* can be achieved (except for 2 and 12) in more than one way, e.g. an overall value of 4 can be achieved from the values of the two dice in three ways:

$$1 + 3 \quad \text{or} \quad 2 + 2 \quad \text{or} \quad 3 + 1$$

(Note that the outcome $1 + 3$ is different from $3 + 1$ – swapping the values on the two dice.)

The overall probability of getting a total of 4 from two dice is therefore:

$$p(\text{Sum of 4}) = p(1 \text{ AND } 3) + p(2 \text{ AND } 2) + p(3 \text{ AND } 1)$$

Q7.25 When rolling two dice, calculate the probabilities of scoring a *total* of exactly:

(i) 2   (iii) 7

(ii) 3   (iv) 12

Q7.26 On average, 90% of seeds of a given type of seed have been found to germinate successfully.

Calculate the probabilities that:

(i)   one selected seed will germinate

(ii)   one selected seed will not germinate

(iii) every one of four selected seeds will germinate

(iv)   none of four selected seeds will germinate

(v)   just one of four selected seeds will germinate

   (Hint: consider how many different ways it is possible to have one seed germinating and three seeds not germinating.)

(vi)   three of four selected seeds will germinate

7.3.7 Conditional probability

The probability of a given outcome (e.g. getting wet, W), may be dependent on (*conditional* on) an existing condition (e.g. it is raining, R). Such a conditional probability would be written as $p(W|R)$, i.e. the probability of 'W' occurring, given that condition 'R' exists.

Quite clearly, the probability of getting wet if it is NOT raining, $p(W|\bar{R})$, will generally be considerably less than the probability of getting wet if it IS raining, $p(W|R)$,

A good example of conditional probability occurs with diagnostic testing for a particular disease. The probability of getting a positive (P) result from a diagnostic test for a particular illness will normally depend on whether the person being tested has (D), or does not have ($\bar{D}$), the disease.

The probability of a:

true positive result (has the disease) would be given by $p(P|D)$,

false positive result (but does not have the disease) would be given by $p(P|\bar{D})$,

true negative result (does not have the disease) would be given by $p(\bar{P}|\bar{D})$,

false negative result (but has the disease) would be given by $p(\bar{P}|D)$.

The use of conditional probability can be illustrated using the following Example.

E7.15

It is known that the conditional probabilities for the results for a new diagnostic screening test for a male disease are:

	Man does have disease	Man does NOT have disease
Probability of positive result:	$p(P|D) = 0.9$	$p(P|\bar{D}) = 0.2$
Probability of negative result:	$p(\bar{P}|D) = 0.1$	$p(\bar{P}|\bar{D}) = 0.8$
Total probabilities:	$p(P|D) + p(\bar{P}|D) = 1.0$	$p(P|\bar{D}) + p(\bar{P}|\bar{D}) = 1.0$

Assuming that 5% (0.05) of men of age 60 do have the disease, calculate, for this age group:

(i) the probability that a man selected at random would record a *true* positive result, (he must both have the disease and then record positive).

(ii) the probability that a man selected at random would record a *false* positive result, (he must be both free of the disease and then record positive).

(iii) the probability that a man selected at random would record a positive result, (this will include the possibility of a true positive as well as false positive).

(iv) the probability that a man selected at random, who tests positive, would actually have the disease.

See text for calculation

The calculations for E7.15:

(i) For a man, selected at random, to record a *true positive* result requires that the man must have the disease AND he must then test positive.
  - Probability that a man, selected at random, has the disease, $p(D) = 0.05$.
  - Probability that a man, with the disease, will test positive, $p(P|D) = 0.9$.

  Hence, the probability that a man, selected at random, will record a true positive result

$$p(\text{True positive}) = p(P|D) \times p(D) = 0.05 \times 0.9 = 0.045$$

(ii) For a man, selected at random, to record a *false positive* result requires that the man does NOT have the disease AND he must then test positive.
  - Probability that a man, selected at random, does not have the disease,

$$p(\bar{D}) = 1 - p(D) = 1 - 0.05 = 0.95$$

  - Probability that a man, free of the disease, will test positive, $p(P|\bar{D}) = 0.2$.

  Hence, the probability that a man, selected at random, will record a false positive result

$$p(\text{False positive}) = p(P|\bar{D}) \times p(\bar{D}) = 0.95 \times 0.2 = 0.19$$

(iii) For a man, selected at random, to record a positive result (*either* true or false) requires that either a true positive OR a false positive is recorded:

$$p(\text{Positive}) = p(\text{True positive}) + p(\text{False positive}) = 0.045 + 0.19 = 0.235$$

(iv) For a man, *selected at random*, who then records positive, the probability that he does *in fact have the disease* will be given by:

$$p(\text{Does have disease}) = \frac{p(\text{True positive})}{p(\text{True positive}) + p(\text{False positive})} = \frac{0.045}{0.235} = 0.191 \qquad [7.30]$$

This example shows the care that must be taken with random screening tests. The possibility of false positives occurring in the majority of people who do not have the disease can make the probability of a false diagnosis quite high. In the above example, a man tested at random who gives a positive result has less than a 20% chance of actually being ill.

Screening tests are normally conducted with people identified as already being 'at risk', and are then followed up by more accurate diagnostic techniques.

Q7.27 Using the data from E7.15, calculate the probability that a man, selected at random, who then tests negative for the disease, could actually have the disease.

7.3.8 Probability trees

In complex conditional probability problems, as in E7.15, it is often useful to describe the dependent probabilities using a probability tree – Figure 7.13.

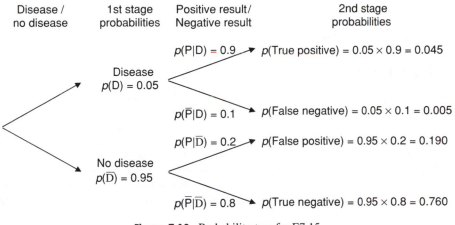

Figure 7.13. Probability tree for E7.15.

Probability trees are particularly useful when the probabilities in the second stage of the analysis are conditional on the outcomes from the first stage of the analysis.

Q7.28 A bag initially contains three black sweets and four red sweets. In this problem two sweets are taken from the bag one at a time, and the first sweet is *not* returned to the bag. Note that the probabilities for selection of the second sweet depend on what sweets are left after the first selection has been made. Use a probability tree to work out the answers.

Calculate the probabilities of drawing:

(i)   a black sweet followed by another black sweet

(ii)   a black sweet followed by a red sweet

(iii)   a red sweet followed by another red sweet

(iv)   a red sweet followed by a black sweet

(v)   any one of the above four combinations

Q7.29 What is the probability that, out of five people, at least two people have their birthdays in the same month (assume that every month $= 1/12$ of a year)?

Hint 1: it is easiest to first calculate the probability that no one has their birthday in the same month as anyone else.

Hint 2: pick *one* person and then calculate the probability that the *next* person to be picked will have a birthday in a *different* month, and then the probability that the *next* person will *also* have a birthday in a *different* month to the previous two, etc., etc.

7.4 Bayesian odds

7.4.1 Introduction

The basic ideas of probability have been introduced in Unit 7.3, leading to the concept of conditional probability. We now introduce the Bayesian approach to probability that uses a 'likelihood ratio' to quantify the way in which new information can change the expectation, or 'odds', that a particular condition is true. In forensic terms, a court would be interested in whether the evidence, E, is sufficiently strong to persuade a jury to accept the statement, S, of events proposed by the prosecution, or to accept the proposition from the defence that the statement is false, $\bar{S}$. We introduce the concepts of 'prior odds' and 'posterior odds' as the 'odds' before and after introducing a new piece of evidence. Bayesian statistics have particular advantages due to their simplicity in combining the cumulative 'odds' from a range from different contributing factors.

7.4.2 Conditional and joint probabilities

It is important to be very clear about our definitions of probabilities.

$p(A|B)$ is the **conditional probability** (7.3.7) that outcome A will occur, given that condition B already exists.

We now define $p(A \cap B)$ as the **joint probability** that both outcomes A and B are true simultaneously.

The joint probability of A occurring with B must be the same as the joint probability of B occurring with A, and we can deduce that:

$$p(A \cap B) = p(B \cap A) \qquad [7.31]$$

E7.16

Using the data from E7.15, calculate the *joint* probability that a man, selected at random, will *both* have the particular disease AND test positive, $p(P \cap D)$, i.e. the probability of recording a *true positive*.

The probability that a man, selected at random, will have the disease, $p(D) = 0.05$. Given that he has the disease, the man must then test positive to satisfy the joint probability. The *conditional* probability that he will test positive, given that he has the disease, is $p(P|D) = 0.9$.

Hence the *joint* probability of having the disease AND testing positive is given by:

$$p(P \cap D) = p(P|D) \times p(D) = 0.9 \times 0.05 = 0.045$$

We can write the general relationship:

$$p(A \cap B) = p(A|B) \times p(B) \qquad [7.32]$$

E7.17

In a group of 10 people, one person, G, was responsible for breaking a window. The probability that G will have glass on his/her clothing is estimated to be $p(glass|G) = 0.99$.

The probability that a different person, $\bar{G}$, may have glass on his/her clothing is estimated to be $p(glass | \bar{G}) = 0.2$.

Calculate the probabilities that a person selected at random from the group of 10:

(i) has glass on his/her clothing and is the person (G) who broke the window
    $= p(glass \cap G)$

(ii) has glass on his/her clothing and is *not* the person ($\bar{G}$) who broke the window
    $= p(glass \cap \bar{G})$

Answer:

(i) Probability that the person G will be selected at random, $p(G) = 1/10 = 0.1$.
    Given that $p(glass|G) = 0.99$ we can use [7.32] to write:

$$p(glass \cap G) = p(glass|G) \times p(G) = 0.99 \times 0.1 = 0.099$$

(ii) Probability that a person selected at random will *not* be G,

$$p(\bar{G}) = 1 - p(G) = 1.0 - 0.1 = 0.9$$

Given that $p(glass|\bar{G}) = 0.2$ we can use [7.32] to write:

$$p(glass \cap \bar{G}) = p(glass|\bar{G}) \times p(\bar{G}) = 0.2 \times 0.9 = 0.18$$

Q7.30 A population of 200 000 possible suspects is screened using a DNA test to match a specimen collected from a crime scene. It is estimated that the DNA results would produce a random, false positive, match for 1 person in 1.0 million. Assume that the person, K, who left the specimen, would definitely provide a DNA match.

Calculate the probabilities that a person selected at random from the 200 000 possible suspects:

(i)   has a DNA match and is the person who left the DNA specimen,

(ii) has a DNA match but is *not* the person who left the DNA specimen.

7.4.3 Bayes' rule

From [7.32] we have:

$$p(A \cap B) = p(A|B) \times p(B)$$

Similarly using [7.32] we could write:

$$p(B \cap A) = p(B|A) \times p(A)$$

and, using [7.31], we can combine the two equations above, to give:

$$p(A|B) \times p(B) = p(B|A) \times p(A)$$

which then gives **Bayes' rule**:

$$p(A|B) = p(B|A) \times \frac{p(A)}{p(B)} \qquad [7.33]$$

The above equation is very important because it '*transposes the conditional*'. It changes from the probability of B, given that A is true, to the probability of A, given that B is true. This is a subtle, but very significant, change.

For example, in a diagnostic test (e.g. in E7.15) the scientist is primarily concerned with the *accuracy of the test*, i.e. 'what is the probability that the test will give a positive result, given that the patient does have the disease?' However, *from the patient's point of view*, the more important question (with the *transposed conditional*) is 'what is the probability that I have the disease, given that I have just tested positive?'

The probability, $p(B)$, of getting outcome B will be the sum of the probability of getting B with A true plus the probability of getting B with A false ($\bar{A}$ or not A):

$$p(B) = p(B \cap A) + p(B \cap \bar{A}) = p(B|A) \times p(A) + p(B|\bar{A}) \times p(\bar{A}) \qquad [7.34]$$

We can substitute $p(B)$ from [7.34] into [7.33] and obtain:

$$p(A|B) = \frac{p(B|A) \times p(A)}{\{p(B|A) \times p(A) + p(B|\bar{A}) \times p(\bar{A})\}} \qquad [7.35]$$

This is a very useful rearrangement of Bayes' rule, as we will see in the following example.

E7.18

Using the data again from E7.15, we know that the probability that a man with the disease, will test positive in a diagnostic test is given by, $p(P|D) = 0.9$. i.e. there is a 90% chance that a man with the disease will test positive. But what is the probability, $p(D|P)$, that a man selected at random, who then tests positive, will actually have the disease? (Note the change of the conditional in this problem.)

To answer this we use [7.35]:

$$p(D|P) = \frac{p(P|D) \times p(D)}{\{p(P|D) \times p(D) + p(P|\bar{D}) \times p(\bar{D})\}}$$

We have the necessary data from E7.15:

$$p(P|D) = 0.9, \ p(P|\bar{D}) = 0.2, \ p(D) = 0.05 \quad \text{and} \quad p(\bar{D}) = 1 - p(D) = 0.95$$

Combining these gives:

$$p(D|P) = \frac{0.9 \times 0.05}{\{0.9 \times 0.05 + 0.2 \times 0.95\}} = 0.191$$

This is the same result as was achieved in [7.30].

E7.19

Using the data from E7.17, calculate the probability, $p(G|glass)$, that a person, selected at random, who is found to have glass on his/her clothing, is the person, G, who broke the window.

Using [7.35] and data from E7.17:

$$p(G|glass) = \frac{p(glass|G) \times p(G)}{\{p(glass|G) \times p(G) + p(glass|\bar{G}) \times p(\bar{G})\}} = \frac{0.099}{0.099 + 0.18} = 0.355$$

Q7.31 Using the data from example Q7.30, calculate the probability that a person, selected at random from the 200 000, and who is found to have a DNA match, will be the person, K, who was responsible for leaving the specimen.

7.4.4 Bayesian odds

If the possible outcomes of a statistical trial are either A or $\bar{A}$, then: $p(A) = 1 - p(\bar{A})$.

The **odds of** A, $O(A)$, are an alternative way of expressing the likelihood of a particular event, A, occurring, and are defined:

$$O(A) = \frac{p(A)}{p(\bar{A})} = \frac{p(A)}{1 - p(A)} \qquad\qquad [7.36]$$

This equation can be rearranged to give:

$$p(A) = \frac{O(A)}{1 + O(A)} \qquad [7.37]$$

The relationships between probabilities, 'p', and odds, 'O', are illustrated in Table 7.3.

Table 7.3. Odds and probabilities.

$p(A)$	0.01	0.1	0.5	0.9	0.99
$p(\bar{A})$	0.99	0.9	0.5	0.1	0.01
Odds, $O(A)$	0.0101	0.111	1	9	99
Odds *for*	—	—	Evens	9:1	99:1
Odds *against*	99:1	9:1	Evens	—	—

E7.20

Using the data again from E7.15:

(i) Calculate the odds, *before* being tested, that a man, selected at random, has the particular disease, given that $p(D) = 0.05$.

(ii) Use the result from E7.18 to calculate the *conditional* odds, $O(D|P)$, that a man who *has tested* positive does indeed have the particular disease.

(iii) How *significant* is a positive test result to a man selected at random?

Answer:

(i) Using [7.36], we can calculate the odds:

$$O(D) = p(D)/\{1 - p(D)\} = 0.05/0.95 = 0.0526$$

(ii) From E7.18, $p(D|P) = 0.191$

$$O(D|P) = p(D|P)/\{1 - p(D|P)\} = 0.191/0.809 = 0.236$$

(iii) As far as the man is concerned, the positive result from the test has made it $0.236/0.053 = 4.5$ *times more likely* that he does have the disease.

From E7.20:
Odds *before* the test are called **prior odds**

$$O(D) = p(D)/\{1 - p(D)\} = 0.0526 \qquad [7.38]$$

Odds *after* testing positive are called **posterior odds**

$$O(D|P) = p(D|P)/\{1 - p(D|P)\} = 0.236 \qquad [7.39]$$

Using Bayes' rule from [7.33], we can write:

$$p(D|P) = p(P|D) \times \frac{p(D)}{p(P)}$$

and

$$p(\bar{D}|P) = p(P|\bar{D}) \times \frac{p(\bar{D})}{p(P)}$$

and from [7.36]

$$O(D|P) = \frac{p(D|P)}{p(\bar{D}|P)} = \frac{\left\{ p(P|D) \times \dfrac{p(D)}{p(P)} \right\}}{\left\{ p(P|\bar{D}) \times \dfrac{p(\bar{D})}{p(P)} \right\}} = \frac{p(P|D)}{p(P|\bar{D})} \times \frac{p(D)}{p(\bar{D})}$$

giving

$$O(D|P) = \frac{p(P|D)}{p(P|\bar{D})} \times O(D) \qquad [7.40]$$

We can now define the **likelihood ratio** by using the equations:

$$\text{Posterior odds} = \text{Likelihood ratio} \times \text{Prior odds} \qquad [7.41]$$

$$\text{Likelihood ratio, } LR = \frac{p(P|D)}{p(P|\bar{D})} \qquad [7.42]$$

The *effect of the test* has been to change the odds of the man having the disease by multiplying the prior odds by the likelihood ratio.

7.4.5 Calculating likelihood ratios

The likelihood ratio (LR) can be calculated directly by using a quantitative assessment of the evidence, i.e. calculating the alternative probabilities of a given outcome, P, given two exclusive conditions, D and $\bar{D}$.

In a courtroom situation, a jury may be faced with the problem of deciding whether a statement describing a situation or an event, is *true*. The prosecution may claim that the statement is true, S, while the defence claims that the statement is not true, $\bar{S}$. Evidence, E, may be presented which changes the odds, $O(S|E)$, of the statement being true, S, given the evidence.

A 'new' piece of evidence, E, will change the odds that the statement is true. The likelihood ratio (LR) for this new piece of evidence is given by the ratio of the probabilities of the 'evidence' occurring given that the statement is either true, $p(E|S)$, or not true, $p(E|\bar{S})$:

$$\text{Likelihood ratio, } LR = \frac{p(E|S)}{p(E|\bar{S})} \qquad [7.43]$$

Then using [7.40] we get:

$$O(S|E) \quad = \quad \frac{p(E|S)}{p(E|\bar{S})} \quad \times \quad O(S)$$

[7.44]

Posterior odds	=	Likelihood ratio	×	Prior odds
supporting statement,		due to new		supporting
given the new evidence		evidence		statement

An important aspect of the above equation, [7.44], for the forensic scientist, is the fact that the data required to calculate the likelihood ratio is in a form that is produced by scientific investigation. It is necessary only to calculate the probabilities of observing the evidence, E, given the two conditions that the statement is either true or not true.

The equation then provides the posterior odds, $O(S|E)$, with the *transposed conditional*, i.e. the probability of the statement being true, given the new evidence. The posterior odds are presented in a form appropriate for consideration in court.

E7.21

Use the data from E7.15 to calculate the likelihood ratio for having the disease, given a positive result from the diagnostic test.

From E7.15, $p(P|D) = 0.9$ and $p(P|\bar{D}) = 0.2$.

Using [7.42]

$$\text{Likelihood ratio, } LR = \frac{p(P|D)}{p(P|\bar{D})} = \frac{0.9}{0.2} = 4.5$$

This agrees with the increased odds calculated in E7.20.

E7.22

Assuming that the proportion of left-handed people is 10%, and that it is known that the person, G, who committed a crime was *definitely* left-handed, calculate the likelihood ratio for the evidence of left-handedness.

The probability of being left-handed for the culprit, $p(L|G) = 1.0$.

For a randomly selected person, the probability of being left-handed,

$$p(L|\bar{G}) = 1/10 = 0.1$$

Hence, using [7.43] the likelihood ratio for the evidence of left-handedness becomes:

$$LR = \frac{p(L|G)}{p(L|\bar{G})} = \frac{1.0}{0.1} = 10$$

Q7.32 For the problem outlined in Q7.31 and Q7.30, calculate

(i)   Prior odds that a randomly selected person is the person, K, who left the sample, $O(K)$.

(ii) Likelihood ratio due to the evidence, E, of a DNA match.

(iii) Posterior odds that a randomly selected person, who gives a DNA match, is the person, K, who left the sample, $O(K|E)$.

(iv) Does the result in (iii) agree with the result in Q7.31?

Q7.33 A particular blood group is present in only 5% of the population. If it is known that the person who committed a particular crime had this blood group, calculate the likelihood ratio for this evidence when used to assess whether a suspect committed the crime. Assume that the experimental measurement of the blood group is 100% accurate.

The advantage of using the Bayesian, 'odds', approach is that it is easy to combine different factors that lead to an overall probability, by simply multiplying the relevant likelihood ratios. However, for combination by simple multiplication, it is necessary that the probabilities of the two tests are *independent* of each other.

E7.23

In E7.17, a witness states that the person, G, who broke the window was left-handed. Assuming that the proportion of left-handed people is 10%, calculate the posterior odds that a person from the group of 10, who is both left-handed *and* has glass on his clothing is the person, G.

Probability of selecting the person, G, at random from a group of 10, $p(G) = 0.1$. Hence the prior odds for randomly selecting, G, before considering the evidence:

$$O(G) = \frac{p(G)}{1 - p(G)} = \frac{0.1}{1.0 - 0.1} = \frac{0.1}{0.9} \approx 0.11$$

Using data from E7.17, the likelihood ratio for discovery of glass on clothing,

$$LR(glass) = \frac{p(glass|G)}{p(glass|\bar{G})} = \frac{0.99}{0.2} = 4.95 \approx 5$$

From E7.22, the likelihood ratio for 'left-handedness',

$$LR(left) = \frac{p(left|G)}{p(left|\bar{G})} = \frac{1.0}{0.1} = 10$$

The posterior odds, taking both tests into account (we assume that the probability of glass on the clothing is independent of whether the culprit is left-handed or not),

$$O(G|glass,left) = LR(glass) \times LR(left) \times O(G) = 5 \times 10 \times 0.11 = 5.5$$

It is not possible to use simple multiplication of likelihood ratios if the two pieces of evidence are not independent. For example, if the pieces of evidence were that a person was *believed* to be male and that he (or she) was over 6 feet tall, then the likelihood ratio for the height will be different depending on whether the person is actually male or female. In these situations it is necessary to investigate the probabilities of the different options using a form of 'probability tree'. Such calculations are heavily dependent on the particular situation involved and are not covered in this book.

7.5 Factorials, permutations and combinations

7.5.1 Introduction

This Unit develops the statistics for calculating the number of ways of obtaining particular arrangements of objects and events. These calculations have particular relevance in a range of probability calculations, from the likelihood of winning the National Lottery to the distribution of plants across a field.

7.5.2 Factorial

The **factorial**, $n!$, of a number, n, is the number multiplied by every integer between itself and one

$$n! = n \times (n-1) \times (n-2) \times \cdots \times 3 \times 2 \times 1 \qquad [7.45]$$

Scientific calculators can calculate the factorial of a number directly – look for the $x!$ function. The factorial can be calculated in Excel using the function FACT(x).

It is useful to note some particular values:

$$1! = 1 \qquad\qquad [7.46]$$

$$n! = n \times (n-1)! \qquad\qquad [7.47]$$

$$0! = 1 \qquad \text{(This is not obvious, but should be remembered.)} \qquad [7.48]$$

E7.24

$1! = 1$

$2! = 2 \times 1! = 2 \times 1 = 2$

$3! = 3 \times 2! = 3 \times 2 = 6$

$4! = 4 \times 3! = 4 \times 3 \times 2 = 24$

etc.

$0! = 1$

$0! / 3! = 1/6$

$20! = 20 \times 19!$

Division using factorials   Using a simple numerical example first:

E7.25

$$\frac{8!}{5!} = \frac{8 \times 7 \times 6 \times 5 \times 4 \times 3 \times 2 \times 1}{5 \times 4 \times 3 \times 2 \times 1} = \frac{8 \times 7 \times 6}{1} = 336$$

In E7.25, the 5! in the denominator cancels with all the terms from 5 to 1 in the numerator, leaving only $8 \times 7 \times 6$.

Hence, we can write a general formula for the division of factorials, where $n > m$:

$$\frac{n!}{m!} = n \times (n-1) \times \cdots \times (m+2) \times (m+1) \qquad [7.49]$$

E7.26 🖥

Using equation [7.49]:

(i) $6! \,/\, 4! = 6 \times 5 = 30$

(ii) $46! \,/\, 41! = 46 \times (46-1) \times \ldots \times (41+2) \times (41+1) = 46 \times 45 \times 44 \times 43 \times 42$

Q7.34   Work out the values of each of the following WITHOUT using a calculator. (Check the results afterwards using a calculator.)

(i)   5!

(ii) $(4-3)!$

(iii) $(3-3)!$

(iv) $3! - 3!$

(v) $\dfrac{7!}{5!}$

(vi) $\dfrac{101!}{99!}$

7.5.3 Permutations

A mathematical *permutation* is the *number of ways of arranging 'r' items in order* when selected from '*n*' items.

Arranging *n* items in order   The simplest problem is to find out how many ways it is possible to arrange *n* items into different sequences (or order).

E7.27

If we have three items: X, Y and Z, we can produce six possible arrangements which differ in the ordering of the items:

XYZ, XZY, YXZ, YZX, ZXY, ZYX

We can calculate the number of ways mathematically by considering how many 'choices' we have when fill the *first*, *second* and *third* places in the order. Initially we have *three* choices for which letter to put in first place (e.g. Y), but we will then only have a choice of *two* letters left (e.g. X and Z) for the second place, and then only *one* letter for the third place:

We have a choice of three items for the first place (X, Y or Z):	We now have two choices left for the second place:	We only have one 'choice' left for the final place:	Total number of different ways:
3	$\times 2$	$\times 1$	$= 3! = 6$

Thus the total number of possible arrangements was $3 \times 2 \times 1 = 3!$

In general:

$$\textbf{Number of ways} \text{ of arranging } r \text{ items in order} = r! \qquad [7.50]$$

Q7.35 In how many ways can the four nitrogen bases, A, T, G, and C, be arranged in different orders?

Arranging *r* items in order when selected from *n* possible items

We can illustrate the problem using the following example:

E7.28 🖳

As a specific example, consider selecting 3 $(= r)$ items, *in a specific order*, from 7 $(= n)$ items (e.g. TUVWXYZ). The number of ways can be calculated as follows:

We have a choice of $n = 7$ items for the first place:	We now have $(7 - 1)$ choices left for the second place:	We now have $(7 - 2)$ choices left for the third place:	Total number of different ways:
7	$\times 6$	$\times 5$	$= 7 \times 6 \times 5$

In this example the total number of ways of ordering 3 $(= r)$ from 7 $(= n)$ items $= 7 \times 6 \times 5 = 210$.

We can rewrite the number as follows:

$$7 \times 6 \times 5 = \frac{7 \times 6 \times 5 \times \cancel{4} \times \cancel{3} \times \cancel{2} \times 1}{\cancel{4} \times \cancel{3} \times \cancel{2} \times 1} = \frac{7!}{4!} = \frac{7!}{(7-3)!} = \frac{n!}{(n-r)!}$$

The general formula for arranging *r* items *in order* when selected from *n* possible items is given by the **permutation function**:

$$_nP_r = \frac{n!}{(n-r)!} \qquad [7.51]$$

This function can also be calculated directly in most scientific calculators.

$_nP_r$ can be calculated in Excel using the function PERMUT(n, r).

E7.29 🖥️

Calculate the number of ways of selecting the first, second and third horses (order important) in a horse race with 10 horses:

$$_{10}P_3 = 10!/(10 - 3)! = 10 \times 9 \times 8 = 720$$

Q7.36 A sportsman has eight different trophies, but only has room for four trophies in a display cabinet.

How many different ways is it possible to display just four trophies out of 8, assuming that the display order is important?

7.5.4 Combinations

A mathematical *combination* is the number of different *ways of selecting* 'r' items from 'n' items when the *order is NOT important.*

E7.30 🖥️

Calculate the number of ways of selecting two items from four possible items.

There are 12 ($=_4P_2$) different ways of *ordering* two letters from the four letters W, X, Y and Z:

WX, WY, WZ, XW, XY, XZ, YW, YX, YZ, ZW, ZX, ZY

However, if the *order of selection* is *not* important, then there are *no differences* between each of the following six pairs of selections:

WX and XW, WY and YW, WZ and ZW, XY and YX, XZ and ZX, YZ and ZY

In this case there will then be only six *different* ways of selecting two items from four items with the order *not* important:

WX, WY, WZ, XY, XZ, YZ

The number of ways of *selecting r* items from *n* items when the order is not important is given by the function $_nC_r$.

We know that the number of ways that each *selection* of the r items can be ordered amongst themselves $= r!$.

Hence, from the previous two statements, we see that the number of ways of first *selecting* and then *ordering* r items from n items, $_nP_r$, is given by the equation:

$$_nP_r = {}_nC_r \times r!$$
[7.52]

The number of combinations can therefore be calculated using:

$$_nC_r = \frac{_nP_r}{r!}$$
[7.53]

which is the same as:

$$_nC_r = \frac{n!}{r! \times (n-r)!}$$
[7.54]

This function can also be calculated directly in most scientific calculators.

$_nC_r$ can be calculated in Excel using the function COMBIN(n, r).

The example E7.30 can be solved using [7.54], to give the number of ways of selecting ($r =$) two items from ($n =$) four items (W, X, Y and Z) where the order is NOT important:

$$_4C_2 = \frac{4!}{2! \times (4-2)!} = \frac{24}{2 \times 2} = 6$$

E7.31 🖥

Calculate the number of ways of selecting the first, second and third horse in a horse race with 10 horses, where order is NOT important:

$$_{10}C_3 = {}_{10}P_3 \, / \, 3! = 10!/\{3! \times (10-3)!\} = 720 \, / \, 6 = 120$$

E7.32 UK National Lottery (Lotto) 🖥

Calculate the probability of getting the jackpot in the UK National Lottery (Lotto), where six different numbers must be selected from 49 numbers.

The number of *arrangements* of six numbers selected from 49 numbers is:

$$_{49}P_6 = 1.0068 \times 10^{10}$$

However, many of the above arrangements will have selected the *same numbers but in a different order*. For each set of 6 numbers, there are 6! ways of rearranging them into a different order.

To win the lottery, the order of the six numbers is *not* important. The number of ways of selecting six numbers *in any order* from 49 numbers is

$$_{49}C_6 = {}_{49}P_6 / 6! = 1.0068 \times 10^{10} / 6! = 13\,983\,816 \approx 14\,\text{million}$$

The aim, when buying a Lottery ticket, is to guess which one of the 14 million possible selections of the six numbers will occur.

Thus the probability of selecting the *correct* six numbers is 1 in 14 million.

Q7.37 A football manager has a squad of 20 players.

How many different teams of 11 players could be selected from the squad?

8
Distributions and Uncertainty

Overview

The concept of a continuous 'distribution' in statistics was introduced in 7.2.8, and is a method of describing how the values of a data variable may be spread over a given range. The distribution may record the variability of data that has been recorded in the *past* (frequency distribution) or predict the variability with which data values might be expected to be recorded in the *future* (probability distribution).

This chapter describes three distributions important in science – the normal, binomial, and Poisson distributions. These have importance in modelling different systems in science:

- The normal distribution can be used to model a very wide range of systems where the distribution follows the familiar 'bell-shaped' curve. A particularly common application in science occurs in relation to experimental uncertainty.

- The binomial distribution applies to systems where, at the fundamental level, the system can be in *one of two* possible states – hence the prefix 'bi-'. A classic example of binomial statistics might calculate the probability of getting eight heads when a coin is tossed 10 times.

- The Poisson distribution is a special case of the binomial distribution, which occurs when the individual probability of a given event is low. This distribution can be used to estimate the probabilities for the occurrences of such events, for example, the probability that a particular number of cases of a randomly occurring disease may appear in a specific locality.

The most important parameters used to describe a distribution – mean, standard deviation, variance – have already been introduced in 7.1.4 and 7.1.6, and are used to describe the *location* and *spread* of the distribution. An additional parameter, the coefficient of dispersion, *CD*, has also been introduced in 7.1.6 to describe the particular *shape* of a distribution. Other parameters that are used to describe the characteristic shape of distributions are given in the associated web site.

Different types of scientific mechanisms may produce experimental data with different distributions. If it is possible to confirm that the data falls into one or other type of distribution, then it is then possible to deduce that one mechanism is more likely to be operating than a possible alternative. In this respect, 'goodness of fit' (Unit 11.3) is a useful hypothesis test that can be used to assess whether an experimental distribution might be different from a particular theoretical model.

The impact of experimental uncertainty on scientific measurements has been introduced in Unit 1.2. The normal distribution has particular value in dealing with experimental uncertainty, as it closely models the variability in the results of many types of experimental measurement. In particular, the random uncertainty can be quantified by defining a *confidence interval (CI)* at X%, which has an X% probability of including the true value being measured.

For example if an experimental result is presented as:

$$4.67 \pm 0.05 \, (95\% \, \text{CI})$$

then it can be stated with 95% probability that the true value of the variable being measured lies between 4.62 (= 4.67 − 0.05) and 4.72 (= 4.67 + 0.05).

The ways in which the uncertainties in separate variables combine in a calculation of overall uncertainty is discussed in 8.3.3.

8.1 Normal distribution

8.1.1 Introduction

The bell-shaped normal distribution is the most widely used distribution in science. Of particular importance is the fact that many experimental variations can be described effectively by using the normal distribution.

8.1.2 Experimental uncertainty

There is always some uncertainty when making any experimental measurement. If we are trying to measure some experimental property, whose true value is μ, then we are more likely to record a value close to μ, and less likely to record a value a long way away from μ.

The probability density, $pd(x)$ (7.2.8), of recording a particular value, x, is largest when x is close to μ, but then the probability falls off as a 'bell-shaped' curve on either side of μ:

E8.1

Figure 8.1 shows the probabilities of getting an experimental result, x, in specific ranges, when measuring a **true** blood-alcohol level of 80 mg $(100 \, \text{mL})^{-1}$, the *standard deviation* uncertainty in the measurement process being used is 2.0 mg $(100 \, \text{mL})^{-1}$.

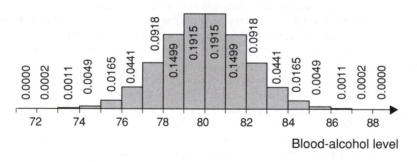

Figure 8.1. Histogram of experimental results.

For example, the histogram shows that the probability of recording a value between 81 and 82 is 0.1499 (or about 15%).

The mathematical name for this 'bell-shaped' distribution is the **normal distribution**. Most *random* experimental errors can be considered to have a frequency spread of results that follow a normal distribution.

Q8.1 Figure 8.1 shows the probabilities of a single experimental result, with a standard deviation of 2.0, falling within unit ranges about a mean value of 80.

Calculate, *using the data* in Figure 8.1, the probabilities that a single experimental result will fall within the following ranges in mg $(100 \text{ mL})^{-1}$:

  (i)   80–81

  (ii) 81–83

  (iii) >80

  (iv) 78–82 (i.e. 80 ± 1 standard deviation)

  (v)   76–84 (i.e. 80 ± 2 standard deviation)

  (vi) 74–86 (i.e. 80 ± 3 standard deviation)

  (vii) 72–88 (i.e. 80 ± 4 standard deviation)

8.1.3 Normal distribution

The **normal distribution** (also called the Gaussian distribution) is defined by the probability density equation, $pd(x)$, as a function of the variable, x (with mean value, μ, and standard deviation, σ):

$$pd(x) = \frac{1}{\sqrt{2\pi\sigma^2}} \exp\left\{-\frac{(x-\mu)^2}{2\sigma^2}\right\} \qquad [8.1]$$

Any particular normal distribution can be identified using the shorthand: $N(\mu, \sigma^2)$. Remember that it is necessary to take the square root of the variance, σ^2, to get the standard deviation, σ. For example, $N(34,9)$ describes a normal distribution with a mean value, $\mu = 34$ and a variance, $\sigma^2 = 9$. The standard deviation for the distribution, $\sigma = \sqrt{9} = 3$.

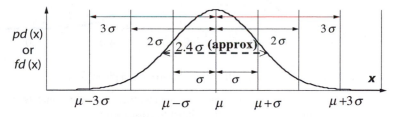

Figure 8.2. Normal curve with mean, μ, and standard deviation, σ.

The normal curve in Figure 8.2 gives the probability density, $pd(x)$, or frequency density, $fd(x)$, of recording a particular value, x, when the values are distributed with a standard deviation, σ, about a mean, μ.

E8.2

Make a careful note of the 'width' of the curve in Figure 8.2 compared to the size of the standard deviation, σ:

- There are *about* six 'standard deviations' between the two extreme 'tails' on either side.

- The distribution *width at half its maximum height* is approximately equal to $2.4 \times \sigma$.

There are very many different normal curves, each defined by a different *mean* and *standard deviation*. However, two important facts are that

- they all have the same overall 'bell' *shape*, and

- they all have the same *area* (in the case of the probability distribution, this area is 1.0).

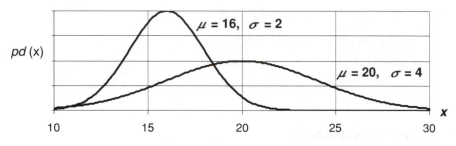

Figure 8.3. Distributions $N(16,4)$ and $N(20,16)$.

Figure 8.3 shows normal distributions, $N(16,4)$ and $N(20,16)$, with means 16 and 20 and standard deviations 2 and 4 respectively. Note that they have the same probability area ($= 1.0$), with one curve having twice the height of the other but half the width.

Q8.2

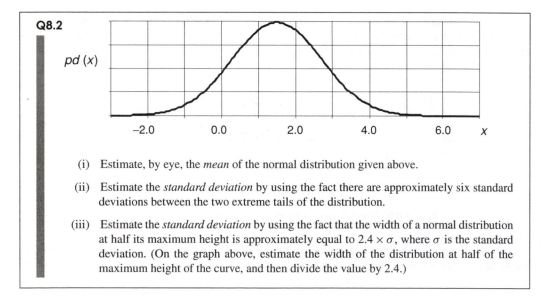

(i)   Estimate, by eye, the *mean* of the normal distribution given above.

(ii) Estimate the *standard deviation* by using the fact there are approximately six standard deviations between the two extreme tails of the distribution.

(iii) Estimate the *standard deviation* by using the fact that the width of a normal distribution at half its maximum height is approximately equal to $2.4 \times \sigma$, where σ is the standard deviation. (On the graph above, estimate the width of the distribution at half of the maximum height of the curve, and then divide the value by 2.4.)

8.1.4 Standard normal distribution

The **standard normal distribution**, $N(0,1)$, is a specific normal distribution, which has a *mean value* of **0.0**, a *standard deviation* of **1.0** – see Figure 8.4. The variable on the abscissa (horizontal axis) of the standard normal distribution is usually written as z.

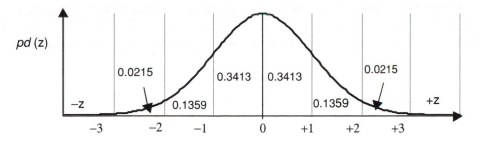

Figure 8.4. Some probability areas for a standard normal distribution.

The probabilities of recording a value that falls between specific z-values can be calculated from standard Tables – see Appendix II.

⌨ Study Resources > Statistical Tables

In particular, Figure 8.4 shows the probabilities (areas) of recording values between *integer* z-values.
The probability of recording a value between z_1 and z_2 can be written as $p(z_1 < z < z_2)$.
The *probability*, $p(z_1 < z < z_2)$, is equal to the *area* under the standard normal distribution between values z_1 and z_2.
The *total probability* (*area*) for *all* values of z, $p(-\infty < z < +\infty) = 1.0$, and the probability area for each 'half' of the distribution, $p(-\infty < z < 0) = p(0 < z < +\infty) = 0.5$.
The normal distribution is symmetrical – the areas on the negative side of the distribution are the same as the equivalent areas on the positive side.

E8.3

Using the probability areas in Figure 8.4, we can calculate the probabilities of recording z values within specific ranges (it is often necessary to add or subtract areas to get the desired final area):

(i) $p(0 < z < +1) = 0.3413 \approx 34\%$

(ii) $p(-1 < z < +1) = 0.3413 + 0.3413 = 0.6826 \approx 68\%$

(iii) $p(+1 < z < +2) = 0.1359$

(iv) $p(0 < z < +2) = p(0 < z < +1) + p(+1 < z < +2) = 0.3413 + 0.1359 = 0.4772 \approx$ 47.7%

(v) $p(-2 < z < +2) = 2 \times 0.4772 = 0.9544 \approx 95\%$

(vi) $p(-\infty < z < 0) = 0.5 = 50\%$

(vii) $p(-\infty < z < 2) = p(-\infty < z < 0) + p(0 < z < +2) = 0.5 + 0.4772 = 0.9772$

(viii) $p(2 < z < \infty) = 1.0 - p(-\infty < z < +2) = 1.0 - 0.9772 = 0.0228 \approx 2.3\%$

Note in this case, how the area in the right-hand 'tail' ($z > 2$) is calculated by taking the area between $z = -\infty$ and $z = 2$ away from the total area (1.0).

Q8.3 Use the data in Figure 8.4 to calculate the following probability areas for the standard normal distribution, $N(0,1)$:

(i) $p(+2 < z < +3)$ (iv) $p(+3 < z < \infty)$

(ii) $p(0 < z < +3)$ (v) $p(-\infty < z < +3)$

(iii) $p(-3 < z < +3)$

8.1.5 Calculating z probability areas

The fractional areas under the *standard normal distribution* are commonly published as a function of the *abscissa variable*, which is usually written as z.

The **z-area table** (Appendix II) gives the total probability (= area), of readings between $z = -\infty$ (the extreme left of the distribution) and the selected value of z.

For example, the probability area, $A_2 (= 0.7517)$, shown in Figure 8.5(a) gives the probability of recording a value between $z = -\infty$ and $z_2 = +0.68$, and the probability area, $A_1 (= 0.0735)$ in Figure 8.5(b) gives the probability of recording a value between $z = -\infty$ and $z_1 = -1.45$:

By taking the *difference* between the areas in Figures 8.5(a) and 8.5(b), we can then calculate the probability area, $A = A_2 - A_1 (= 0.6782)$ in Figure 8.5(c), which gives the probability of recording a value between $z_1 = -1.45$ and $z_2 = 0.68$.

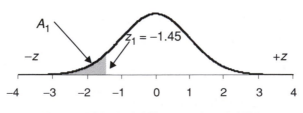

Figure 8.5(a). Probability area, $A_2 = 0.7517$.

Figure 8.5(b). Probability area, $A_1 = 0.0735$.

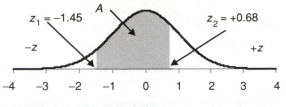

Figure 8.5(c). Probability area $A = A_2 - A_1$.

To calculate the probability (area) between any two specific values of z

(For example: from $z_1 = -1.45$ to $z_2 = +0.68$):

1   Look up the areas, A_1 and A_2, from the z-Area Table for z_1 and z_2. For example, for $z_2 = 0.68$, the area, $A_2 = p(-\infty < z < z_2) = 0.7517 -$ see Figure 8.5(a).

2   If the z-value is **negative**, look up the area for the equivalent *positive* z-value, and **subtract the area from 1.0000**. For example, for $z_1 = -1.45$, look up the probability area for $z = +1.45$, $p(-\infty < z < +1.45)$, which gives 0.9265, and then the area, $A_1 = p(-\infty < z < -1.45) = 1.0000 - 0.9265 = 0.0735 -$ see Figure 8.5(b).

3   Calculate the difference between the two values, $A = A_2 - A_1$:

$$A = A_2 - A_1 = 0.7517 - 0.0735 = 0.6782 \text{ (see Figure 8.5(c))}.$$

4   The probability of randomly selecting a value between $z = -1.45$ and $z = +0.68$ from the standard normal distribution is given by the probability area, A, shown in Figure 8.5(c). The probability, $A, = 0.6782$.

It is useful (as in Figure 8.5) to draw small sketches of the areas being calculated from the z-area table, and use common sense to check correct addition or subtraction.

Areas in the positive (or upper) 'tail' of the distribution can be calculated by *subtracting the z-area from 1.0*. For example, the probability area for $z > 2.0$ can be written as $p(2.0 < z)$, and is calculated as follows:

$$p(2.0 < z) = p(2.0 < z < +\infty) = 1.0000 - p(-\infty < z < 2.0) = 1.0 - 0.9772 = 0.0228$$

Areas in the lower tail are equal (symmetrical) to the equivalent upper tail value, i.e. $p(z < -2.0) = p(2.0 < z) = 0.0228$.

The probability areas for the standard normal distribution can also be calculated using the function NORMSDIST in Excel.

💻 Computer Techniques > Excel Tutorial: Distributions

E8.4 💻

Use statistical Tables to calculate the following probability areas for the standard normal distribution, $N(0,1)$. (It may be helpful to draw small sketches of the areas on a normal distribution.)

(i) $p(-\infty < z < 0) = 0.5000$

(exactly half of the distribution)

(ii) $p(-\infty < z < +0.9) = 0.8159$

(value directly from Tables)

(iii) $p(0 < z < +0.9) = p(-\infty < z < +0.9) - p(-\infty < z < 0) = 0.8159 - 0.5000 = 0.3159$

(difference between (ii) and (i))

(iv) $p(-0.9 < z < 0) = 0.3159$

(area is symmetrical with (iii))

(v) $p(-0.9 < z < +0.9) = 0.3159 + 0.3159 = 0.6318$

(addition of (iii) and (iv))

(vi) $p(+0.9 < z < +\infty) = 1.0000 - p(-\infty < z < +0.9) = 1.0000 - 0.8159 = 0.1841$

(gives the upper tail area to the right of $z = 0.9$)

(vii) $p(-\infty < z < -0.9) = 1.0000 - p(-\infty < z < +0.9) = 0.1841$

(the lower tail area is found by calculating the equivalent upper tail area)

(viii) $p(-\infty < z < -1.62) = 1.0000 - p(-\infty < z < +1.62) = 1.0000 - 0.9474 = 0.0526$

(taking value from Tables)

(ix) $p(-1.62 < z < +0.9) = p(-\infty < z < +0.9) - p(-\infty < z < -1.62) = 0.8159 - 0.0526 = 0.7633$

(difference between (ii) and (viii))

Q8.4 Use statistical Tables to calculate the following probability areas for the standard normal distribution, $N(0,1)$:

(i) $p(-\infty < z < +0.75)$ (v) $p(1.31 < z < 2.0)$

(ii) $p(0 < z < +0.75)$ (vi) $p(-1.56 < z < -0.42)$

(iii) $p(-0.75 < z < 0)$ (vii) $p(2.5 < z) = p(2.5 < z < +\infty)$

(iv) $p(-0.22 < z < 1.31)$ (viii) $p(z < -1.78) = p(-\infty < z < -1.78)$

8.1.6 Reverse z calculations

There are occasions when it is necessary to carry out a z calculation *backwards*, i.e. starting with a probability area, A, and then looking up the equivalent z-value from the Tables.

It is possible to calculate z-values from probabilities for the standard normal distribution by using the function NORMSINV in EXCEL.

💻 Computer Techniques > Excel Tutorial: Distributions

E8.5 🖳

Find the z-value such that the probability of getting a value greater than this value is 5%.

In this case, we need to find a value, z', such that $p(z' < z) = 0.0500$.

If $p(z' < z) = p(z' < z < \infty) = 0.0500$, we can deduce that $p(-\infty < z < z') = 0.9500$.

Hence we need to look in the z-area table to find a value of z' such that the probability area, $A = 0.9500$.

We find that $z = 1.64$ gives $A = 0.9495$ and $z = 1.65$ gives $A = 0.9505$.

By interpolation, we infer that the value of z' must lie midway between these values, giving $z' = 1.645$.

Q8.5 Use statistical Tables to calculate the following values of z' for the standard normal distribution, $N(0,1)$. (It would be useful to draw the probability areas on small sketch graphs – this will help work out the additions and subtractions required.)

(i) Value of z', such that $p(0 < z < z') = 0.3944$

Hint: $p(0 < z < z') = p(-\infty < z < z') - 0.5000$.

(ii) Value of z', such that $p(z' < z < 0) = 0.3944$

Hint: by symmetry $p(z' < z < 0) = p(0 < z < z'')$ where $z' = -z''$, then use result from (i).

(iii) Value of z', such that $p(z' < z) = p(z' < z < +\infty) = 0.025$

Hint: $p(z' < z < +\infty) = 1.0000 - p(-\infty < z < z')$.

(iv) Value of z', such that $p(z' < z) = p(z' < z < +\infty) = 0.005$

Hint: as for (iii).

(v) Value of z', such that $p(z < z') = p(-\infty < z < z') = 0.025$

Hint: by symmetry $p(z < z') = p(z'' < z)$ where $z' = -z''$, then use result from (iii).

8.1.7 Common probability areas

We can use the probability areas from the standard normal distribution to work out some common probabilities associated with *all* normal distributions:

- 68.3% ($\approx$ 2/3) of all data points lie within $\pm$ONE standard deviation from the mean.

- 95.4% ($\approx$ 95%) of all data points lie within $\pm$TWO standard deviations from the mean.

- 99.7% of all data points lie within $\pm$THREE standard deviations from the mean.

- 90% of all data points lie within ± 1.64 standard deviations from the mean.

- 95% of all data points lie within ± 1.96 ($\approx \pm 2$) standard deviations from the mean.

- 99% of all data points lie within ± 2.58 standard deviations from the mean.

Q8.6 Use statistical Tables (or the Excel function NORMSINV), to calculate the following values of z' in the standard normal distribution such that

(i) 2.5% of the probability area lies in the region defined by: $z' < z$

(ii) 0.5% of the probability area lies in the region defined by: $z' < z$

(iii) 95% of the probability area lies in the region defined by: $-z' < z < z'$

(iv) 99% of the probability area lies in the region defined by: $-z' < z < z'$

8.1.8 Using z-values for ANY normal distribution

The probability areas (between x_1 and x_2) for *any* normal distribution (with mean, μ, and standard deviation, σ) can be calculated by comparing it with the *identical* probability areas between equivalent points (z_1 and z_2) of the standard normal distribution.

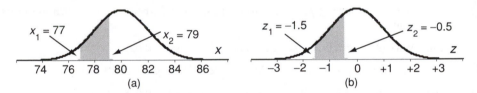

Figure 8.6. Identical probability areas in x- and z-distributions.

For example, we will compare the probability area for the distribution $N(80,4)$ in Figure 8.6(a) with the same probability area in the standard normal distribution, $N(0,1)$ in Figure 8.6(b).

We can see from Figure 8.6 that the probability area between $x_1 = 77$ and $x_2 = 79$ in the normal distribution, $N(80,4)$, **is the same** as the probability area between $z_1 = -1.5$ and $z_2 = -0.5$ in the standard normal distribution, $N(0,1)$.

The 'z-value', z_i, in Figure 8.6(b) that is equivalent to the 'x-value', x_i, in Figure 8.6(a) is given by the important equation:

$$z_i = \frac{(x_i - \mu)}{\sigma} \qquad\qquad [8.2]$$

We outline, in E8.6 below, the general procedure for addressing this type of problem.

E8.6 🖳

Using the distribution in Figure 8.6 with a normal distribution of data values with mean $= 80.0$ and standard deviation $= 2.0$, calculate the probability of recording a data value between 77.0 and 79.0.

1   Calculate the z_i values equivalent to the particular x_i values using [8.2] with $\mu = 80$ and $\sigma = 2$:

$$z_i = \frac{(x_i - 80)}{2}$$

For $x_1 = 77$, the equivalent $z_1 = (77 - 80)/2 = -1.5$.

For $x_2 = 79$, the equivalent $z_2 = (79 - 80)/2 = -0.5$.

2   Calculate the probability areas from the z-values:

Probability area for $z_1 = -1.5$, $A_1 = 0.0668$

Probability area for $z_2 = -0.5$, $A_2 = 0.3085$

3   Calculate the area between the two values:

$$A = A_2 - A_1 = 0.3085 - 0.0668 = 0.2417$$

This is the probability of recording a data value between -1.5 and -0.5 in the standard normal distribution.

4   The probability areas in Figures 8.6(a) and (b) are the same. Hence, we can also say that there is a probability of 0.2417 ($= 24.17\%$) of recording a data value between 77.00 and 79.00 for a normal distribution of data values with mean $= 80.0$ and standard deviation $= 2.0$.

Answer $= 0.2417$.

Q8.7   A normal probability distribution along the variable axis x is described by $N(5,9)$. Calculate

   (i)   mean value of the distribution

   (ii)   standard deviation of the distribution

   (iii) z-value equivalent to the mean value, $x = 5$

   (iv) z-value equivalent to $x = 8$

   (v) z-value equivalent to $x = 2$

   (vi) z-value equivalent to $x = 11$

   (vii) z-value equivalent to $x = -1$

   (viii) probability area, $p(5 < x < 11)$, between the mean value, $x = 5$, and $x = 11$

   (ix)   probability area, $p(2 < x < 8)$, between $x = 2$ and $x = 8$

   (x)   probability area, $p(-1 < x < 11)$, between $x = -1$ and $x = 11$

   (xi)   probability area, $p(11 < x < \infty)$, greater than $x = 11$

The cumulative probability areas for *any* normal distribution can also be calculated using the function NORMDIST in Excel.

⌨ Computer Techniques > Excel Tutorial: Distributions

E8.7 ⌨

The distribution of (integer) marks for a particular examination is described by a normal distribution with

$$\text{Mean value, } \bar{x} = 53.50$$

$$\text{Standard deviation, } s = 11.47$$

For a group of 200 students, use Excel to estimate the numbers of students falling into each of the ranges given in the table below.

– Take the boundaries between each range at values 9.5, 19.5, etc.

– Use the function NORMDIST to calculate the *cumulative probability* up to each range boundary, e.g. NORMDIST(39.5,53.5,11.47,TRUE) calculates the cumulative probability up to 39.5.

– Calculate the *probability for each range* by taking the difference between successive values for the cumulative probability.

– Multiply the probability for each range by 200.

– The results are given in the table below (with probabilities to three decimal places and the final estimated numbers given to two decimal places):

Range	0–9	10–19	20–29	30–39	40–49	50–59	60–69	70–79	80–89	90+
Boundary	9.5	19.5	29.5	39.5	49.5	59.5	69.5	79.5	89.5	100
Cumulative Probability	0.000	0.001	0.018	0.111	0.364	0.700	0.918	0.988	0.999	1.000
Probability	0.000	0.001	0.017	0.093	0.253	0.336	0.219	0.070	0.011	0.001
Number	0.01	0.29	3.34	18.58	50.50	67.18	43.79	13.96	2.17	0.16

See also E11.8.

8.1.9 Reverse calculations for ANY normal distribution

When performing 'reverse' calculations (i.e. finding values of x from z-values) it is necessary to calculate x_i from z_i, and this requires a rearrangement of [8.2]:

$$x_i = \mu + (z_i \times \sigma) \tag{8.3}$$

It is also possible to calculate x-values from probabilities for any normal distribution by using the function NORMINV in Excel.

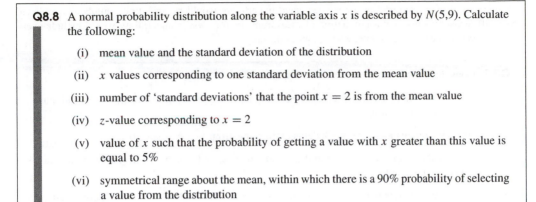

Q8.8 A normal probability distribution along the variable axis x is described by $N(5,9)$. Calculate the following:

   (i)   mean value and the standard deviation of the distribution

   (ii) x values corresponding to one standard deviation from the mean value

   (iii) number of 'standard deviations' that the point $x = 2$ is from the mean value

   (iv) z-value corresponding to $x = 2$

   (v)   value of x such that the probability of getting a value with x greater than this value is equal to 5%

   (vi) symmetrical range about the mean, within which there is a 90% probability of selecting a value from the distribution

   (vii) symmetrical range about the mean, within which there is a 95% probability of selecting a value from the distribution

8.2 Uncertainties in measurement

8.2.1 Introduction

We have seen in Unit 1.2 that a typical experimental measurement will have a random uncertainty, which may be due to, either

- a *variation in the measurement procedure* itself (for example, uncertainty in the response of the pH electrode), or

- a *natural variation in the subjects* being measured (for example, different plants of the same crop grow at different rates).

The problems of experimental uncertainty are fundamental to all experimental science, and a more extensive discussion of experimental variation is given in Unit 13.1.

8.2.2 Experimental variation and true value

In a typical experimental measurement, we aim to try to identify the *unknown value*, μ, of some scientific parameter (e.g. *pH* of a particular solution, or the average growth rate of crop plants).

The **true value** (Unit 1.2) means something slightly different in each of the two types of experimental uncertainty:
- Where the uncertainty is due entirely to the measurement process itself, the true value is the *single* actual value, μ, of the subject that is being measured (e.g. the true *pH* of the solution).

- Where the major uncertainty lies in the subjects that are being measured, the true value is the mean value, μ, of ALL the possible measurements (population) that could be made (for example, the mean growth rate of all the plants of a given crop).

Fortunately, we can apply the *same statistics* to both of the above situations. We see that the mean value, $\bar{x}$, of a *sample* of n experimental measurements is used as a 'best estimate' for true mean value, μ, that would be obtained if a full *population* of measurements could be made.

A population of measurements for *measurement* uncertainty would be an infinite number of replicate (repeated) measurements, and a population of measurements for *subject* uncertainty would include every possible measurement (for example, every plant in the crop).

Q8.9 Consider the two experimental investigations:

(A) A metallurgist makes four repeated (replicate) analyses for the concentration of cadmium in a piece of iron.

(B) A biologist measures the lengths of 30 stems from a crop of soya bean.

Both of these scientists will record a mean value, $\bar{x}$, and a standard deviation, s, for their experimental results.

(i) Explain, in both cases, the probable sources of the variation, s, in the experimental results.

(ii) What will be the relevance of the value, $\bar{x}$, to the metallurgist in (A)?

(iii) What will be the relevance of the value, $\bar{x}$, to the biologist in (B)?

The effect of experimental variation can be described by using the probability density, $pd(x)$, of recording a particular experimental value, x.

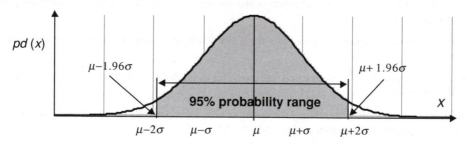

Figure 8.7. Distribution of individual results, x, for an experimental uncertainty which has a normal distribution with a mean, μ, and standard deviation, σ.

Figure 8.7 shows that the *true value* being measured is μ, but experimental uncertainty means that *probability* of getting an experimental *value*, x, is often given by a *bell-shaped* normal distribution.

The distribution shows that it is more likely that the experimental result, x, will be close to μ, but there is also a decreasing probability that a result, x might be recorded a long way from μ.

From 8.1.7 we can see that, for *individual* experimental results, x,

$$90\% \text{ of } x \text{ would fall in the range: } \mu - 1.64 \times \sigma \quad \text{to} \quad \mu + 1.64 \times \sigma$$
$$95\% \text{ of } x \text{ would fall in the range: } \mu - 1.96 \times \sigma \quad \text{to} \quad \mu + 1.96 \times \sigma \qquad [8.4]$$
$$99\% \text{ of } x \text{ would fall in the range: } \mu - 2.58 \times \sigma \quad \text{to} \quad \mu + 2.58 \times \sigma$$

N.B. It is worth remembering the numbers: 1.64, 1.96 and 2.58 – they occur frequently.

Q8.10 In a particular experiment to measure the concentration of lead in drinking water, the experimental procedure has a standard deviation uncertainty of 2.0 ppb. A number of analysts each make a *single* experimental measurement.

If the true value of the concentration is 23 ppb, estimate the range of values within which 95% of the analysts' results are likely to fall.

8.2.3 Replicate measurements, sample means and standard error

The uncertainty in a *single* experimental measurement is given directly by the standard deviation, σ, of the experimental variation – see 8.2.2.

However, it is good scientific practice to take several repeat (or *replicate*) measurements, n, and then take the average (mean) of that sample of n measurements.

When we take an increasing number of replicate measurements, the mean, $\bar{x}$, of our replicates will tend to become closer to the 'true' value, μ. This is a consequence of a statistical theory called the Central Limit Theorem.

Another important consequence of the Central Limit Theorem is that the distribution of *sample mean* values will tend towards a *normal* distribution, even if the distribution of the individual measurements is *not normal* (e.g. a Poisson distribution). By taking the mean values of data samples, we are more likely to ensure that our experimental uncertainties do follow normal distributions – this is important for the validity of much of the statistics used to analyse such data. In practice, a sample size of at least 30 data values is sufficient.

The reduced uncertainty (or *improved* accuracy) when taking the *mean* of n *replicate measurements* is given by:

$$\text{Standard error of the mean, } SE = \frac{\sigma}{\sqrt{n}} \qquad\qquad [8.5]$$

(Note that the equivalent statistic when using the sample standard deviation, s, is called the standard uncertainty – see 8.2.5.)

According to the Central Limit Theorem, the distribution of the possible mean values, $\bar{x}$, of n experimental measurements follows a normal distribution, as in Figure 8.8, with a mean (of mean values) $= \mu$ and a standard deviation $= SE = \sigma/\sqrt{n}$.

Figure 8.8 gives the probability density, $pd(\bar{x})$, for recording mean values, $\bar{x}$, of different sample sizes, n. The sample size, $n = 1$, represents a single measurement as in Figure 8.7.

As the sample size, n, increases:

- the uncertainty in the mean value gets smaller,

- the probability curve gets narrower, and

- the experimental result, $\bar{x}$, is likely to be closer to the 'true' value, μ.

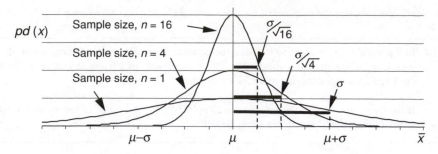

Figure 8.8. Distribution of mean values, $\bar{x}$, for samples of different sizes.

Comparing with [8.4], we can see that, for the *mean values*, $\bar{x}$, of n experimental measurements (note the effect of the number of replicate measurements, n):

90% of $\bar{x}$ values would fall in the range: $\mu - \left\{ 1.64 \times \dfrac{\sigma}{\sqrt{n}} \right\}$ to $\mu + \left\{ 1.64 \times \dfrac{\sigma}{\sqrt{n}} \right\}$

95% of $\bar{x}$ values would fall in the range: $\mu - \left\{ 1.96 \times \dfrac{\sigma}{\sqrt{n}} \right\}$ to $\mu + \left\{ 1.96 \times \dfrac{\sigma}{\sqrt{n}} \right\}$ [8.6]

99% of $\bar{x}$ values would fall in the range: $\mu - \left\{ 2.58 \times \dfrac{\sigma}{\sqrt{n}} \right\}$ to $\mu + \left\{ 2.58 \times \dfrac{\sigma}{\sqrt{n}} \right\}$

Q8.11 In the same experimental procedure as in Q8.10 the standard deviation uncertainty of measurement is 2.0 ppb. In this case, each analyst makes four measurements and then calculates the mean value, $\bar{x}$, of his/her four measurements.

   (i)   Calculate the standard error of the mean values, $\bar{x}$.

   (ii) If the true value of the concentration is 23 ppb, estimate the range of values within which 95% of the experimental means, $\bar{x}$, are likely to fall.

For problems that cannot be solved directly using [8.6], it is necessary to use the general statistics developed in 8.1.8 for performing calculations based on the normal distribution of experimental mean values.

Q8.12 An experimental measurement process gives a result described by a normal distribution, $N(15.0, 0.25)$, i.e. a mean value of 15.0 and a standard deviation of 0.5.

Single measurements are made. Calculate:

   (i)   The probability of recording a result greater than 16.0.

   (ii) The probability of recording a result between 14.0 and 14.5.

   (iii) The value, x, such that there is a 5% probability of recording a result greater than x.

Ten *replicate* measurements are made. Calculate:

(iv) The probability that the *mean* of these 10 replicates will be greater than 15.2.

(v) The range of values within which there is a 95% probability of recording the *mean* of the 10 replicates.

(vi) The value, x, such that there is a 5% probability of recording a *mean value* greater than x.

8.2.4 Experimental perspective

The theory, given in sections 8.2.2 and 8.2.3 above, presents the statistics on the basis that the values of the true mean, μ, and the standard deviation, σ, of the experimental variation *are both known*.

For example, [8.6] states, with 95% confidence, that the mean value, $\bar{x}$, of n measurements will fall within the range:

$$\mu - \left\{ 1.96 \times \frac{\sigma}{\sqrt{n}} \right\} \quad \text{to} \quad \mu + \left\{ 1.96 \times \frac{\sigma}{\sqrt{n}} \right\} \qquad [8.7]$$

However, the *true value* of μ will *not be known*.

The aim of most *actual* experimental investigations is to obtain a 'best estimate' for an *unknown* 'true' value, μ, by using the value of the *sample* mean, $\bar{x}$.

Equation [8.7] can be rearranged to give the range of the *probable* values of μ, based on the actual measured value of $\bar{x}$. We can claim with 95% confidence that the 'true' value, μ, lies within the range:

$$\bar{x} - \left\{ 1.96 \times \frac{\sigma}{\sqrt{n}} \right\} \quad \text{to} \quad \bar{x} + \left\{ 1.96 \times \frac{\sigma}{\sqrt{n}} \right\} \qquad [8.8]$$

where $\bar{x}$ is the mean of n measurements.

Equation [8.8] allows us to *predict* the possible value of μ, assuming that we know the value of $\bar{x}$.

It is important to understand the difference between the probability ranges for $\bar{x}$ and for μ as given above in [8.7] and [8.8] respectively.

E8.8

A set of $n = 6$ replicate measurements are made, giving experimental results: 25, 28, 23, 20, 25, 24. In this example, assume that the standard deviation uncertainty in the measurement procedure is already known from extensive previous measurements, and has the value: $\sigma = 3.0$. Calculate, with 95% confidence, the range of values within which the unknown true value, μ, lies.

The mean of the six results, $\bar{x} = 24.17$.

We have the values: $\sigma = 3.0$ and $n = 6$, hence we can claim, with 95% confidence, that the true value, μ, lies within the range (substituting into [8.8]):

$$24.17 - \left\{ 1.96 \times \frac{3.0}{\sqrt{6}} \right\} \quad \text{to} \quad 24.17 + \left\{ 1.96 \times \frac{3.0}{\sqrt{6}} \right\}$$

which equals

$$24.17 - 2.40 \quad \text{to} \quad 24.17 + 2.40$$

giving the range (rounded to 1 dp):

$$21.8 \quad \text{to} \quad 26.6$$

We do not know the true value of μ, but we are 95% confident in stating that it is somewhere between 21.8 and 26.6.

8.2.5 Confidence intervals and *t*-values

Section 8.2.4 explained that the objective of a *typical experiment* is to find the best estimate for an *unknown value*, μ, on the basis of the mean, $\bar{x}$, of n experimental results. Equation [8.8] gives the range within which we would be 95% confident of finding the true value, μ, *assuming that we knew the experimental standard deviation, σ*.

However, for many experimental measurements, the standard deviation, σ, is *not known*, and we must take the sample standard deviation, s, as our 'best estimate' (see 7.1.6) for the population standard deviation value, σ.

If we do not already know the true standard deviation, σ, of the experimental procedure, then we must use the measured value of the sample standard deviation, s, to *estimate* the standard deviation in the sample mean, $\bar{x}$, of a variable, x.

We now define **standard uncertainty** of a sample of values (see also 8.3.2) as the *best estimate* of the standard error of the mean of the population [8.5]:

$$\text{Standard uncertainty } u(x) = \frac{s}{\sqrt{n}} \qquad [8.9]$$

Standard uncertainty gives the standard deviation of repeated *sample* mean values. It shows that the uncertainty in the mean of n measurements is reduced by a *factor* '$\sqrt{n}$' compared to the uncertainty of a *single* ($n = 1$) measurement.

For a single measurement, $n = 1$, the standard uncertainty in the value, $\bar{x}$,

$$u(x) = s$$

For a sample size, $n = 4$, the standard uncertainty in the mean value, $\bar{x}$,

$$u(x) = \frac{s}{\sqrt{4}} = \frac{s}{2}$$

For a sample size, $n = 16$, the standard uncertainty in the mean value, $\bar{x}$,

$$u(x) = \frac{s}{\sqrt{16}} = \frac{s}{4}$$

E8.9 🖥

Seven students each made a measurement of the *pH* (acidity) of the same sample of water, giving the results:

$$pH: \quad 7.81 \quad 8.02 \quad 7.93 \quad 7.99 \quad 7.96 \quad 7.93 \quad 7.89$$

Calculate

(i)   the mean value

(ii)   the measurement standard deviation

(iii)   the standard uncertainty in measurement

Answer:

(i)   Using AVERAGE in Excel: Mean value, $\bar{x} = 7.93$

(ii)   Using STDEV in Excel: Sample standard deviation, $s = 0.069$

(iii)   Standard uncertainty, $u(\bar{x}) = s/\sqrt{n} = 0.069/\sqrt{7} = 0.0261$

Using s instead of σ will *increase the uncertainty* in our estimation of the true value, μ, and we must modify the probability statement given in [8.8].

We now claim with 95% confidence that the 'true' value, μ, lies within the range:

$$\bar{x} - \left\{ t_{2,0.05,(n-1)} \times \frac{s}{\sqrt{n}} \right\} \text{ to } \bar{x} + \left\{ t_{2,0.05,(n-1)} \times \frac{s}{\sqrt{n}} \right\} \qquad [8.10]$$

where $\bar{x}$ is the mean of n measurements.

Notice that, when we use the sample standard deviation, s, instead of the population standard deviation, σ, we replace the value '1.96' in our 95% probability range with a *t-value*, $t_{2,0.05,(n-1)}$.

The value for $t_{2,0.05,(n-1)}$ is greater than 1.96, which results in a *wider range of uncertainty* for μ. This increased uncertainty is a result of the extra uncertainty in using the estimated standard deviation, s, instead of the true value, σ.

The specific value for $t_{2,0.05,(n-1)}$ depends on the sample size, n. A table of values is given in Appendix III, but some example values are given in Table 8.1. Some points to note from Table 8.1:

- For very *small* samples ($n = 2$, 3 etc.), the 't-value' is *large* to reflect the greater uncertainty in estimating the standard deviation.

- For very *large* samples, the 't-value' approaches the same value of 1.96 that is used when the standard deviation is accurately known.

- There is very little change in 't-value' (less than 5%) once the sample size, n, is more than about 30.

Table 8.1. Example values for $t_{2,0.05,(n-1)}$.

Sample size, n	2	3	4	6	10	20	30	∞
$t_{2,0.05,(n-1)}$	12.71	4.30	3.18	2.57	2.26	2.09	2.05	1.96

The t-value, $t_{2,0.05,(n-1)}$ is a special case of the general t-value, $t_{T,\alpha,df}$, which is discussed more fully in section 8.2.7.

8.2.6 General equation for the confidence interval of the mean

We now define the **confidence deviation, Cd (μ, X %)**, as the range, *on either side* of the experimental mean value, $\bar{x}$, where we are $X\%$ confident in stating that the true value, μ, lies:

$$Cd(\mu, X\%) = t_{2,\alpha,df} \times \frac{s}{\sqrt{n}} \qquad [8.11]$$

where (see 8.2.7)

$$\alpha \text{ is the significance level, with } \alpha = 1 - X\%/100, \text{ and}$$
$$df \text{ is the number of degrees of freedom}, \textbf{\textit{df}} = \textbf{\textit{n}} - \textbf{1} \qquad [8.12]$$

The **confidence interval, $CI(\mu, X\%)$,** is defined as the *total range* within which we are $X\%$ confident in stating that the true value, μ, lies:

$$CI(\mu, X\%) = \bar{x} \pm Cd(\mu, X\%) = \bar{x} \pm \left\{ t_{2,\alpha,df} \times \frac{s}{\sqrt{n}} \right\} \qquad [8.13]$$

which is equivalent to the range:

$$\bar{x} - Cd(\mu, X\%) \quad \text{to} \quad \bar{x} + Cd(\mu, X\%)$$

or

$$\bar{x} - \left\{ t_{2,\alpha,df} \times \frac{s}{\sqrt{n}} \right\} \quad \text{to} \quad \bar{x} + \left\{ t_{2,\alpha,df} \times \frac{s}{\sqrt{n}} \right\} \qquad [8.14]$$

N.B. The Excel function CONFIDENCE should **not be used** for *sample* confidence intervals. The function assumes that the population standard deviation, σ, is known, which is not normally true in an experimental context.

E8.10

We use the same data as the previous example, E8.8, with a set of $n = 6$ replicate experimental results: $25, 28, 23, 20, 25, 24$. Calculate the range of values, within which we can be 95% confident that the unknown true value, μ, lies. However in this example, we assume that the uncertainty in the measurement procedure is NOT already known, and it is necessary to use the sample standard deviation, s.

The mean of the six results, $\bar{x} = 24.17$.

The calculated sample standard deviation of the six replicates, $s = 2.64$. (This is our best estimate for the unknown population standard deviation, σ, and, compared to the previous example, the 'estimate' is lower than the 'true' value of 3.0.)

From Table 8.1, a sample size, $n = 6$, gives a value of $t_{2,0.05,5} = 2.57$.

$$CI(\mu, 95\%) = \bar{x} \pm \left\{ t_{2,0.05,(n-1)} \times \frac{s}{\sqrt{n}} \right\} = 24.17 \pm \left\{ 2.57 \times \frac{2.64}{\sqrt{6}} \right\} = 24.17 \pm 2.77$$

We can then say with 95% confidence that, on the basis of six replicate measurements, which have a mean of 24.17 and a sample standard deviation of 2.64, the true value being measured lies between (rounded to 1 dp):

$$21.4 \quad \text{and} \quad 27.0$$

Comparing this with the previous example (where the standard deviation, σ, was known), we can see that, when we calculate s from the experimental results themselves, the *range of uncertainty in the final result has become greater*:

$$21.4 \quad \text{to} \quad 27.0 \quad \text{instead of} \quad 21.8 \quad \text{to} \quad 26.6.$$

Q8.13 A student develops a new quick technique to measure the concentration of cadmium (Cd) in river water, and uses the technique to record the values given below (in ppb) for an unknown sample. In each of the cases below the student wishes to report the true concentration for the unknown sample, with a 95% probability of being correct.

(i) The student makes three replicate measurements: 18.2, 20.2, 19.6.
What can the student say about the true value?

(ii) The student makes six replicate measurements: 18.2, 20.2, 19.6, 17.6, 18.5 19.5.
What can the student say about the true value?

(iii) The student records only one value: 18.2.
What can the student say about the unknown true value?

8.2.7 *t*-Values

The theory underlying the t-values (see also 10.2.1) is based on the use of a 't-distribution' for *sample* data, $\bar{x}$ and s, which is analogous to the use of the normal distribution for *population* data, μ and σ.
The value of $t_{T,\alpha,df}$, depends on three factors (which are further discussed below):

- whether the problem involves both 'tails' of the distribution or not, i.e. 2-tailed or 1-tailed: $T = 1$ or 2.

- significance level, α, or confidence level, $X\%$.

- number of degrees of freedom, df.

The 'number of tails' in the problem depends on whether it is a symmetrical (2-tailed) or a one-sided (1-tailed) problem. For a confidence Interval, CI, calculation, where we are considering probabilities on both sides of the mean value, we use the 2-tailed (symmetrical) value, $t_{2,\alpha,df}$, This is further discussed in 8.2.8.

The **confidence level, $X\%$,** is the confidence (as a percentage probability) in being *right* when quoting a confidence interval. However, statisticians will also use **significance level, α** (discussed further in 9.3.4) which is the maximum probability of being *wrong*. Confidence level, $X\%$, and significance level, α, are related as follows:

$$\alpha = 1 - \frac{X\%}{100} \qquad\qquad [8.15]$$

i.e. a confidence level of $X\% = 95\%$ is equivalent to a significance level, $\alpha = 0.05$.

'Degrees of freedom' is a common terminology used by statisticians to define the amount of information that is available to carry out a particular calculation. We will not discuss the theory in this text, but we will show, for particular cases, how the 'number of degrees of freedom', df, can be calculated.

For a confidence interval, CI, calculation:

$$\text{Degrees of freedom, } \boldsymbol{df = n - 1} \text{ where } n \text{ is the size of the sample.} \qquad [8.16]$$

The values for 2-tailed, $t_{2,\alpha,df}$, and 1-tailed, $t_{1,\alpha,df}$, 't-values' are given in statistical Tables in Appendix III.

The 1-tailed value with a significance level, α, can be calculated from the 2-tailed value for significance level, 2α, using:

$$t_{1,\alpha,df} = t_{2,2\alpha,df} \qquad\qquad [8.17]$$

The Excel function TINV can also be used to find the *2-tailed value, $t_{2,\alpha,df}$* for any given level of confidence and degrees of freedom.

⌨ Computer Techniques > Excel Tutorials: Parametric Tests

Q8.14 Use the Table in Appendix III, and/or the Excel function, TINV, to obtain the following critical t-values:

	Tails	Significance/confidence level	Degrees of freedom	t-Value
(i)	2	0.05	6	
(ii)	1	0.05	6	
(iii)	2	99%	6	
(iv)	1	0.01	6	
(v)	2	95%	6	
(vi)	2	95%	10	
(vii)	2	95%	20	
(viii)	2	95%	50	
(ix)	2	95%	Infinity	
(x)	1	0.025	6	
(xi)	1	99.5	6	

(xii) Explain the relationship between the answers for (i) and (v).

(xiii) Explain the progression of values from (v) to (ix).

(xiv) Where, in the Table of t-values does the t-value become equal to the equivalent z-value?

(xv) Explain the relationship between the answers for (i) and (x).

Q8.15 Calculate the confidence intervals, CI, for the following sample data sets:

	Sample size	Mean	Standard deviation	Confidence level		Confidence interval	
(i)	5	6.02	0.3	95%		to	
(ii)	10	46.2	5.6	90%		to	
(iii)	2	13.6	0.4	95%		to	

8.2.8 Two-tailed and one-tailed

Using the data from the E8.10, we have 95% confidence in saying that the unknown value, μ, lies somewhere within the range 21.4 to 27.0. We illustrate this in Figure 8.9.

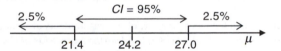

Figure 8.9. Probability ranges for the true value, μ.

- The regions 'outside' the confidence intervals are the **'tails'** of the probability distribution.
- The combined, *2-tailed*, probability that μ is outside the CI is $100\% - 95\% = 5\%$
- The uncertainty distribution is symmetrical.
- The probability that μ is greater than 27.0 is the same as the probability that it could be less than 21.4.
- Hence the *1-tailed* probability that μ could be greater than 27.0 is 2.5%, and the *1-tailed* probability that μ is less than 21.4 is 2.5%.

See also 9.3.3 for the concept of 'tails' in hypothesis testing.

Q8.16 A set of experimental measurements give the following results:

550   553   546   568   548

Calculate the following and answer (v):

(i) the mean of the experimental results;

(ii) the standard deviation of the experimental results;

(iii)   the standard uncertainty of the mean;

(iv)   the confidence interval (at 95%) of the mean value.

(v)   On the basis of the result in (iv), decide whether the mean value is significantly greater than 545 at 97.5% confidence.

8.3 Experimental uncertainty

8.3.1 Introduction

The management of uncertainty in experimental measurements is a recurrent theme throughout this book. In mathematical terms, the aim of a typical measurement is to determine the true value, μ (population mean), of a given variable, for which the scientist makes n replicate measurements of the variable, x, to give an experimental mean value, $\bar{x}$, with a sample standard deviation, s.

Within this Unit, we start by introducing some of the principal terms that are used by scientists to *present* this experimental uncertainty. We then develop the methods that can be used to *combine* the separate uncertainties of individual measurements into a single uncertainty in a final experimental result.

8.3.2 Terminology

It is useful to review some the terms that we have for describing the uncertainty in a particular experimental measurement.

Standard uncertainty, $u(x)$, in the mean value, $\bar{x}$, of n measurements of a variable, x, is defined by the International Organization for Standardization (ISO) *Guide to the Expression of Uncertainty in Measurement* as the standard deviation, s, of the measurements divided by the square root of n:

$$u(x) = \frac{s}{\sqrt{n}} \qquad [8.18]$$

The standard uncertainty (see also 8.2.5) is the best estimate of the standard error of the mean, SE (8.2.3) of the population of measurements from which the sample of experimental measurements was drawn.

We define the new term, **relative uncertainty,** $Ru(x)$, in the mean value, $\bar{x}$, as the percentage ratio of the standard uncertainty to the mean value. Relative uncertainty is the percentage uncertainty in a *particular* experimental *result*.

$$Ru(x) = 100 \times \frac{u(x)}{\bar{x}} = 100 \times \frac{s}{\bar{x} \times \sqrt{n}} \qquad [8.19]$$

We have defined, in 8.2.6, a new term, **confidence deviation** (*Cd*), which is the range *in either direction from the mean value*, $\bar{x}$, within which the true value, μ, can be expected to lie, with a confidence of $X\%$ ($\alpha = 1 - X\%/100$).

$$Cd(\mu, X\%) = t_{2,\alpha,(n-1)} \times \frac{s}{\sqrt{n}} = t_{2,\alpha,(n-1)} \times u(\bar{x}) \qquad [8.20]$$

Confidence interval (CI) is the range of values (8.2.6) within which the true value, μ, can be expected to lie, with a confidence of X% ($\alpha = 1 - X\%/100$).

$$CI(\mu, X\%) = \bar{x} \pm Cd(\mu, X\%) = \bar{x} \pm \left\{ t_{2,\alpha,(n-1)} \times \frac{s}{\sqrt{n}} \right\}$$

$$[8.21]$$

$$CI(\mu, X\%) = \bar{x} \pm \{ t_{2,\alpha,(n-1)} \times u(\bar{x}) \}$$

It is also useful to note two terms that are used to describe the *precision* of an experimental *process* rather than the uncertainty in a particular measurement. **Relative standard deviation (RSD)** and **coefficient of variation (CV)** both give the ratio of the population standard deviation, σ, to the mean value, i.e. the *fractional* standard deviation. The coefficient of variation is the 'percentage' equivalent of relative standard deviation.

$$RSD = \frac{\sigma}{\bar{x}}$$

$$[8.22]$$

$$CV = 100 \times \frac{\sigma}{\bar{x}} = 100 \times RSD$$

When presenting results, the experimental method used should be made explicit, and sufficient information (e.g. sample size, n) should be given to enable the reader to interpret the uncertainty in any of the formats given. The final result should be given with an appropriate number of significant figures (2.1.7).

E8.11 💻

Nine experimental measurements give a mean value of 56.26 with a sample standard deviation of 5.7.

(i)   Calculate values for the standard uncertainty and relative uncertainty.

(ii) Express the final result using a 95% confidence interval.

(i)
$$u(x) = \frac{5.7}{\sqrt{9}} = 1.9$$

$$Ru(x) = 100 \times \frac{1.9}{56.26} = 3.38\%$$

(ii) Degrees of freedom, $df = n - 1 = 9 - 1 = 8$

t-value for 2-tailed, 95%, 8 degrees of freedom $= 2.31$

$$CI(95\%) = 56.26 \pm 2.31 \times 1.9 = 56.26 \pm 4.389$$

which would be expressed as

$$CI(95\%) = 56.3 \pm 4.4$$

Q8.17 A particular measurement procedure is quoted as having a coefficient of variation, CV, of 5%. Calculate the standard uncertainty that would be expected for the mean of eight replicate measurements of a true value of 4.5 ppm.

8.3.3 Combining uncertainties

There are many occasions in science when a final result, r, will be given by the combination of two or more factors, e.g. a, b. The uncertainty in the result, $u(r)$, will depend on a combination of $u(a)$ and $u(b)$.

We will see that the combination of uncertainties requires the addition of the *variances* of either the standard uncertainties or the relative uncertainties.

The methods of combination can be illustrated in Table 8.2.

Table 8.2. Combination of uncertainties.

	Form of equation	Combining uncertainty	
Multiply by a constant, k	$r = k \times a$	$u(r) = k \times u(a)$	[8.23]
Addition or subtraction	$r = a + b$ or $r = a - b$	$u(r) = \sqrt{u(r)^2} = \sqrt{u(a)^2 + u(b)^2}$	[8.24]
Multiplication or division	$r = k\,a\,b$ or $r = k \times \dfrac{a}{b}$	$Ru(r) = \sqrt{Ru(r)^2} = \sqrt{Ru(a)^2 + Ru(b)^2}$	[8.25]
Powers	$r = k \times a^n$	$Ru(r) = n \times Ru(a)$	[8.26]

Some points to note from Table 8.2.

- Multiplying by a constant simply multiplies the standard uncertainty.

- In addition or subtraction, it is necessary to *add* the *squares* of the standard uncertainties (variances) to get the *square* of the final uncertainty.

- In multiplication or division, it is necessary to *add* the squares of the *relative* uncertainties, $Ru()$, to get the square of the *relative uncertainty* in the final result.

- A power term multiplies the *relative* uncertainties, $Ru()$.

In performing calculations, it is often necessary to change between standard uncertainty and relative uncertainty using the equations:

$$Ru(r) = 100 \times \frac{u(r)}{r} \quad \text{and} \quad u(r) = \frac{r \times Ru(r)}{100} \qquad [8.27]$$

E8.12 🖥

Two variables a and b have values and standard uncertainties:

$$a = 5.0 \text{ with } u(a) = 0.3 \quad \text{and} \quad b = 3.0 \text{ with } u(b) = 0.2$$

The table below sets out the calculations for the uncertainties (standard and relative) for various expressions involving a and/or b, using the equations from Table 8.2. Note that in the last three examples, it is necessary to calculate $Ru()$ first and then calculate $u()$. The flows of the calculations are shown by the symbols: $\Rightarrow$ and $\Leftarrow$.

Expression		Value	Standard uncertainty $u()$	$\Rightarrow$	Relative uncertainty $Ru()$
	$a =$	5.0	0.3	$\Rightarrow$	$[8.27] = 100 \times 0.3/5.0 = 6.0\%$
	$b =$	3.0	0.2	$\Rightarrow$	$[8.27] = 100 \times 0.2/3.0 = 6.7\%$
(i)	$r = 5a$	25.0	$[8.23] = 5 \times 0.3 = 1.5$	$\Rightarrow$	$[8.27] = 100 \times 1.5/25 = 6.0\%$
(ii)	$r = a + b$	8.0	$[8.24] = \sqrt{0.3^2 + 0.2^2} = 0.36$	$\Rightarrow$	$[8.27] = 100 \times 0.36/8 = 4.5\%$
(iii)	$r = a - b$	2.0	$[8.24] = \sqrt{0.3^2 + 0.2^2} = 0.36$	$\Rightarrow$	$[8.27] = 100 \times 0.36/2 = 18\%$
(iv)	$r = a \times b$	15.0	$[8.27] = 15 \times 9.0/100 = 1.35$	$\Leftarrow$	$[8.25] = \sqrt{6.0^2 + 6.7^2} = 9.0\%$
(v)	$r = a \div b$	1.67	$[8.27] = 1.67 \times 9.0/100 = 0.15$	$\Leftarrow$	$[8.25] = \sqrt{6.0^2 + 6.7^2} = 9.0\%$
(vi)	$r = a^3$	125.0	$[8.27] = 125 \times 18/100 = 22.5$	$\Leftarrow$	$[8.26] = 3 \times 6 = 18\%$

Note particularly that relative uncertainties *increase dramatically* when the result is:

- a small difference between larger values – (iii),
- due to a positive power – (vi).

Q8.18 Calculate (with the combined uncertainty) the density, ρ, of a piece of compact bone which has been measured to have mass, $m = 4.3$ g with $u(m) = 0.3$ and volume, $V = 2.3$ cm³ with $u(V) = 0.2$. Use the equation:

$$\rho = \frac{m}{V}$$

Use the table below to set out the results as in E8.12.

Variable	Value, x	$u(x)$	$Ru(x)$
Mass, m (g)			
Volume, V (cm³)			
Density, ρ (g cm⁻³)			

8.3.4 Final result

The standard uncertainty, $u(r)$, produced by combining various factors represents a 68% confidence interval for the final result. It is common practice to multiply this by a **coverage factor**, k, to derive a value for a more inclusive confidence interval, *CI*. Typical values for k range from $k = 2 (\approx 95\%\ CI)$ to $k = 3$ $(\approx 99.75\%\ CI)$.

The value for the coverage factor used must be reported with the result.

8.3.5 Deriving standard uncertainties

Before it is possible to 'combine' uncertainties (as in 8.3.3), it is necessary to get them in the same form, preferably as 'standard uncertainties'. However, uncertainties are often described in a variety of different forms, and the conversion required for some common forms is given below.

Confidence interval   When the uncertainty is given as a 95% confidence interval, then the confidence deviation, *Cd*, should be divided by 1.96 (for 99% confidence divide by 2.58), for example:

$$u(x) = \frac{Cd(\mu, 95\%)}{1.96} \qquad [8.28]$$

Standard deviation of small samples   If the uncertainty is given as the standard deviation, *s*, of a small sample (*n* is small), then, to be consistent with other uncertainties, we recommend that the uncertainty is first converted into a confidence interval and then treated as above. Effectively this would modify equation [8.18] to give:

$$u(x) = \frac{t_{2,\alpha,n-1}}{1.96} \times \frac{s}{\sqrt{n}} \qquad [8.29]$$

Limiting uncertainty   Uncertainty is often given as a *limiting* uncertainty that defines the *absolute* range of possible values. For example, a volumetric flask may have a volume certified as 25 mL $\pm$ 0.03 mL, which implies that the volume may lie anywhere within the range 24.97 to 25.03 but will *not* lie outside this range. In effect this is a rectangular distribution.

When combining a limited uncertainty range $V \pm v$, it is first necessary to convert to an effective *standard uncertainty* by dividing by $\sqrt{3}$, i.e. $u(V) = v/\sqrt{3}$.

E8.13

A 100 mL grade A pipette has a limiting uncertainty of $\pm$ 0.08 mL. A technician has a standard deviation uncertainty, $u(V) = 0.09$ mL, when setting the fluid level to the calibrated mark. The technician uses the pipette to deliver a volume of fluid, $V = 100$ mL.

(i)   Calculate the standard uncertainty, $u(V)$, in the volume.

(ii)   State the volume of fluid with an appropriate expression of uncertainty.

Answer:

(i)   Uncertainty in the volume is an additive combination of the uncertainty in the calibration mark and the uncertainty in the setting of the level. The uncertainty in the calibration mark is given by the limiting uncertainty $\pm$ 0.08 mL, which is converted to a *standard uncertainty* by dividing by $\sqrt{3}$:

$$u(V_{\mathrm{CAL}}) = 0.08/\sqrt{3} = 0.046$$

The *standard uncertainty* in setting the fluid level, $u(V_{\mathrm{SET}}) = 0.09$ mL.

The combined *standard uncertainty* is given by:

$$u(V) = \sqrt{u(V_{CAL})^2 + u(V_{SET})^2} = \sqrt{0.046^2 + 0.09^2} = 0.10$$

(ii)   Using a coverage factor of 2:

$$\text{Final volume, } V = 100.00 \pm 0.20^*\text{mL}$$

*Reported uncertainty gives a confidence level of approximately 95%.

8.3.6 Complicated equations

When the final experimental result depends on a complex relationship between a number of separate terms, then it is useful to develop the calculation in the form of a table as given in E8.14.

E8.14 🖥

A body of mass m (g) gives out an amount of heat Q (J) as the temperature falls from T_1 to T_2 (°C).

The specific heat capacity, c, of the body is given by

$$c = \frac{Q}{m(T_1 - T_2)} \text{J g}^{-1}{}^\circ\text{C}^{-1}$$

The value of each variable, x, and its uncertainty, $u(x)$, are given in the table below.

Calculate the value for c, together with the combined uncertainty, $u(c)$.

	Value	Uncertainty	Relative uncertainty
	x	$u(x)$	$Ru(x)$
Heat loss, $Q =$	14.8	0.3	➔ 2.027
Mass, $m =$	0.68	0.01	➔ 1.471
Initial temperature, $T_1 =$	45.6	0.1	
Final temperature, $T_2 =$	40.4	0.1	
Temperature fall, $(T_1 - T_2) =$	5.2	0.141	➔ 2.720
Specific heat capacity, $c =$	4.19	0.155 ◄	3.697

First calculate the temperature difference, $(T_1 - T_2) = 5.2°\text{C}$, and then calculate the specific heat capacity, c, using the above equation: $c = 4.19 \text{ J g}^{-1} {}^\circ\text{C}^{-1}$

The uncertainty in $(T_1 - T_2)$ uses [8.24] $= \sqrt{u(T_1)^2 + u(T_2)^2} = \sqrt{0.1^2 + 0.1^2} = 0.141$ (see data in dashed circle above).

The *relative* uncertainties in all the terms are calculated using [8.19].

The *relative* uncertainty in c uses [8.25] $= \sqrt{2.027^2 + 1.471^2 + 2.720^2} = 3.697$ (see data in dashed rectangle above).

The uncertainty in c is calculated using [8.27]: $u(c) = Ru(c) \times c / 100 = 0.155$.

Specific heat capacity, $c = 4.19 \text{ J g}^{-1} \text{ }^\circ\text{C}^{-1}$ with standard uncertainty 0.16.

Q8.19   An athlete is recorded as running 100 m in 10.15 s. The uncertainty in recording the start time is 0.02 s, the uncertainty in recording the end time is 0.03 s, and the uncertainty in the distance travelled is 0.1 m.

Calculate the average speed of the athlete, together with the uncertainty in the speed.

8.4 Binomial distribution

8.4.1 Introduction

The binomial distribution applies to a physical situation where there are only *two possible results* (hence the prefix 'bi-') for each particular event or *trial*.

8.4.2 Binomial distribution

We will use the following example as an introduction to the distribution.

E8.15   🖥

The parents of four offspring both carry the gene for albinism, which gives each offspring a 1 in 4 probability (0.25) of being an albino.

We wish to calculate the probability, $p(r)$, that, within the group of $n = 4$ offspring, there is a specific number, r, of albinos.

See text for calculation.

In E8.15 there are only two possible states for each offspring – albino (Y) or non-albino (N). Hence this is a *binomial* problem.

Probability that a *particular* offspring is an albino, $p(Y) = p = 0.25$.

Probability that a *particular* offspring is NOT an albino, $p(N) = q = (1 - p) = 0.75$.

The calculation of the probability that r offspring are albinos is an example of a 'multiple outcome' problem discussed in 7.3.6, i.e. there may be several separate outcomes that give the same *overall* outcome. For example, if we calculate the probability that $r = 2$ offspring are albinos (Y), then there are six possible *separate outcomes*, which can be represented by putting the four offspring in a row as follows:

$$YYNN \quad \text{or} \quad YNYN \quad \text{or} \quad YNNY \quad \text{or} \quad NYYN \quad \text{or} \quad NYNY \quad \text{or} \quad NNYY$$

N.B. The number '6' can also be calculated from the combination problem of selecting two from four in any order: $_4C_2 = 6$ – see 7.5.4.

We need to calculate the probabilities of *each separate outcome* (e.g. $p(YYNN)$) and add them together to get the probability of the overall target outcome, $p(2)$:

$$p(2) = p(YYNN) + p(YNYN) + p(YNNY) + p(NYYN) + p(NYNY) + p(NNYY)$$

The probability, $p(YYNN)$, for getting exactly two albinos and two non-albinos requires an AND combination of probabilities:

$$p(YYNN) = p(Y) \times p(Y) \times p(N) \times p(N) = p \times p \times q \times q = 0.25^2 \times 0.75^2$$
$$p(YNYN) = p(Y) \times p(N) \times p(Y) \times p(N) = p \times q \times p \times q = 0.25^2 \times 0.75^2$$
$$p(YNNY) = p(Y) \times p(N) \times p(N) \times p(Y) = p \times q \times q \times p = 0.25^2 \times 0.75^2$$
$$p(NYYN) = p(N) \times p(Y) \times p(Y) \times p(N) = q \times p \times p \times q = 0.25^2 \times 0.75^2$$
$$p(NYNY) = p(N) \times p(Y) \times p(N) \times p(Y) = q \times p \times q \times p = 0.25^2 \times 0.75^2$$
$$p(NNYY) = p(N) \times p(N) \times p(Y) \times p(Y) = q \times q \times p \times p = 0.25^2 \times 0.75^2$$

All the separate probabilities will have the *same* probability – the only difference between them is the *order* of multiplying the p's and q's.

The overall probability of getting two albino offspring is then given by:

$$p(2) = \{6\} \times \{0.25^2 \times 0.75^2\} = 6 \times 0.03516 = 0.211$$

The '6' in the equation is the number of different ways of selecting two albinos out of four, $_4C_2 = 6$.

Using the above example for $p(2)$, we can now develop a general equation for the binomial probability for recording r items $p(r)$.

$$p(r) = \{\text{No. of ways of selecting } r \text{ from } n\} \times \{\text{Probability for each selection}\} \qquad [8.30]$$

giving

$$p(r) = {_nC_r} \times p^r \times (1 - p)^{(n-r)} \qquad [8.31]$$

The binomial distribution applies to a situation where each member of a group (size $= n$) will be in one of two states (e.g. Y or N) with the probability of being in one state (e.g. Y) being equal to p.

The binomial distribution gives the probability, $p(r)$, that, out of the total of n members, exactly r will be in state Y and $(n - r)$ will be in state N.

The combination function, $_nC_r$, is often also called the *binomial coefficient*.

Using, $q = 1 - p$, which is the probability that a specific group member will in state, N, the binomial distribution may be rewritten as:

$$p(r) = {}_nC_r \times p^r \times q^{(n-r)}$$ [8.32]

The Excel function BINOMDIST(r, n, p, FALSE) can be used to calculate values for the probabilities of a binomial distribution. Entering 'TRUE', instead of 'FALSE', in the last argument value would give the *cumulative* probability of $p(r)$.

E8.16 🖥

Calculate the probability, in a class of five students, that exactly two students will be left-handed. Assume that 10% of all students are left-handed, and that each student is either left-handed or right-handed.

In this example, the Y state will be equivalent to a student being left-handed and the N state equivalent to a student being right-handed.

We are required to calculate $p(r)$, when $r = 2$ and $n = 5$.

The first step is to find the value for p, which is the probability that any student, selected at random, will be left-handed.

As 10% of all students are left-handed we can estimate that the probability of one particular student being left-handed is 1 in 10 or $p = 1/10 = 0.1$.

We can now substitute into the distribution equation [8.31]:

$$p(r) = {}_nC_r \times p^r \times (1 - p)^{(n-r)}$$
$$p(2) = {}_5C_2 \times 0.1^2 \times (1 - 0.1)^{(5-2)} = 10 \times 0.1^2 \times 0.9^3 = 0.0729$$

Q8.20 Repeat the calculation in E8.16 for each of the probabilities for 0, 1, 2, 3, 4 and 5 students being left-handed.

Calculate the total of all of these probabilities.

Q8.21 A certain drug treatment cures 70% of people having a particular disease. If 10 people suffering from the disease are treated, calculate the probability (to three significant figures) that

(i)   no one will be cured

(ii) only one person will be cured

(iii) exactly eight people will be cured

(iv) exactly nine people will be cured

(v) exactly ten people will be cured

(vi) at least eight people will be cured

(vii) no more than seven people will be cured

When trying to interpret a particular problem and map the data in the question to the binomial equation it is sometimes useful to use the following information:

In the binomial distribution, we investigate the probability of getting r 'positive' outcomes out of a total of n possible trials.

Hence:

- minimum possible value for $r = 0$

- maximum possible value for $r = n$

The probability, p, is the probability that one particular of the n trials will result in a 'positive' outcome.

We can illustrate the use of these ideas in the following Example E8.17.

E8.17 🖥

It is known that there are 1500 plants randomly distributed over an area of 1000 m^2.

(i) Use the binomial distribution to derive the probability of finding exactly r plants in a quadrat of area 1.0 m^2.

(ii) Calculate the probabilities, $p(r)$, for $r = 0, 1, 2, 3, 4, 5, 6$.

(iii) Plot the results as a histogram.

(i) For the binomial distribution, we need to know the values of n and p. In this problem, the number of plants in the 1.0 m^2 quadrat, could be anything from zero plants ($r = 0$) to (theoretically) all 1500 plants.

The maximum value for r equals n, hence $n = 1500$.

The value of the probability, p, is the probability that a *particular* plant will appear in the quadrat. This will be given by the area of the quadrat divided by the total area:

$$p = 1.0/1000 = 0.001$$

The values for $p(r)$ are given by: $= {}_nC_r \times p^r \times (1 - p)^{(n-r)}$.

(ii) The specific values for, $p(r)$, are calculated in the following table:

Number, r	0	1	2	3	4	5	6
Probability (binomial), $p(r)$	0.223	0.335	0.251	0.126	0.047	0.014	0.004

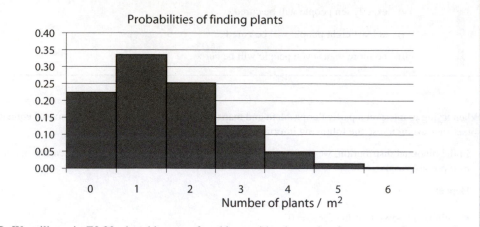

N.B. We will see in E8.23, that this type of problem, with a low value for p, approximates to the Poisson distribution.

8.4.3 Parameters of the binomial distribution

Mean value, μ, of the binomial distribution is given by:

$$\mu = n \times p \qquad [8.33]$$

When making experimental measurements of a binomial distributions, the *experimental mean value, $\bar{r}$,* would be a *best estimate* for the true mean, μ. The *best estimate* for the individual probability, p, would then be given by

$$p \approx \frac{\bar{r}}{n} \qquad [8.34]$$

Standard deviation, σ, of the binomial distribution is given by:

$$\sigma = \sqrt{n \times p \times q} = \sqrt{n \times p \times (1 - p)} \qquad [8.35]$$

Coefficient of dispersion, CD (see 7.1.6), is given by

$$CD = \frac{\text{Variance}}{\text{Mean}} = \frac{\sigma^2}{\mu}$$

which for the binomial distribution becomes:

$$CD = \frac{n \times p \times q}{n \times p} = q \qquad\qquad [8.36]$$

E8.18

The aim of this exercise is to use the results for $p(r)$ from E8.17 to calculate the mean value, $\bar{r}$, of the number of plants found per square metre.

Using the probabilities, $p(r)$, for $r = 0, 1, 2, 3, 4, 5, 6$:

(i)   Calculate the sum of the probabilities over this range, $\Sigma_r p(r)$.

(ii) Calculate the products, $r \times p(r)$, for each value of r, and the sum of these values, $\Sigma_r [r \times p(r)]$.

(iii) Calculate the mean value of r, $\bar{r} = \{\Sigma_r r \times p(r)\} / \{\Sigma_r p(r)\}$ (same method of calculation as used in 7.2.6 to calculate means using frequency data).

(iv) Compare the result with the product, $n \times p$.

The calculations for (i) and (ii) are completed in the following table:

Number, r	0	1	2	3	4	5	6	Total
Probability (binomial), $p(r)$	0.223	0.335	0.251	0.126	0.047	0.014	0.004	0.999
$r \times p(r)$		0.000	0.335	0.502	0.377	0.188	0.070	0.021	1.493

Note that, in the table above, the sum of probabilities, $\Sigma_r p(r)$, is not exactly equal to 1.000. This is because there is still a very small probability that seven or more plants may be recorded, and we have not included this in our total.

The experimental mean value can be calculated,

$$\bar{r} = \{\Sigma_r r \times p(r)\}/\{\Sigma_r p(r)\} = 1.493/0.999 \approx 1.50$$

The product, $n \times p = 1500 \times 0.001 = 1.500$, which agrees with the result calculated from the values of $p(r)$.

Q8.22 Due to a mechanical problem in an egg packing area, one in five eggs are cracked accidentally before they are packed into boxes of six. Fifty boxes are packed before the problem is spotted. Use the table below to calculate the number of boxes with specific numbers of broken eggs.

No. of cracked eggs per box	0	1	2	3	4	5	6
Probability of occurrence							
Number of boxes							

Assume that the packing process is random:

(i) Calculate the probability that any box, chosen at random from the 50 boxes, will have no cracked eggs.

(ii) Repeat the calculation in (i) for the probabilities of finding each of exactly one to six cracked eggs in any given box.

(iii) Using the data from (ii), predict the numbers of boxes, out of the batch of 50, that will have the specific numbers of cracked eggs. Give the answers to a relevant number of decimal places!

8.4.4 Cumulative binomial distribution

The binomial expression for $p(r)$ calculates the probability that exactly r occurrences of a particular condition (Y) will occur.

In practice, many problems are concerned with the *combined* probability of several possible values of r occurring.

The cumulative probability (see 7.2.9) for all values of r from 0 to r' is given by:

$$cp(r \leq r') = p(0) + p(1) + \ldots + p(r' - 1) + p(r')$$

E8.19

Using the data from E8.17, we can calculate the probabilities of finding

(i) two or less plants per quadrat

(ii) three or more plants per quadrat

Answer:

(i) $cp(r \leq 2) = p(0) + p(1) + p(2) = 0.223 + 0.335 + 0.251 = 0.809$

(ii) $cp(r \geq 3) = 1 - cp(r \leq 2) = 1 - 0.809 = 0.191$

Using the Excel function BINOMDIST, it is possible to select the *cumulative probability* as the result by entering 'TRUE' into the last argument.

8.4.5 Approximations for the binomial distribution

There are two important approximations for the binomial distribution, which occur when:

- $p \ll 1$ (i.e. p is very small)

  the probabilities of the binomial distribution become approximately the same as the *Poisson distribution*. Figure 8.10(a) compares the binomial distribution ($n = 200$, $p = 0.02$) with the Poisson distribution ($\mu = n \times p = 4$) shown as a continuous line.

- $np(1-p) \gg 1$ (typically $np(1-p) \geq 5$)

the probabilities of the binomial distribution become approximately the same as the *normal distribution*. Figure 8.10(b) compares the binomial distribution ($n = 20$, $p = 0.6$) with the normal distribution ($\mu = n \times p = 12$, $\sigma = \sqrt{np(1-p)} = 2.19$) shown as a continuous line.

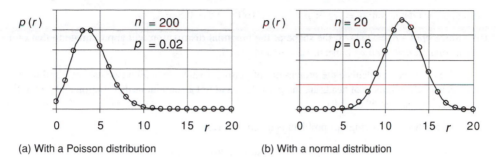

(a) With a Poisson distribution   (b) With a normal distribution

Figure 8.10. Comparisons of binomial distribution (circle points).

The approximation, $p \ll 1$, to the Poisson distribution is considered in Unit 8.5.

The approximation, $np(1-p) \gg 1$, to the normal distribution is useful because we can then use the equations developed for the normal distribution. For example, using [8.13] and [8.35], it is possible to develop the 2-tailed confidence interval, *CI*, for the true mean of the binomial distribution:

$$CI(\mu, X\%) \approx \bar{r} \pm \left\{ z_{2,\alpha} \times \sqrt{np(1-p)} \right\} \approx \bar{r} \pm \left\{ z_{2,\alpha} \times \sqrt{\bar{r} \times \left(1 - \frac{\bar{r}}{n}\right)} \right\} \qquad [8.37]$$

where $z_{T,\alpha}$ is the z-value equivalent to $t_{T,\alpha,\infty}$ (i.e. the t-value for $df = \infty$, infinity), and $\bar{r}$ is the experimentally measured mean value for r.

Useful values for $z_{T,\alpha}$ at 95% and 99% confidence:

Table 8.3.

Significance level, α:	0.05	0.01
Tails, *T*: $\begin{cases} \text{1-tailed} \\ \text{2-tailed} \end{cases}$	1.65 1.96	2.35 2.58

The test for proportion, given in Unit 10.4, is the hypothesis test equivalent to the above confidence interval.

E8.20

A coin is tossed 100 times. Calculate

(i) the standard deviation in the probability distribution for the number of heads, r,

(ii) the probability that the number of heads will be 60 or more.

Answer:

(i) This is a binomial distribution with $n = 100$, $p = q = 0.5$.

Using [8.35]:

Standard deviation: $\sigma = \sqrt{n \times p \times q} = \sqrt{100 \times 0.5 \times 0.5} = 5$.

(ii) Since $np(1-p) = 25 \gg 1$, the shape of the binomial distribution will approximate to that of a *normal distribution* with mean, $\mu = 50$ and $\sigma = 5$.

We now need to calculate the probability of getting a value, $x > 59.5$ (we use 59.5 and not 60 because the numbers of heads are integer values and not a continuous distribution and 59.5 falls midway between 59 and 60)

The method for solving this problem is given in 8.1.8:

$$z = (59.5 - 50)/5 = 1.9$$
$$p(z > 1.9) = 1.0000 - 0.9713 = 0.0287$$

E8.21 🖥

A country is due to have a referendum with only two choices – Yes or No.

In anticipation of the referendum, 1000 people are selected at random in an 'opinion poll' and it is found that 520 will vote Yes and 480 will vote No.

Using this information, calculate whether it is possible to be confident at a 95% level that the result of the actual referendum will be 'Yes'.

From the sample, $n = 1000$ and $\bar{r} = 520$.

A *1-tailed* confidence deviation (see [8.37]) for the true mean value will be:

$$Cd(\mu, X\%) \approx \left\{ z_{1,\alpha} \times \sqrt{\bar{r} \times \left(1 - \frac{\bar{r}}{n}\right)} \right\} = 1.65 \times \sqrt{520 \times \left(1 - \frac{520}{1000}\right)} = 26$$

Thus the true mean value for a 'Yes' vote *could* be as low as $520 - 26 = 494$.

Hence it is not possible to be 95% confident that the vote will return a 'Yes'.

Q8.23 A random sample of 50 frogs is taken from a lake, and it is found that 37 are female and 13 are male. Estimate, with a 95% confidence interval, the true proportion of female frogs in the lake's frog population.

(See also Q10.13 and E11.2.)

8.5 Poisson distribution

8.5.1 Introduction

The Poisson distribution is an *approximation* of the binomial distribution when the *probability of a particular event, p, is very low*, for example, describing the uniformity in spread of plants across a field.

8.5.2 Poisson distribution

The Poisson distribution gives the probability, $p(r)$, of finding r occurrences of a particular outcome, and is defined by

$$p(r) = \frac{e^{-\mu} \times \mu^r}{r!}$$
[8.38]

where μ is the *mean number* of such outcomes that would normally be expected to occur.

The Excel function POISSON (r, μ, False) can be used to calculate values for the probabilities of a Poisson distribution. Entering 'True', instead of 'False', in the last argument value would give the *cumulative probability* of $p(r)$.

The following example illustrates that, when the probability, p, is small, the results from the binomial and Poisson distributions are virtually identical.

E8.22 💻

A service engineer is responsible for responding to emergency breakdowns from 2500 customers. He normally receives four calls per day.

Calculate the probability that he would receive eight calls in a particular day (assume no customer has more than one breakdown per day).

Use *both* the binomial and Poisson distributions to answer the question.

This is a binomial-type problem because any one customer will only have a breakdown (Y) or no breakdown (N).

Calculation assuming a binomial distribution:

We need to calculate values for n and p to use in [8.31].

The value of n equals the maximum possible value for r, which is the maximum number of calls per day, $n = 2500$.

The mean number of calls is known to be $\mu = 4$.

From [8.33], the probability, p, that any one customer will have a breakdown,

$$p = \mu/n = 4/2500 = 0.0016$$

We need to calculate $p(r)$, where $r = 8$:

$$p(8) = {}_{2500}C_8 \times 0.0016^8 \times (1 - 0.0016)^{2492} = 0.029\,72$$

Calculation assuming a Poisson distribution:

For the Poisson distribution we already have the mean value, $\mu = 4$. Hence

$$p(8) = \frac{e^{-4} \times 4^8}{8!} = 0.029\,77$$

The two results are virtually identical.

Looking at E8.22, we can see that:

- for the binomial distribution it is necessary to know values for n and p, but

- for the Poisson distribution it is only necessary to know the value of μ.

There are many problems with inherently low values of p, for which only the average value, μ, is known, and it is then appropriate to use the Poisson distribution. Typical examples include situations where relatively few events are being counted in a small section of a much larger environment, e.g. counting:

- plants in a small area of a large field,

- radioactive decay events occurring in a small time interval out of an overall decay period,

- occurrences of illness in a town within a large country.

In each of the above examples, the probability, p, that a particular event (plant, decay event, ill person) will be found in the particular situation (area, time interval, town) will be small.

The other condition for the Poisson distribution to be valid is that each event occurs randomly with the same probability, p, i.e. it is independent of any other event.

E8.23 🖳

It is known that, on average, a random spread of particular rare plants gives an average of 1.5 plants per square metre. Use the Poisson distribution to derive the probability of finding exactly r plants in a quadrat of area 1.0 m^2.

Calculate the probabilities, $p(r)$, for $r = 0, 1, 2, 3, 4, 5, 6$

Compare the results in this example with those in E8.17.

For the Poisson distribution, we need to know the mean value, μ.

In fact, we already know that the mean value, $\mu = 1.5$.

The values for $p(r)$ are given by: $p(r) = \dfrac{e^{-\mu} \times \mu^r}{r!}$

which can be calculated to give:

Number, r	0	1	2	3	4	5	6
Probability (Poisson), $p(r)$	0.223	0.335	0.251	0.126	0.047	0.014	0.004
**Probability (binomial), $p(r)$	0.223	0.335	0.251	0.126	0.047	0.014	0.004

**Binomial probabilities taken from E8.17.

Note that for this particular type of problem (p is very low), the results obtained by the Poisson distribution are virtually the same as for the binomial distribution.

Q8.24 The number of deaths due to a specific disease per year per 100 000 people is on average 50. Assuming that the occurrence of each incidence of the disease is random, calculate the probability that in a particular group of 1000 people there may be

(i) no deaths   (iii) two deaths

(ii) one death   (iv) more than two deaths

8.5.3 Parameters of the Poisson distribution

Mean value of the Poisson distribution is given directly by μ [8.39]

Since p is very small,

$$q = 1 - p \approx 1 \tag{8.40}$$

Standard deviation, σ, of the Poisson distribution is given by:

$$\sigma = \sqrt{n \times p \times q} \approx \sqrt{n \times p} \tag{8.41}$$

Coefficient of dispersion, CD (see 7.1.6), for the Poisson distribution is given by

$$CD = \frac{\text{Variance}}{\text{Mean}} = \frac{s^2}{\bar{x}}$$

$$CD \approx \frac{n \times p}{n \times p} \approx 1 \tag{8.42}$$

A good example, in the use of the coefficient of dispersion, CD, is in the distribution of random 'events', e.g. plants in a field.

- If the events (plants) are individually distributed at random, then the number of plants in a small area will be given by the Poisson distribution (see E8.23). For the Poisson distribution, $CD \approx 1$.

- If there is a supportive 'interaction' between the events (plants), then the events may group together (plant clumping). Another example would be the development of 'clusters' of an otherwise random disease, e.g. possible increase in the incidence of disease cases due to local pollution. In these cases, the increased probability of 'high count' groups would give $CD > 1$.

- If the events compete for resources, then they will tend to spread out and give a more even distribution. This will give a smaller dispersion of values around the mean value, resulting in $CD < 1$.

Q8.25 🖥️

In assessing the distribution of a species of plant, the two data sets, A and B, in the table record the frequencies with which specific numbers were recorded in 100 quadrats. The means and standard deviations of the two distributions are given.

Number of plants	0	1	2	3	4	5	6	7	8	9	10	Mean	Standard deviation
Frequency, A	5	13	21	22	18	11	6	3	1	0	0	3.13	1.76
Frequency, B	10	14	17	18	16	12	7	4	1	1	0	3.13	2.04

The frequency distributions are also recorded in the line graph shown in Figure 8.11.

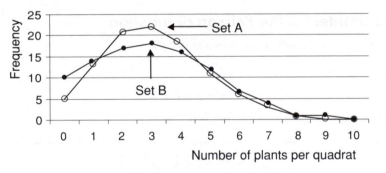

Figure 8.11. Plant distributions

(i)   Calculate the coefficients of dispersion for the two data sets.

(ii) On the basis of (i) decide which data set shows a Poisson distribution.

(iii) For the data set that is not Poisson, decide whether the distribution of plants shows possible clumping of plants or a more even spread.

9

Scientific Investigation

Overview

Science is concerned with understanding the way in which the natural world works. This includes the biology of the living world, the chemistry of atomic and molecular structures, the physics of forces and systems, but also includes the more applied sciences, e.g. sports science, forensic science, the environmental sciences, etc.

In this chapter we see that science is used to investigate a wide range of different systems, and that there are many different types of 'scientific investigation'.

We also see that the power and effectiveness of science, as a way of accurately describing nature, grew out of the Renaissance in the sixteenth century with the development of the 'scientific method'.

As an essential feature of the 'scientific method', we concentrate on the issues involved with the setting and testing of hypotheses. The statistical implementation of hypothesis tests for a range of different system parameters is developed in the following chapters using both parametric and non-parametric tests.

The final chapter in the book takes hypothesis testing into more complex scientific investigations, and highlights the interdependence between the experiment design and the statistics involved in assessing the validity of the proposed hypotheses.

9.1 Scientific systems

Any system that is being investigated can normally be characterized by:

- **Outcomes**   observable changes in the system.
- **Factors**   'inputs' to the system that may affect the outcomes.
- **Mechanisms**   the actual processes (often hidden) within the system that link the 'input' factors to the 'observed' outcomes.
- **Subjects**   representative examples of the 'scientific system' that are used for the particular experiment.

Each 'factor' will have two, or more, 'levels', and a change of 'level' may cause a particular 'outcome' from the system. 'Outcomes' and 'levels' will be measured by suitable experimental variables.

E9.1

It is claimed that omega-3 oils, found in many fish, reduce 'clogging' of the arteries. In this case the 'system' is the human body, the 'outcome' is reduced deposits in the arteries, the 'factor' relates to different 'levels' of intake of omega-3 oils, and the 'mechanism' is the way in which the body metabolizes these oils. In an experiment, suitable 'variables' must be used to measure the amount of arterial deposit and the intake ('levels') of omega-3 oils.

There are three main levels in which science can normally understand such a system such as that described in E9.1. It may be possible to:

1   Confirm a **correlation (or association)** between the input factor(s) and the system outcome(s), e.g., an outcome may change following a change in a factor. However, a correlation between factor and outcome does *not necessarily* mean that the change in the factor is the *reason* for the change in the outcome.

2   Confirm a **causal relationship** between the input factor(s) and the system outcome(s). In this case, a change in a factor (cause) can be demonstrated to be the *reason* for the change in outcome (effect).

3   Explain the **mechanism** by which the factor causes the outcome to change. The understanding of this mechanism must give more than a description of *past* observations; it must be sufficiently good to be able to *predict* how the system will work under modified conditions.

In the case of the 'omega-3' example (E9.1):

* At level 1, the variable that records arterial deposits and the variable that measures oil intake may show close correlation, but the oil intake might not be the 'cause' of the reduction in deposits. For example, the reduction in artery deposits may be due to the fish diet for reasons not related to the omega-3 oils. The fish diet may be the 'cause' that both reduces arterial deposits (an effect) and, coincidentally, increases the intake of omega-3 oils (now an effect and not a cause).

* Alternatively at level 2, a 'cause and effect' relationship could be shown to exist between 'factor' and 'outcome' (i.e. the omega-3 oils actually reduce arterial deposits), but the mechanism by which this works might not be fully understood.

* Ideally, at level 3, science should be able to demonstrate, within the constraints of the 'scientific method' (Unit 9.2), a robust understanding of the mechanism involved in the process. This understanding must continue to be essentially correct as further discoveries are made about the system.

Scientific 'systems' can be recognized in all branches of science – from an understanding of the properties of sub-atomic particles to the behaviour of the universe itself. In all cases, the development of understanding identifies the same three levels: – *correlation, causation, mechanism*.

9.2 The 'scientific method'

At a minimum level, any work in science must be characterized by an accuracy and objectivity in the collection of data, and a commitment to base conclusions solely on experimental results.

An important distinction should now be made between the processes used to reach conclusions:

* **Induction**: *comparative* conclusions about one system are made on the basis of what is known to happen in a *different* system, and

* **Deduction**: *new* conclusions are drawn as direct *logical* consequences of known facts.

Inductive reasoning may be considered to be equivalent to an *intelligent guess* that might prove to be wrong. Inductive reasoning is often a good starting point to consider how a new system might work and suggest possible hypotheses, but we will see that deduction, and not induction, is the fabric of true science.

E9.2

In trying to understand the mechanism of elasticity in the common elastic band, we may compare the effect with elasticity in other materials. For example, we know that when we pull on a metal wire it will stretch and when we pull on an elastic band it will also stretch. We may make an *induction* that the mechanisms are basically similar in the two cases, although different in magnitude.

If our *induction* is correct, we might expect that, when we heat up an elastic band, it will stretch in the same way that a metal wire will stretch on heating. However, when we perform the experiment, we find that a stretched elastic band will tend to *contract* on heating. Our *induction* about similar mechanisms must be wrong in this case!

Deductive reasoning starts from known facts, and then proceeds to work out the inevitable consequences of those facts. The process of deduction is faultless, but the accuracy of the conclusions still depends on the accuracy of the original facts on which the deduction was based.

The **'scientific method'** itself is based on a cycle of hypothesis, experiment, deduction and prediction.

- *Observations* or *inductive reasoning* are often the starting point where observations begin to suggest a possible **mechanism** that might explain the behaviour of a system.

  (In E9.2 it is thought that the mechanism for elasticity in the elastic band could be similar to that in the metal wire.)

- A **hypothesis** is developed which describes an aspect of the proposed mechanism in the scientific system. The choice between the truth or otherwise of this hypothesis is often written as two alternatives:
  - H_O: Null hypothesis – the proposed mechanism is NOT correct.
  - H_A: Alternative hypothesis – the proposed mechanism IS correct.

  (In E9.2 the alternative hypothesis is that the mechanism in the elastic band is similar to that in the metal wire.)

- An **experiment** is devised, whose outcome should give evidence to support, or not support, the alternative hypothesis, H_A.

  (In E9.2 the alternative hypothesis predicts that an elastic band should expand on heating.)

- The experiment is performed and a decision is made as to whether the new understanding of the system is likely to be true. The conclusion is a **deduction** based on the observed facts.

  (In E9.2, the observed facts do *not* support the alternative hypothesis. A new mechanism must be found to explain the behaviour of the elastic band.)

- If the alternative hypothesis is supported, then the proposed mechanism should be used to *predict* other observable behaviour patterns for the system. If these predictions are tested using the 'hypothesis-testing' approach, and continue to confirm the new explanations, then this understanding begins to be accepted as a good description of the system.

The process of the 'scientific method' is also referred to as the 'hypothetico-deductive' method, emphasizing the two key elements of the concept.

When a hypothesis continues to satisfy more and more experimental tests, then it will gradually be accepted as a scientific fact. However, it must be remembered that the process of the 'scientific method' does not end with the discovery of 'fact', just because it satisfies all current experimental data. Any scientific 'fact' may be modified in the future if further experimental results fail to conform to the accepted theory.

E9.3

An excellent example of the 'scientific method' occurred in the developed of the theory of relativity by Albert Einstein. The new theory, for which there were many doubters, predicted that light from a star would be *deflected* as it passed close to our sun.

This prediction would test the theory of relativity, and in 1919 during an eclipse of the sun the apparent positions of stars close to the sun were found to shift according to Einstein's prediction. Since that time, the theory of relativity has continued to predict events accurately, and is now an accepted 'fact' of science.

However, as part of the 'cycle' of the scientific method, it is always probable that the current theory of relativity will be superseded by new 'facts' as more precise measurements demand a revised or alternative explanation.

9.3 Hypothesis testing

9.3.1 Introduction

Many investigations in science seek to answer a simple 'Yes/No' question. For example:

- Is a new vaccine effective in preventing an infection?

- Do glass fragments at the scene of a crime come from different sources?

- Does a particular training regime improve athletic performance?

The statement of the initial 'Yes/No' question is a hypothesis statement, which is normally presented in two parts:

- 'No' is the null hypothesis and

- 'Yes' is the alternative hypothesis.

However, any scientific investigation is subject to errors and uncertainties, and the results do not give an *absolute* answer to the 'Yes/No' question – this is where the statistics come in!

The statistical analysis in a hypothesis test is based on the experimental data recorded, and normally calculates the probability (p-value) that it would be wrong to choose the 'Yes' (alternative hypothesis) answer to the question. There is never an absolute certainty in the final answer, but at least it is possible to assess the probability of making at least one type of error.

9.3.2 Hypothesis statement

It is essential, *before* performing any experimental measurements, to *state the hypothesis* being tested.

The hypothesis is presented using two *mutually exclusive* statements (often written as H_O and H_A) that *cover all possible outcomes*, for example:

Null hypothesis, H_O: Smoking does not increase the probability of lung cancer

Alternative hypothesis, H_A: Smoking increases the probability of lung cancer

The NULL hypothesis usually demonstrates 'no factor effect' or the 'status quo'. The ALTERNATIVE hypothesis usually demonstrates that a factor (Unit 9.1) does indeed cause some 'effect' or 'change of value'.

A scientific experiment must be precise about what it is actually measuring, and will normally measure some physical, chemical or biological variable, which will then be used as an *indicator* in choosing which hypothesis is believed to be correct. For scientific use, it is also important that any 'vague' terminology should be accurately defined – for example, 'smoking' in the above example may be defined (for instance) as five cigarettes per day.

E9.4

As part of a monitoring procedure, an analyst wishes to test whether the lead content of a water supply *exceeds* 50 ppb. If the true lead content is μ, the appropriate hypotheses for this test would be:

Null hypothesis, H_O: Lead content does not exceed 50 ppb: $\mu \leq 50$ ppb

Alternative hypothesis, H_A: Lead content does exceed 50 ppb: $\mu > 50$ ppb

---

9.3.3 Tails

In defining a hypothesis statement in respect of a possible factor 'effect', there are two possibilities (see also 8.2.8):

- **2-tailed**: the factor effect may operate in *either direction*, or

- **1-tailed**: the factor is only being tested for an effect in *one particular direction*.

The hypothesis statement in E9.4 is '1-tailed', because the alternative hypothesis only proposes an *increase* in lead concentration. Example E9.5 gives an example of a '2-tailed' hypothesis in a similar context. In this case the alternative hypothesis proposes either an increase *or* a decrease.

E9.5

As part of a monitoring procedure, an analyst wishes to test whether the lead content of a water supply is *not equal* to 50 ppb. If the true lead content is μ, the appropriate hypotheses for this test would be:

Null hypothesis, H_O: Lead content equals 50 ppb: $\mu = 50$ ppb

Alternative hypothesis, H_A: Lead content does not equal 50 ppb: $\mu \neq 50$ ppb

---

9.3.4 Significance level

It is also essential, *before* performing any experimental measurements, to decide on the *probability of error* that would be acceptable when making the decision.

Before starting the experiment, the significance level, α (see also 8.2.7) must be stated:

> **Significance level**, α. This is the *largest probability of error* that is acceptable when choosing the alternative hypothesis, H_A.

If a conclusion is claimed to be *'significant'*, then the results normally indicate that H_A is correct with less than a 0.05 (5%) chance of being wrong.

If a conclusion is claimed to be *'highly significant'*, then the probability of H_A being wrong is normally less than 0.01 (or 1%).

Note that the term **confidence level** is often used to express the percentage probability of being *right*, e.g., confidence levels of 95% and 99% for being right are equivalent to significance levels of 0.05 and 0.01 respectively for being wrong! (See also 8.2.7.)

The choice of a value for the significance level, α, depends on the context of the problem, and the consequences that would arise if H_A were chosen in error. The default level for most scientific research is taken as $\alpha = 0.05$.

9.3.5 Errors

With a 'Yes/No' type of question, there are two ways of being wrong:

* deciding 'Yes' when 'No' is correct (type I error) and

* deciding 'No' when 'Yes' is correct (type II error).

In terms of the hypotheses, the type I and type II errors could occur as shown in Table 9.1.

Table 9.1. Errors in a hypothesis test.

Conclusion	True situation	
	H_O – actually true	H_A – actually true
Do not reject H_O	Correct result	Type II error (β)
Accept H_A	Type I error (p)	Correct result

The *probability* of a type I error for a *given experiment* is given by the *p*-value, which can be calculated from the experimental results – see 9.3.6.

The *probability* of a type II error is denoted by β (beta), which is usually much more difficult to calculate – see 9.3.7.

9.3.6 Significance level, α, and p-value

The **p-value** is the *estimated probability of a type I error* for that particular set of results, and it can be calculated from the experimental values themselves.

The significance level, α (alpha), is the maximum probability of a type I error that would be acceptable. Its value must be decided as part of the *initial* experiment design process.

On the basis of the relative values of α and p, it is then possible to make the decision:

If $p \leq \alpha$ Accept alternative hypothesis, H_A, at a significance level of α

[9.1]

If $p > \alpha$ Insufficient evidence to accept alternative hypothesis

---

Most software programs (e.g. Excel, MINITAB) will calculate a p-value for the given statistical test. This makes performing a statistical test very easy, as the decision is based only on the relative sizes of p and α.

It is useful to remember that, for the same data set:

$$p\text{-value for a 2-}tailed\text{ test} = 2 \times p\text{-value for a 1-}tailed\text{ test} \qquad [9.2]$$

9.3.7 Power of the experiment

The **power** of an experiment is a measure of how successful the experiment will be in confirming the existence of a 'factor' effect, if such an effect actually exists.

It is the 'opposite' of the probability of a type II error:

$$\text{Power} = 1 - \beta \qquad [9.3]$$

Ideally the power of the experiment should be as close to '1.0' as possible. In practice, the power of an experiment is often limited by many experimental uncertainties that may make the existence of the 'factor' effect difficult to detect.

It is only possible to calculate the value of *power* (and β) for a particular experiment if all the effects of any random uncertainties and external bias are known and understood. This is not normally the case.

9.3.8 Statistic and critical value method

Using the ability of a computer program to calculate the p-value is a quick and convenient way of reaching a decision in most statistical tests. However, an alternative method exists which predates the use of computers, and it is important to be aware of the procedure involved:

1   For each test it is possible to calculate the value of a **test statistic**, e.g. F-statistic for the F-test (Unit 10.1), t-statistic for the t-test (Unit 10.2). Each test has its own formula for calculating the value of its statistic directly from the experimental results. A large value for the test statistic normally implies a large 'factor' effect.

2   For each test, there are tables of **critical values**, e.g. F-critical for the F-test, t-critical for the t-test. The choice of the appropriate critical value for a given experiment will depend on the significance level, α, required and the degrees of freedom (see the *individual tests*) available in the experimental data.

3   The alternative hypothesis, H_A, will normally be accepted at a significance level of α, if the value of the test statistic is *greater than* the critical value. Contrast this with the p-value where H_A is accepted if the p-value is *less than* the significance level, α.

E9.6

Two of the following sets of results from a t-test are valid. Two of the sets have inconsistent data values. In this particular t-test, the alternative hypothesis is accepted if $t_{STAT} > t_{CRIT}$.

Identify the two result sets that have inconsistencies:

$$A \quad t_{STAT} = 2.24, \quad t_{CRIT} = 1.86, \quad \alpha = 0.05, \quad p = 0.028$$
$$B \quad t_{STAT} = 2.46, \quad t_{CRIT} = 2.68, \quad \alpha = 0.05, \quad p = 0.015$$
$$C \quad t_{STAT} = 2.45, \quad t_{CRIT} = 2.45, \quad \alpha = 0.05, \quad p = 0.02$$
$$D \quad t_{STAT} = 2.61, \quad t_{CRIT} = 3.00, \quad \alpha = 0.01, \quad p = 0.017$$

In set A,

$$t_{STAT} > t_{CRIT} - \text{leads to 'accept the alternative hypothesis'}$$
$$p < \alpha - \text{leads to 'accept the alternative hypothesis'}$$

Set A is consistent.

In set B,

$$t_{STAT} < t_{CRIT} - \text{leads to 'do not reject the null hypothesis'}$$
$$p < \alpha - \text{leads to 'accept the alternative hypothesis'}$$

Set B is *not* consistent - data gives conflicting conclusions.

In set C,

$$t_{STAT} = t_{CRIT} - \text{exactly on the choice boundary, and}$$

we would then expect that

$$p = \alpha, \text{ but in the data } p < \alpha.$$

Set C is *not* consistent.

In set D,

$$t_{STAT} < t_{CRIT} - \text{leads to 'do not reject the null hypothesis'}$$
$$p > \alpha - \text{leads to 'do not reject the null hypothesis'}$$

Set D is consistent.

Q9.1 The *critical value* for a particular F-test at a significance level of 0.05 is $F_{CRIT} = 9.60$.

    (i)   If the measured *test statistic* is found to be $F_{STAT} = 8.70$, should H_A be accepted?

   (ii)   If the measured *test statistic* is found to be $F_{STAT} = 10.20$, would the equivalent p-value be greater than, or less than, 0.05?

  (iii)   If the measured *test statistic* is found coincidentally to have exactly the *same* value as the critical value: $F_{STAT} = 9.60$, what is the p-value that would be calculated for this test?

9.3.9 Stating results

In hypothesis testing it is usually possible to calculate the probability of a type I error (see 9.3.6) when accepting the alternative hypothesis, H_A. Thus, an *acceptable* statement of the 'yes' conclusion would then be:

Accept the alternative hypothesis, H_A, at a significance level of (e.g.) 5%

However, it is *not* usually possible to calculate the probability of the Type II Error for accepting the null hypothesis, H_O. In other words, it is often difficult to 'prove' a negative.

Do NOT therefore use the statement: "Accept the null hypothesis"

An *acceptable* statement for the 'no' result would therefore be either:

Insufficient evidence to reject the null hypothesis, H_O, at a significance level of (e.g.) 5%

or

Insufficient evidence to accept the alternative hypothesis, H_A, at a significance level of (e.g.) 5%

Q9.2 The results of a statistical test give a *p*-value, $p = 0.03$.

Which of the following statements are correct statements?

(i) The alternative hypothesis, H_A, should be accepted at a significance level of 0.01   True / False

(ii) The alternative hypothesis, H_A, should be accepted at a significance level of 0.05   True / False

(iii) The null hypothesis, H_O, should be accepted at a significance level of 0.01   True / False

(iv) The null hypothesis, H_O, should be accepted at a significance level of 0.05   True / False

---

9.3.10 Choice of test

The choice of an appropriate test depends on a range of factors. It is necessary to decide whether a particular test addresses the type of problem that is being investigated (e.g. two-sample test for difference), and then whether the data that is available satisfies the criteria for the test (e.g. normally distributed).

Some parametric tests are more robust than others when dealing with data that is not normally distributed, and, for some types of data, it is possible to apply a transformation to produce near-normal data.

Further information on the choice and application of different tests is given on the web site.

⌨ Study Resources > Additional Material

10

Parametric Tests

Overview

We saw in Chapter 9 that the validity of modern science is based on the use of the 'scientific method' of experimental investigation. Integral to this method is the proposing and testing of *hypotheses*.

This chapter introduces a number of different hypothesis 'tests', which test for different properties in the system being investigated:

- *F*-test test for differences in variance (spread) between data samples.

- *t*-test tests for differences in mean value (location) between data samples.

- Correlation tests for existence of a linear relationship between data variables.

- Test for differences in proportion.

The 'tests' in this chapter are categorized as 'parametric' because the actual values of the data are used directly in the calculations. In non-parametric tests the data values are used only to *rank* the values in order; the tests are then based on the ranking of data rather than actual values.

There are some features that are common to all of these hypothesis tests. In Chapter 9 we introduced the idea of a system with 'input factor' and 'outcomes'. Using that model, we can now describe a hypothesis test as an experiment that is designed to test whether a change in the 'level' of an input 'factor' has a measurable 'outcome'. For example, we may test to see whether:

- an increase in temperature accelerates the growth rate of bacteria (in this case the 'factor' is temperature, the 'levels' are different values of temperature, and the 'outcome' is a change in growth rate), or

- the manganese concentration is different between two rivers (in this case the 'factor' is the choice of river, the 'levels' identify particular rivers, and the 'outcome' is a difference in manganese concentration).

In a hypothesis test (9.3.2), two *mutually exclusive* statements (often written as H_O and H_A) are stated that *cover all possible outcomes*, for example:

Null hypothesis, H_O:	Represents a 'no effect' result.
Alternative hypothesis, H_A:	Represents the detection of a factor 'effect'.

There are two possible ways (9.3.5) in which the *wrong conclusion* might be reached:

Type I error:	The test claims to have detected a factor 'effect' when no such condition actually exists.
Type II error:	The test fails to detect a factor 'effect' when such condition actually does exist.

The probability of getting a type I error is the *p-value*. This can normally be calculated from the statistics of the data recorded.

The *significance level, α* (9.3.4), for the test is the maximum acceptable value for a type I error, and it is normal practice to decide on this value before conducting the test.

$$\text{The alternative hypothesis will be accepted if } p \leq \alpha \qquad [10.1]$$

A type II error is an indication that the experiment was *not powerful enough* to detect a true factor 'effect'. The usual reason for a 'lack of power' in an experiment is that the experimental uncertainties swamp the factor effect that is being measured.

Good experimental design can be used to make the various uncertainties more explicit and, by separating their effects, enable a more focused analysis of the factor effect itself. The techniques of experiment design are developed in Unit 13.3.

10.1   Sample variances: *F*-test

10.1.1   Introduction

It is expected that the reader is familiar with the application and terminologies of the *hypothesis test* as set out in Unit 9.3. The *F*-test is a test for significant differences between the *variances* of two data samples that are drawn from normal distributions, and was named after the statistician Sir Ronald A. Fisher. The *F*-test is also used *within* other tests, e.g. in analysis of variance (13.2.6).

Where the data may not have been drawn from a normal distribution it is possible to use Levene's test to compare variances. See web site:

⌨ Study Resources > Additional Materials

10.1.2   Hypotheses

Consider two data samples, A and B:

- Data sample, A, is drawn from a population of values which has a *population variance*, $\sigma_A{}^2$.

  The measured *sample variance*, $s_A{}^2$, of the group A is a 'best-estimate' of the population variance.

- Data sample, B, is drawn from a population of values which has a *population variance*, $\sigma_B{}^2$.

  The measured *sample variance*, $s_B{}^2$, of the group B is a 'best-estimate' of the population variance.

In a practical situation, an analyst may have two samples of data and can calculate the variances, $s_A{}^2$ and $s_B{}^2$, but does not known whether the two samples had been drawn from the *same* population of data values. There are then two possibilities that can be described by the *2-tailed* hypotheses:

Null hypothesis, H_O:	The source populations are the same: $\sigma_A{}^2 = \sigma_B{}^2$.
Alternative hypothesis, H_A:	The source populations are different: $\sigma_A{}^2 \neq \sigma_B{}^2$.

In some cases, the test may need to check whether specifically one variance is greater than the other, giving the *1-tailed* hypotheses, e.g.:

Null hypothesis, H_O: Variance of A is not greater than variance, B: $\sigma_A^2 \leq \sigma_B^2$.

Alternative hypothesis, H_A: Variance of A is greater than variance, B: $\sigma_A^2 > \sigma_B^2$.

10.1.3 *F-statistic and F-critical*

The *F*-statistic is defined as:

$$F_{STAT} = \frac{s_A^2}{s_B^2} \qquad [10.2]$$

where s_A^2 and s_B^2 are the two *sample* variances being compared.

In Excel, calculate F_{STAT} using the equation '= VAR(*arrayA*)/VAR(*arrayB*)', where '*arrayA*' and '*arrayB*' identify the data for sample sets *A* and *B*, and the function VAR calculates the sample variances s_A^2 and s_B^2.

The *F*-critical value, $F_{CRIT} = F_{(T,\alpha, dfA, dfB)}$, depends on:

- T – the number of 'tails' for the problem: 1-tailed or 2-tailed,

- α – the significance level required (default, $\alpha = 0.05$),

- *dfA* and *dfB* – the degrees of freedom of the variances in the numerator and denominator respectively:

$$dfA = n_A - 1 \quad \text{and} \quad dfB = n_B - 1 \qquad [10.3]$$

where n_A are n_B the sample sizes for the two data samples.

In principle, a difference between σ_A^2 and σ_B^2 could go either way: $\sigma_A^2 > \sigma_B^2$ or $\sigma_A^2 < \sigma_B^2$. Hence values for F could be significantly greater than 1.0 or significantly less than 1.0. There are therefore two critical values for F: an *upper* critical value for $\sigma_A^2 > \sigma_B^2$ and a *lower* critical value for $\sigma_A^2 < \sigma_B^2$.

Typically, many published tables give the *upper* 1-tailed value $F_{(1,\alpha, dfA, dfB)}$ for $F > 1.0$. Tables for critical *F*-values for $\alpha = 0.05$ are given in Appendix IV. For example, the 1-tailed critical value, $F_{(1,0.05,5,5)} = 5.05$.

In Excel, the function FINV(α, *dfA*, *dfB*) gives the upper 1-tailed critical value. The *2-tailed* value of $F_{(2,\alpha, dfA, dfB)}$, for a significance level of α, equals the *1-tailed* value, $F_{(1,\alpha/2, dfA, dfB)}$, for a significance level of $\alpha/2$:

$$F_{(2,\alpha, dfA, dfB)} = F_{(1,\alpha/2, dfA, dfB)} \qquad [10.4]$$

For example, the 2-tailed value, $F_{(2,0.05,5,5)} = F_{(1,0.025,5,5)} = 7.15$.

Q10.1 Find the values for F_{CRIT} for:

  (i) A 1-tailed test for a significance level of 0.05 and a numerator with 10 data values and denominator with 8 data values.

  (ii) A 2-tailed test for a significance level of 0.05 and a numerator with 16 data values and denominator with 10 data values.

Making the decision   Accept the alternative hypothesis if: $F_{\text{STAT}} \geq F_{(T,\alpha,\,dfA,\,dfB)}$.

1-tailed test

Use a 1-tailed test for testing that one specific sample, A, has a *greater* variance than the other sample, B.

The variance of A must be placed in the *numerator* to calculate $F_{\text{STAT}} = s_A{}^2/s_B{}^2$.

Use the *1-tailed* value, $F_{(1,\alpha,\,dfA,\,dfB)}$ for F_{CRIT}.

2-tailed test

Use a 2-tailed test when testing for a *difference* between the variance of samples, A and B. This assumes that *it is not known beforehand* which would be the larger.

Calculate the variances first, and place the *larger* sample variance in the *numerator* when calculating F_{STAT}. This will ensure that F_{STAT} is greater than 1.

Use the *2-tailed* value of $F_{(2,\alpha,\,dfA,\,dfB)}$ for F_{CRIT}.

E10.1 ⌨

In order to assess whether two examinations, P and Q, had a difference in their ability to discriminate between students, it was decided to compare the variances of the results for two similar groups of 21 students each. Recorded variances were as follows: $s_P{}^2 = 85.0$ and $s_Q{}^2 = 195.5$.

Apply:

(i)   a 2-tailed test, with an alternative hypothesis which states that there is a difference between the variances of the two exams.

(ii) a 1-tailed test, with an alternative hypothesis which states that the variance of exam P would be less than that of exam Q.

Answers:

(i)   For the 2-tailed test, calculate the ratio of the variances, with the largest value in the numerator:

$$F_{\text{STAT}} = \frac{s_Q^2}{s_P^2} = \frac{195.5}{85.0} = 2.3$$

The 2-tailed critical value with $n_A = n_B = 21$, $F_{\text{CRIT},} = 2.46$.

Since $F_{\text{STAT}} < F_{\text{CRIT}}$, we do not reject the null hypothesis – there is not sufficient evidence for a difference between the examinations.

(ii) For the 1-tailed test, F_{STAT} is calculated with $s_Q{}^2$ in the numerator, giving the same value as in (i).

The 1-tailed critical value with $n_A = n_B = 21$, $F_{\text{CRIT},} = 2.12$.

In this case, $F_{\text{STAT}} > F_{\text{CRIT}}$, we accept the alternative hypothesis – there IS sufficient evidence that exam Q gives greater variance than exam P.

Important note. The choice of 1-tailed or 2-tailed test *must* be made before the data is calculated. It would not be correct to apply the 1-tailed test after it is found that one variance is much bigger than the other.

Q10.2 In a forensic analysis, the refractive indices of two samples of glass fragments are measured as part of a t-test (E10.4), and it is necessary to test whether the variances of the two samples are significantly different.

- Sample A gave seven fragments with standard deviation $s_A = 37.8$.

- Sample B gave six fragments with standard deviation $s_B = 75.5$.

Perform a F-test to test for a possible difference in variance.

10.1.4 Using the p-value in Excel

There is a confusing variety of ways of conducting an F-test using Excel. The most *straightforward* method is:

1 Follow the steps outlined in 10.1.3 to calculate F_{STAT}: use the function VAR to calculate the variances of each data array, and the equation '= VAR(*arrayA*)/VAR(*arrayB*)' to give the value of F_{STAT}.

2 Then, for a *1-tailed* test, use the Excel function FDIST(F_{STAT}, *dfA*, *dfB*) to give the p-value directly.

3 For a *2-tailed* test, it is necessary to *double* the value [9.2] given by the function FDIST(F_{STAT}, *dfA*, *dfB*) to get the *2-tailed* p-value:

$$p\text{-value}(1\text{-}tailed) = \text{FDIST}(F_{STAT}, dfA, dfB)$$

$$p\text{-value}(2\text{-}tailed) = 2 \times \text{FDIST}(F_{STAT}, dfA, dfB)$$

[10.5]

Alternatively, the Excel function FTEST(*array1,array2*) uses the *original sample data arrays* and gives the p-value directly for a *2-tailed* test for difference between the variances of the arrays. Note, however, that the 'Help' facility in Excel, versions 97 and XP, states that the FTEST performs a *1-tailed* test. When using FTEST(*array1,array2*) for a *1-tailed* test:

- the result must be *divided by* 2 to get the true p-value,

$$1\text{-}tailed\ p\text{-value} = \text{FTEST}(array1,array2)/2$$

[10.6]

- and it is necessary to confirm that the expected sample does indeed have the greater variance.

In Excel, **Tools > Data Analysis > F-Test for Two Sample Variances** also uses original sample data arrays to perform a 1-tailed F-test.

Example E10.2 illustrates the use of the various software options for performing an F-test.

E10.2 🖥

Two technicians, X and Y, regularly perform the same type of analysis. However, it is suspected that technician, Y, has become less precise in his procedures, leading to a greater variability in the results that he produces. The data below shows replicate measurements made by each technician on the same material.

Is there evidence to say that the results from Y are significantly (at 95% confidence) more variable than those from X?

									Mean	Standard deviation	Variance
X	23.8	24.7	23.8	24.6	24.2	24.2	24.1	24.7	24.26	0.370	0.137
Y	23.6	25.4	23.5	25.2	24.3	24.5	24.2	25.3	24.50	0.745	0.554

We can state the conditions of the test:

Null hypothesis, H_O:	Variance of Y is not greater than variance, X: $\sigma_Y^2 \leq \sigma_X^2$
Alternative hypothesis, H_A:	Variance of Y is greater than variance, X: $\sigma_Y^2 > \sigma_X^2$
Tails, T:	1 (there is a sense of direction)
Significance level, α:	0.05 (this is the default choice for a 'significant' effect)

Calculate the **F-statistic:**

$$F_{STAT} = \frac{s_Y^2}{s_X^2} = \frac{0.554}{0.137} = 4.04$$

Look up the critical F-value:

$$dfY = 8 - 1 = 7, dfX = 8 - 1 = 7$$

$$F_{(1, 0.05, 7, 7)} = 3.79$$

As $F_{STAT} > F_{CRIT}$ – accept the alternative hypothesis: Y does show *increased variability* compared to X.

p-Method

$$p = FDIST(4.04, 7, 7) = 0.04276$$

We find that $p < \alpha$ – which confirms that we should accept the alternative hypothesis.

Q10.3 Two different methods, P and Q, for measuring the lead content of water are compared. The same sample of water is analysed by each method, giving eight replicate results using method P and seven replicate results using method Q, as given in the table below:

									Mean	Standard deviation	Variance
P	83.3	81.4	81	79.8	82.5	84.7	83.2	83.3	82.40	1.57	2.47
Q	81.5	77.9	76.9	74.7	79.9	84.3	81.3		79.50	3.24	10.47

(i)   Test for a difference in the experimental uncertainty between the two methods.

(ii) As we can see from the data that $s_Q^2 > s_P^2$, would it be acceptable to use a 1-tailed test?

10.2 Sample means: Student's *t*-test

10.2.1 Introduction

It is expected that the reader is familiar with the application and terminologies of the *hypothesis test* as set out in Unit 9.3.

The concepts of the confidence interval of the mean, $CI(\mu, X\%)$, and the confidence deviation, $Cd(\mu, X\%)$, were developed in 8.2.6:

$$CI(\mu, X\%) = \bar{x} \pm \left\{ t_{2,\alpha,(n-1)} \times \frac{s}{\sqrt{n}} \right\} \qquad [10.7]$$

This equation represents the outcome of an experiment designed to measure the *'true'* value, μ, of a variable.

The mean value, $\bar{x}$, of n experimental results is the *best estimate* of the unknown 'true' value, μ, and the standard deviation, s, is a measure of experimental uncertainties.

The value of the '*t*-value', $t_{2,\alpha,(n-1)}$, in the *CI* equation above depends on the number of data values, n, in the data sample, and on the *level of confidence, X%* ($\alpha = 1 - X\%/100$), required for the 'true' value, μ, to be within the *CI* range.

In this Unit we introduce the **t-test**, which uses the same statistics as the *CI* equation. The *t*-test is a *hypothesis test* (see Unit 9.3) that is used to decide:

*   whether an unknown 'true' value, μ, of a SINGLE *sample set* is likely to differ from a *specific value*, μ_0, or

*   whether unknown 'true' values, μ_A and μ_B, of TWO *sample sets* are likely to be *different from each other*.

We also introduce the **paired *t*-test** (10.2.9), which tests for differences between 'paired' data values, for example, for performance differences between values, 'before' and 'after', some treatment.

The theory of both the *t*-test and *t*-value (see also 8.2.7) was developed by William Gossett to analyse the statistics of small data samples in the brewing industry. He published his work under the name of 'Student' to protect confidential industrial information – hence the name of the Student's *t*-test.

10.2.2 Uses of the *t*-test

Experimental variation (Unit 1.2) has two main possible sources:

*   variations due to the measurement process (e.g. response of a pH electrode), and

*   inherent variations between the subjects being measured (e.g. plants in a crop).

However, we see in 8.2.2 that we can use the same statistical processes to manage both types of variation.

*   For variations in the *measurement process*, the *population of data values* would be an *infinite number of replicate* (repeated) measurements.

*   For variations in the *subjects* being measured, the *population of data values* would include the measurement of *every member of the system* (e.g. plants in a crop).

The t-test is a *statistical test* that is applied to one or two *sample* sets of experimental data values.

- In a **one-sample t-test**, the test assesses the *probability* that the 'true' mean value, μ, of the population from which the data sample was taken is different from a specific value, μ_O.

- In a **two-sample t-test**, the test assesses the *probability* that the 'true' mean value, μ_A, of the population for one data sample is different from the 'true' mean value, μ_B, of the population for another data sample.

Remember that the 'true' values, μ, μ_A and μ_B, are never known, and that the mean values $\bar{x}$, $\bar{x}_A$ and $\bar{x}_B$ of the experimental data sets are the *best estimates* for these values.

Q10.4 Decide, for each of the following experiments, whether it is a one-sample or two-sample test.

	Number of samples
(i) Test whether a particular training regime might improve athletic performance by recording the changes in performance of two groups of athletes when only one group adopts the new regime.	One/Two
(ii) Measure the contents of a sample of five cereal packets to test whether the mean contents due to a certain filling machine is below 580 g	One/Two
(iii) Measure the enzyme activity in the livers of two groups (one infected with hepatitis and one normal) to test whether there is a significant difference.	One/Two

---

10.2.3 Hypothesis test

The t-test follows the procedures of a hypothesis test (Unit 9.3), and, *before carrying out the test*, it is necessary to:

- State the null hypothesis, H_O, and alternative hypothesis, H_A.

- Decide on the significance level, α, for the test (usually $\alpha = 0.05$).

We see from [8.15] that significance level, $\alpha = 1-$ (Level of confidence)/100, e.g., a level of confidence of 95% is equivalent to $\alpha = 0.05$.

There are then two ways of carrying out a t-test, either using the *t-statistic* method (see 10.2.4) or using the *p-value* method (10.2.5).

10.2.4 t-statistic method

The *t-statistic*, t_{STAT}, is a statistic that can be calculated (see 10.2.7 and 10.2.8) for a particular set of data, and its value is a measure of the *deviation of the data from the null hypothesis*.

A *large* absolute value for t_{STAT} suggests that it is more likely that the alternative hypothesis is true. However, we need to know 'how large' the t_{STAT} should be before we accept H_A. The answer to 'how large' is given by a *critical t-value*, t_{CRIT}, for the particular problem.

The value of the critical t-value, t_{CRIT}, for the particular problem can be found from **t-tables** in Appendix III.

The *t*-value, $t_{T,\alpha\,df}$ (8.2.7), will depend on:

- number of 'tails', T, 1 or 2, for the problem (see 10.2.5)

- significance level, α (or level of confidence, X%)

- degrees of freedom, *df*, of the problem (see 10.2.7 and 10.2.8)

The decision on whether to accept H_A is based on the *relative values* of the t_{STAT} and t_{CRIT} (see 10.2.7 and 10.2.8).

10.2.5 *p*-value method

It is now possible (and easy) to enter the raw experimental data directly into suitable software packages, which will then calculate a *'p-value'* (9.3.6) for the given data.

The *p*-value is the *probability* that accepting the alternative hypothesis, H_A, would be an *incorrect decision*. The *p*-value is the probability of a type I error (9.3.5). A *small value* for the *p*-value suggests that it is more likely that the alternative hypothesis is true.

> If $p \le \alpha$, accept H_A, otherwise do not reject H_O.   [10.8]

Q10.5 Two students carry out *t*-tests to see whether there is any difference in the mean blood glucose levels between the fish in two fish farm tanks. Each student analyses five different fish from each tank. Student A calculates the *t*-statistic, $t_{STAT} = 3.15$ for his data, and student B calculates the *t*-statistic, $t_{STAT} = 4.09$ for her data.

The two students also note down *p*-values of 0.0078 and 0.0018 calculated from the raw data, but later cannot remember which *p*-value belongs to which student.

(i) Decide which *p*-value would be associated which each *t*-statistic.
(Hint: check the relative magnitudes of the two values of t_{STAT}.)

(ii) Could one or both of the students be at least 95% confident that that there was a difference between the fish in the two tanks?

10.2.6 Tails for a *t*-test

The *'tails'* for the *t*-distribution are also discussed in 8.2.8 in respect of the calculation of confidence interval, and also in 9.3.3.

A **2-tailed test** applies to *t*-tests that are looking for an effect (or difference) that could be in *either 'direction'*.

A **1-tailed test** applies to *t*-tests that are looking for an effect in *only one 'direction'*.

The choice of 'tails' will depend on the wording of the alternative hypothesis, and whether the words carry a sense of *direction* (1-tailed) or just measure *difference* (2-tailed).

Note that the choice of 'tails' *must* be made *before the experiment is carried out* – otherwise it is not possible to state the hypotheses accurately.

Q10.6 Decide, for each of the following statements for alternative hypotheses, which imply a 1-tailed test and which imply a 2-tailed test.

	Number of tails
(i) An athlete's performance is enhanced when using a particular dietary supplement.	1 / 2
(ii) Low-level background music affects people's capacity to memorize a list of names.	1 / 2
(iii) The mean contents of breakfast cereal packets filled by a certain filling machine is less than 580 g.	1 / 2
(iv) An increase in a car driver's blood alcohol level to 40 mg $(100 \text{ mL})^{-1}$ will change the probability of having a road accident.	1 / 2

10.2.7 One-sample *t*-tests

In a one-sample test, the mean, $\bar{x}$, of the set (sample) of data values is the best estimate for the unknown 'true' value, μ. The aim of the one-sample t-test is to decide whether μ is likely to be different from, greater than, or less than, a *specific value*, μ_O.

The t-test uses this information to try to decide whether, or not, the alternative hypothesis, H_A, is likely to be true.

Hypotheses for a **one-sample 2-tailed *t*-test**:

Null hypothesis, H_O:	$\mu = \mu_O$
Alternative hypothesis, H_A:	$\mu \neq \mu_O$
Accept H_A if	$t_{STAT} < -t_{CRIT}$ or $t_{STAT} > +t_{CRIT}$

Hypotheses for the two alternative 'directions' of a **one-sample 1-tailed *t*-test**:

		or	
Null hypothesis, H_O:	$\mu \leq \mu_O$		$\mu \geq \mu_O$
Alternative hypothesis, H_A:	$\mu > \mu_O$		$\mu < \mu_O$
Accept H_A if	$t_{STAT} > +t_{CRIT}$		$t_{STAT} < -t_{CRIT}$

One-Sample *t*-statistic:

$$t_{STAT} = \frac{(\bar{x} - \mu_O)}{s/\sqrt{n}}$$

[10.9]

where s is the sample standard deviation and n is the number of data values in the sample.

When looking up the *critical t-value*, $t_{CRIT} = t_{T, \alpha, df}$, use:

- one or two tails, T, as appropriate,

- significance level $= \alpha$, and

- degrees of freedom, $df = n - 1$.

[10.10]

E10.3 💻

Perform a *t*-test to assess whether the following five data values

$$0.56 \quad 0.62 \quad 0.59 \quad 0.55 \quad 0.67$$

have been drawn from a population with a mean value, μ, that is *not* equal to 0.54.

Answer:

Null hypothesis, H_O: Population mean, $\mu = 0.54$

Alternative hypothesis, H_A: Population mean, $\mu \neq 0.54$

Sample mean, $\bar{x} = 0.598$

Sample size, $n = 5$

Sample standard deviation, $s = 0.0487$

Value for comparison, $\mu_O = 0.54$

Using [10.9]:

$$t_{STAT} = \frac{(\bar{x} - \mu_O)}{s/\sqrt{n}} = \frac{(0.598 - 0.54)}{0.0487/\sqrt{5}} = 2.664$$

No. of tails $= 2$ Significance level: choose default $\alpha = 0.05$ (95%)

Degrees of freedom, $df = 5 - 1 = 4$

$t_{CRIT} = 2.78$

Since $t_{STAT} < t_{CRIT}$ we cannot reject the null hypothesis, H_O.

There is not sufficient evidence at 0.05 significance to conclude that the sample data was not drawn from a population with a mean value of 0.54.

See E10.6 for a computer analysis using *p*-values.

Q10.7 The mean level of pollutant in the wastewater of an industrial company should not exceed 3.50×10^{-8} g L^{-1}. An experiment is conducted to decide whether the wastewater has a mean level of pollutant greater than 3.50×10^{-8} g L^{-1}. If it is concluded that it does, then the probability of error in the conclusion should not be greater than 1 in 20.

Replicate analysis of six samples of the water reveal levels of 3.45, 3.65, 3.58, 3.56, 3.68, 3.59 (all $\times 10^{-8}$ g L^{-1}).

(i) State the hypotheses appropriate to this problem.

(ii) What is the appropriate level of confidence for this test?

(iii) Calculate the mean and standard deviation of the sample.

(iv) Calculate the t-statistic, t_{STAT} appropriate to this problem.

(v) Decide if this is a 1-tailed or 2-tailed t-test.

(vi) Calculate the number of degrees of freedom for t_{CRIT}.

(vii) Look up the value for t-critical t_{CRIT}.

(viii) Comment on whether the level of pollutant is greater than $3.50 \times 10^{-8} \text{ g L}^{-1}$.

10.2.8 Two-sample t-tests

In a two-sample test, the means, $\bar{x}_A$ and $\bar{x}_B$ of two sets (samples) of data values are the best estimates for the unknown 'true' values, μ_A and μ_B. The aim of the two-sample t-test is to decide whether μ_A is likely to be different from, greater than, or less than, the value, μ_B.

The t-test uses this information to try to decide whether, or not, the alternative hypothesis, H_A, is likely to be true.

Hypotheses for a **two-sample 2-tailed t-test**:

Null hypothesis, H_O: | $\mu_A = \mu_B$
Alternative hypothesis, H_A: | $\mu_A \neq \mu_B$
Accept H_A if | $t_{STAT} < -t_{CRIT}$ or $t_{STAT} > +t_{CRIT}$

Hypotheses for the two alternative 'directions' of a **two-sample 1-tailed t-test**:

Null hypothesis, H_O: | $\mu_A \leq \mu_B$ | **or** | $\mu_A \geq \mu_B$
Alternative hypothesis, H_A: | $\mu_A > \mu_B$ | | $\mu_A < \mu_B$
Accept H_A if | $t_{STAT} > +t_{CRIT}$ | | $t_{STAT} < -t_{CRIT}$

Two-sample t-statistic:

$$t_{STAT} = \frac{(\bar{x}_A - \bar{x}_B)}{s' \times \sqrt{\{1/n_A + 1/n_B\}}} \qquad [10.11]$$

where n_A and n_B are the numbers of data values in the data sets A and B respectively, and s' is a 'pooled' standard deviation.

$$s' = \sqrt{\frac{\{(n_A - 1)s_A^2 + (n_B - 1)s_B^2\}}{(n_A + n_B - 2)}} \qquad [10.12]$$

In this case, the assumption has been made that both samples have been drawn from populations that have similar variances, and the 'pooled' standard deviation is a 'best estimate' of the common standard deviation, calculated from the standard deviations of the two samples s_A and s_B.

When looking up the *critical t-value*, $t_{CRIT} = t_{T,\alpha,df}$, use:

- one or two tails, T, as appropriate,
- significance level $= \alpha$, and
- degrees of freedom, $df = n_A + n_B - 2$. [10.13]

The calculation of the *t*-values in cases when the two samples are drawn from populations with DIFFERENT variances is not included in this book. However, two-sample *t*-tests can be performed easily from raw data in both Excel and other statistics software for conditions of 'not equal' variances.

It is good practice, before assuming conditions of 'equal' variance in a *t*-test, to carry out a 2-tailed *F*-test for any significant difference in the variance of the two data samples. Levene's test for the homogeneity of variances performs the same function.

⌨ Study Resources > Additional Material

E10.4 ⌨

In a forensic analysis, the refractive indices of two samples, A and B, of glass fragments are measured. Sample, A, collected from a broken window contains seven fragments and the sample, B, from the clothing of a suspect contains six fragments – the measured values of their refractive indices are given below:

A	1.524184	1.524135	1.524129	1.524095	1.524132	1.524067	1.524152
B	1.524348	1.524255	1.524244	1.524179	1.524250	1.524126	

Assess whether it can be said with 99% confidence that the two samples of glass originated from different sources.

Answer:
This example has small differences between much larger numbers. Provided that both samples are treated in the same way (i.e. multiply each number by 1 000 000 and subtract 1 524 000), it is possible to reduce the problem to more manageable numbers as below – the *t*-test will give the *same conclusion*:

A	184	135	129	95	132	67	152
B	348	255	244	179	250	126	

Null hypothesis, H$_0$:	Means of source populations are equal: $\mu_A = \mu_B$
Alternative hypothesis, H$_A$:	Means of source populations are not equal: $\mu_A \neq \mu_B$
Mean values,	$\bar{x}_A = 127.7$ $\bar{x}_B = 233.7$
Sample standard deviations,	$s_A = 37.8$ $s_B = 75.5$

Using a *F*-test to compare the standard deviations (see Q10.2) we find that there is no significant differences between variances. Hence we assume equal variances and use a pooled standard deviation.

Using [10.12], pooled standard deviation:

$$s' = \sqrt{\frac{\{(n_A - 1)s_A^2 + (n_B - 1)s_B^2\}}{(n_A + n_B - 2)}} = \sqrt{\frac{\{(7 - 1) \times 37.8^2 + (6 - 1) \times 75.5^2\}}{(7 + 6 - 2)}} = 58.05$$

Using [10.11]:

$$t_{STAT} = \frac{(\bar{x}_A - \bar{x}_B)}{s' \times \sqrt{\{1/n_A + 1/n_B\}}} = \frac{(127.7 - 233.7)}{58.05 \times \sqrt{1/7 + 1/6}} = -3.28$$

Number of tails $= 2$ Significance level, $\alpha = 0.01$ (99%)

Degrees of freedom, $df = 7 + 6 - 2 = 11$

$t_{CRIT} = 3.11$

Since $t_{STAT} < -t_{CRIT}$ we can accept the alternative hypothesis, H_A.

There is sufficient evidence at 0.01 significance to conclude that the glass on the suspects clothing did *not* come from the broken window.

See E10.6 for a computer analysis using *p*-values.

Q10.8   Samples of two powders were analysed for their particle diameters. Use a *t*-test to test whether the two powders came from original distributions that had the same, or different, mean diameters. The level of confidence should be 95%.

The results of the measurements (in arbitrary units) were:

Sample A (11 observations): mean $= 6.65$, standard deviation $= 3.91$.

Sample B (16 observations): mean $= 4.28$, standard deviation $= 2.83$.

10.2.9 Paired *t*-tests

A paired *t*-test is a special form of a two sample test, in which each data value in one sample can be 'paired' uniquely with one data value in the other sample.

The use of 'pairing' is an example of 'repeated measures' in experimental design – see 13.3.6.

The procedure for carrying out the paired *t*-test is simple:

1   The difference, d_i, between each pair of responses is calculated: $d_i = a_i - b_i$.

2   The sample of values, d_i, is then treated as a one-sample *t*-test, comparing the unknown mean difference, μ_D, to $\mu_O = 0$.

E10.5 🖥

A student aims to test whether exercise increases systolic blood pressure. Seven subjects measure their systolic blood pressure (in mm Hg) before and after exercise, with the results given in the table below.

Perform a paired *t*-test to investigate whether there is any significant increase (at 0.05) in blood pressure:

Subject	1	2	3	4	5	6	7
BP before	115	122	145	133	147	118	130
BP after	125	127	155	136	142	124	139
Difference	10	5	10	3	−5	6	9

Each subject links unique 'pairs' of data – 'before' and 'after'.

Take the differences, d_i, between the values in each data pair.

Sample standard deviation of the differences, $s = 5.32$.

Mean value of differences $\bar{x} = 5.43$.

Using the one sample t_{STAT} in [10.9], compare $\bar{x}$ with $\mu_O = 0$

$$t_{STAT} = \frac{(\bar{x} - \mu_o)}{s/\sqrt{n}} = \frac{5.43}{5.32/\sqrt{7}} = 2.70$$

Degrees of freedom, $df = n - 1 = 6$.

This is a test for an 'increase' and is therefore a 1-tailed test:

$$t_{CRIT} = t_{1,0.05,6} = 1.94$$

Since $t_{STAT} > t_{CRIT}$, we accept the alternative hypothesis, that the data shows that there is an increase in blood pressure after exercise.

See E10.6 for calculations using Excel.

Q10.9 Seven 'experts' in the taste of real ale have been asked to give a 'taste' score to each of two brands of beer, Old Whallop and Rough Deal. The results, a_i and b_i for each expert are as follows, together with the differences, d_i, between the scores for each expert:

Experts:	A	B	C	D	E	F	G
Old Whallop, a_i	60	59	65	53	86	78	56
Rough Deal, b_i	45	62	53	47	65	80	46
Difference, $d_i = a_i - b_i$	15	−3	12	6	21	−2	10

This is a *paired test*, because the identification of the specific expert is the factor that links the values, a_i and b_i, in the pair of values for each expert.

Assume that the scoring for each expert follows a normal distribution. Complete a paired *t*-test, by carrying out a one-sample *t*-test on the set of seven difference values to test whether the mean of the differences, μ_D, is significantly different (at a level of 0.05) from zero.

The equivalent *non-parametric test*, where the data may not be normally distributed, is given in Q12.2.

10.2.10   Using Excel for *t*-tests

The Excel function TTEST and the Excel Data Analysis Tools will perform either a two-sample *t*-test or a paired *t*-test on two arrays of data.

The results from the *t*-tests are given in the form of the *'p-value'* – see 10.2.5.

If $p \le \alpha$, accept H_A, otherwise do not reject H_O.   [10.14]

However Excel will *not* perform a one-sample *t*-test directly on 'raw' data – it is necessary to calculate the *t*-statistic by entering the appropriate formula.

E10.6   💻

Use appropriate software to solve the problems set in Examples E10.3, E10.4 and E10.5.

E10.3 is a 1-tailed problem and, in Excel, requires the calculation of the t_{STAT} using [10.9]. Using TDIST(t_{STAT},df,tails) it is then possible to calculate the *p*-value: $p = 0.056$. This is consistent with the answer to E10.3 in not rejecting H_O.

E10.4 can be analysed using TTEST(data1,data2,2,2) for a 2-tailed, two-sample, *t*-test using equal variances. TTEST returns the *p*-value, $p = 0.0073$. Since $p < 0.01$, this is consistent with accepting H_A at 0.01 significance.

E10.5 can be analysed using TTEST(data1,data2,1,1) for a 1-tailed, paired *t*-test. TTEST returns the *p*-value, $p = 0.0178$, which implies that there is a significant difference in blood pressure.
Note that if an unpaired *t*-test is used, the *p*-value, $p = 0.206$, which would suggest that there is no evidence of a significant difference; this shows the extra 'power' of using a 'paired' design – see 13.4.6.

Most statistics software packages can be used to perform all three *t*-tests.

Q10.10 A number of students are separately investigating whether two types of growth media, X and Y, give different rates, μ_X and μ_Y, of bacterial growth. They all measure growth rates in several dishes using each of the two media. They start with the hypotheses:

H_0: Choice of media does not affect growth rates: $\mu_X = \mu_Y$,

H_A: Choice of media affects growth rates: $\mu_X \neq \mu_Y$.

Consider the significance of the following student outcomes:

(i) Student A applies a two-sample, *2-tailed*, *t*-test to his data using the Excel function TTEST, which returns the *p*-value: $p = 0.06$. Should the student conclude that there is a difference between the means, μ_X and μ_Y, of the populations from which the samples were drawn?

(ii) Student B decides to set a significance level $\alpha = 0.05$ for his test, and carries out a *2-tailed* test, which returns a value, $p = 0.076$. However, by looking at the mean values of the two samples, the student concludes that media X appears to give a higher growth rate, and so he then restates his hypotheses:

H_0: Media X does not give a higher growth rate than media Y: $\mu_X \leq \mu_Y$,

H_A: Media X does give a higher growth rate than media Y: $\mu_X > \mu_Y$.

He then uses a *1-tailed* test, which returns a value, $p = 0.038$, which he interprets as showing a significant difference, and he accepts the alternative hypothesis. Is the student correct in his analysis? If not, why not?

(iii) Student C uses a two-sample, *2-tailed*, *t*-test on his data using the Excel function TTEST, and a significance level $\alpha = 0.05$. The test returns the *p*-value: $p = 0.03$. What is the probability that the student will be wrong if he accepts the alternative hypothesis?

10.2.11 Statistical constraints for *t*-tests

The statistical theory of the *t*-tests is based on some assumptions:

1 The sampling procedure is truly random.

2 Each item of data is measured independently of all other data values.

3 The population of data values from which the samples are drawn should be a normal distribution.

The experimental procedure should be designed to ensure that the assumptions, 1 and 2, are met – see 'Experimental Design', Unit 13.3.

Assumption 3 is often met because many important distributions in science do follow normal distributions, e.g. experimental uncertainty.

When the data does not follow a normal distribution, it is possible to use non-parametric tests (Chapter 12), although these tests are generally less powerful (9.3.7) than the equivalent parametric tests.

10.3 Linear correlation

10.3.1 Introduction

Correlation is a measure of the extent to which the values of two variables of a system are *related* (or correlated). For example, both the height and weight of a child normally increase as the child gets older – there is a correlation between height and weight.

It must be noted, however, that the fact that two parameters are correlated does NOT *necessarily* mean that one is a *cause* and the other an *effect*. In the example of the height and weight of a child, both increases are the result of an increase in age of the child. The underlying 'cause' (factor) is the age of the child, and the height and weight are *both* 'effects' (outcomes).

If two variables are correlated, the relationship between them could follow very many different mathematical functions, e.g., the numbers of bacteria in a dying population may follow an exponential decay equation. *Linear correlation* is, however, a measure of the extent to which there is a *linear* relationship between two variables, i.e. the extent to which their relationship can be expressed in terms of a *straight line*.

The statistics of linear correlation is covered in greater depth in 13.1.6.

10.3.2 Linear correlation coefficient

The **linear correlation coefficient, r,** between the two variables, x and y, is a measure of the extent to which the data follows a relationship of the form:

$$y = mx + c$$

The variable, r, is also called the 'product-moment correlation coefficient' or the 'Pearson's correlation coefficient'.

The square of the correlation coefficient, r^2, is called the **coefficient of determination** and is also often used as a measure of correlation.

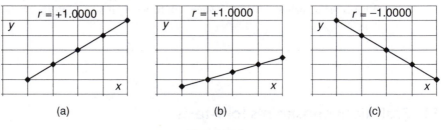

| (a) | (b) | (c) |

Figure 10.1.

- If the data values for x and y fall *exactly* on straight lines as in Figure 10.1, then there is *perfect* linear correlation, $r^2 = 1$, and either $r = +1$ or $r = -1$.

- If the data values do *not* fall exactly on a straight line, then the value of r^2 lies between 0 and 1, i.e. $0 \le r^2 < 1$.

The *sign* of r actually depends on the *direction of slope* of the straight line. If the slope, m, is positive, then the correlation coefficient, r, is also positive. Similarly, if the slope, m, is negative, then the correlation coefficient, r, is negative.

However, the *magnitude* of r does *not* depend on the *magnitude* of the slope, m. The relationship can be summarized in the Table 10.1.

Table 10.1. Ranges of the correlation coefficient.

Relationship	Slope m	Correlation coefficient r	Coefficient of determination r^2
Perfect linear correlation	Positive	$+1$	$+1$
Correlation plus random variation	Positive	$0 < r < +1$	$0 < r^2 < +1$
No correlation	Zero	0	0
Correlation plus random variation	Negative	$-1 < r < 0$	$0 < r^2 < +1$
Perfect linear correlation	Negative	-1	$+1$

It is important to remember that the correlation coefficient ONLY records the significance of a *linear* term in a relationship. It is *always* necessary to plot the data (e.g. in Excel) and *examine the data by eye* to see whether there may be a *non-linear relationship*, before deriving the correlation coefficient.

E10.7

The graphs in Figure 10.2 illustrate some 'correlation' data, together with the 'best-fit' straight line, and the calculated values for the correlation coefficient, r.

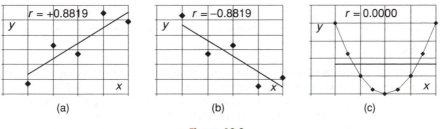

(a) (b) (c)

Figure 10.2.

Graphs (a) and (b) show the same degree of correlation but with opposite slopes. In both cases $r^2 = 0.7777$.

Graph (c) illustrates an extreme situation when a simple calculation on raw data will give zero *linear* correlation, but a quick plot of the data shows very clearly that there is a high degree of *non-linear* correlation between the variables.

10.3.3 Hypothesis test for linear correlation

The hypothesis test for linear correlation tests whether a straight line, which has a non-zero slope, will give a better 'fit' using linear regression (Unit 4.2) than a line with zero slope.

Null hypothesis, H_O: No linear relationship between x and y.
Alternative hypothesis, H_A: Linear relationship between x and y.

The test statistic is the value of the correlation coefficient, r.

In Excel, the correlation coefficient, r, can be calculated using the functions CORREL or PEARSON. The statistics of the correlation coefficient, r, is discussed further in 13.1.6.

The critical value, $r_{CRIT} = r_{T,\alpha,df}$, can be derived from the table of Pearson's correlation coefficients (Appendix V), using appropriate values for

T – number of tails,
α – the significance level required (typically $\alpha = 0.05$),
df – degrees of freedom: $df = n - 2$ (where n = the number of data pairs).

In choosing the number of tails, T, it is necessary to choose:

• 1-tailed for correlation with a particular sign for the slope (for example, it may be expected that m will be positive), or

• 2-tailed for correlation in which a slope of either direction would be acceptable in accepting H_A.

Decision making

$$r \geq +r_{CRIT} \text{ or } r \leq -r_{CRIT} : \text{accept } H_A, \text{ otherwise do not reject } H_O. \qquad [10.15]$$

E10.8 💻

The data below shows the number of 'press-ups' and 'sit-ups' achieved by seven children (subjects). From experience of previous experiments, we devise the hypothesis that a more athletic child who can perform more 'press-ups' will also be able to perform more 'sit-ups'. Hence we investigate whether the data shows a *1-tailed* correlation between performance in the two activities.

Subject:	1	2	3	4	5	6	7
Press-ups:	10	8	2	6	7	3	5
Sit-ups:	24	27	12	14	21	16	22

See text for calculations.

Entering the data from E10.8, into the function CORREL (or PEARSON) in Excel, returns the correlation coefficient, $r = 0.781$.

The 1-tailed critical value for $\alpha = 0.05$ from Appendix V is $r_{CRIT} = 0.669$.

Since $r_S > r_{CRIT}$ we accept the alternative hypothesis.

10.3.4 Using the *p*-value in Statistics Software and Excel

Statistics software can be used to calculate p-values for the correlation coefficient.

If the p-value returned is the 2-tailed value, it is possible to use [9.2] to calculate the 1-tailed p-value by dividing the 2-tailed value by '2'.

The 2-tailed p-value can be obtained in Excel by using:

Tools > Data Analysis > Regression

The 2-tailed correlation p-value appears as the 'P-value' for the 'X-variable', and as 'Significance F' in the ANOVA section of the Regression results – see 13.1.8.

E10.9 🖥

Use statistics software or Excel to derive the p-value for the problem in E10.8.

Statistics software (e.g. MINITAB) gives the 2-tailed value $p = 0.038$.

Excel produces the same value ($p = 0.0383$).

The 1-tailed p-value is then calculated: p(1-tailed) $= p$(2-tailed)/2 $= 0.019$.

Since $p < 0.05$ we again accept the alternative hypothesis.

Q10.11 The following data shows the distances for a standing jump achieved by 10 students, whose heights are recorded:

Student	1	2	3	4	5	6	7	8	9	10
Height (m)	1.59	1.63	1.67	1.7	1.72	1.8	1.74	1.75	1.82	1.84
Jump (m)	1.49	1.45	1.78	1.6	1.9	1.55	1.69	1.81	1.69	1.93

(i) Calculate the Pearson's product moment correlation coefficient between the two data sets.

(ii) Compare the result in (i) with the critical value for a 1-tailed test for positive correlation between jump distance and height.

(iii) Use Excel or statistics software to calculate the appropriate p-value, and compare with conclusion reached in (ii).

10.3.5 Correlation in calibration - checking residuals

When the data is being used for a linear calibration graph (see 4.2.6), then it is normal to expect a high level of linear correlation, with (typically) $r^2 > 0.99$. However, a high value of r^2 can still hide significant non-linear characteristics that might indicate an inherent problem with the 'linear' calibration process.

The **residual** for each data point is the deviation (in the y-direction) between the point itself and the best-fit calibration line (see also 13.1.5). In any linear calibration experiment, it is advisable to check the residuals arising from the best-fit straight line. Example E10.10 illustrates how a simple plot of residual values highlights inherent curvature within a 'linear' calibration.

E10.10

Two 'linear calibration' graphs, A and B, are shown in Figure 10.3(a) and Figure 10.3(b) respectively. Both show good linear correlation with $r^2 = 0.991$. The 'residuals' for every data point in the two lines are shown (magnified) in the graphs underneath: Figure 10.4(a) and Figure 10.4(b) respectively. What information do the 'residual' plots provide?

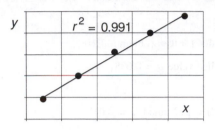 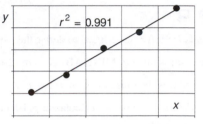

Figure 10.3(a). Calibration A.   **Figure 10.3(b).** Calibration B.

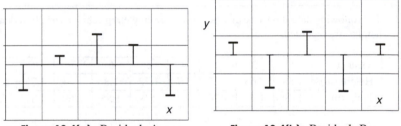

Figure 10.4(a). Residuals A.   **Figure 10.4(b).** Residuals B.

The residuals for line A show a 'systematic' variation along the line, clearly indicating that the 'true' line has a distinct curvature. The 'less than perfect correlation' for line A is due mainly to a curvature of the line and not to measurement uncertainty. The experimenter should investigate the reason for this apparent curvature before proceeding with the experiment.

The residuals for line B show 'random' variations consistent with experimental uncertainty. The 'less than perfect correlation' for line B is due mainly to experimental uncertainty which can then be taken into account (see 13.1.7) when calculating the overall uncertainty in the final result of the experiment.

Information about 'residuals' from linear regression can be obtained using

- Excel Tools > Data Analysis > Regression and then checking the Residuals' and 'Residual Plots' boxes.
- Other statistics software.

Q10.12 Analyse the linear correlation of the following data sets and comment on their use as possible calibration graphs (the dependent variable is in the second row of each data set):

(i)

x	1	3	5	7	9	11
y	0.6	1.77	2.8	3.85	4.77	5.55

(ii)

T	0.1	0.3	0.5	0.7	0.9	1.1
V	0.07	0.18	0.31	0.46	0.57	0.73

(iii)

C	1	3	5	7	9	11
A	0.06	0.18	0.3	0.42	0.52	0.59

10.4 Test for proportion

10.4.1 Introduction

A simple proportion, where the outcome is either Y or N ($= \bar{Y}$), is a direct binary choice. The associated probabilities are determined by the statistics of the binomial distribution (Unit 8.4).

10.4.2 True proportion

The unknown *true* proportion of outcomes, Y, for a system is a *parameter* of the system and we will describe it by the Greek symbol, Π (capital pi). The true value, Π, is the proportion of Y outcomes that would be achieved if the whole *population* (7.1.2) of the system were measured.

The experimentally measured proportion, P, from a *sample* is the *best estimate* for the unknown true proportion, Π. If r outcomes, giving Y, are returned from a sample size, n:

$$P = \frac{r}{n} \qquad [10.16]$$

The unknown *true* mean, μ, of the number, r, in a binomial distribution (where each trial has a probability p of giving the outcome Y) is given by [8.33]:

$$\mu = n \times p \qquad [10.17]$$

The unknown *true* proportion, Π, for the number of 'Y outcomes will be given by:

$$\Pi = \frac{\mu}{n} = p \qquad [10.18]$$

The *true* standard deviation for the number of Y outcomes, r, is given by [8.35]:

$$\sigma = \sqrt{n \times p \times (1 - p)} \qquad [10.19]$$

Using the fact that the experimental proportion, P, is the best estimate for Π, together with [10.18] gives:

$$p \approx P \qquad [10.20]$$

We can then write, combining [10.19] and [10.20]:

$$\sigma \approx \sqrt{n \times P \times (1 - P)} \qquad [10.21]$$

The best estimate of the true standard deviation of the *proportion* is then given by:

$$\sigma_P = \frac{\sigma}{n} \approx \frac{\sqrt{n \times P \times (1 - P)}}{n} = \sqrt{\frac{P \times (1 - P)}{n}} \qquad [10.22]$$

The 95% confidence interval for the value of Π is given by (see [8.37]):

$$CI(\Pi, 95\%) = P \pm 1.96 \times \sqrt{\frac{P \times (1 - P)}{n}} \qquad [10.23]$$

10.4.3 One-sample test

In a one-sample test, the measured proportion, P, is compared with a particular value for the proportion, Π_O.

A test statistic, z, can be calculated:

$$z = \frac{P - \Pi_O}{\sigma_P} = \frac{P - \Pi_O}{\sqrt{\dfrac{P(1 - P)}{n}}} \qquad [10.24]$$

This is equivalent to the t-statistic in [10.9]

The critical value $z_{T,\alpha}$ is equal to $t_{T,\alpha,\infty}$, which is the equivalent t-value evaluated for $df = \infty$ (infinity) – see 8.4.5.

E10.11

A country is due to have a referendum with only two choices – Yes or No. In anticipation of the referendum, 1000 people are selected at random in an 'opinion poll' and it is found that 520 will vote Yes and 480 will vote No. Using this information, calculate whether it is possible to be confident at a 95% level that the result of the actual referendum will be 'Yes'.

The hypotheses for the test become:

Null hypothesis, H_O: True proportion of Yes votes $\leq 50\%$
Alternative hypothesis, H_A: True proportion of Yes votes $> 50\%$

This is a one-sample, 1-tailed, test to compare the proportion, $P = 0.52$ returned in one sample with a specific target proportion, $\Pi = 0.5$. Using [10.24]

$$z = \frac{0.520 - 0.500}{\sqrt{\dfrac{0.52\,(1 - 0.52)}{1000}}} = \frac{0.020}{0.0158} = 1.27$$

The critical value, $z_{T,\alpha}$, can be calculated from the 1-tailed t-value $df = \infty$:

$$z_{1,0.05} = t_{1,0.05,\infty} = 1.64$$

(see Appendix III).

Since

$$z < z_{T,\alpha}$$

we cannot reject the null hypothesis and it is not possible to be 95% confident that the vote will return a 'Yes'.

See also Example E8.20.

Q10.13 A random sample of 50 frogs is taken from a lake, and it is found that 37 are female and 13 are male.

Is this experimental result significantly (at 0.05) different from the expected proportion of 60% females to 40% males?

(See also Q8.23 and E11.2.)

10.4.4 Two-sample test

A two-sample test compares the measured proportions, P_A and P_B. The relevant statistic is:
(this is very similar to the t-statistic [10.11])

$$z = \frac{P_A - P_B}{\sqrt{P'\,(1 - P') \times (1/n_A + 1/n_B)}} \tag{10.25}$$

where P' is a 'pooled' value for the proportions of the two samples:
(this is very similar to pooled standard deviation [10.12])

$$P' = \frac{n_A P_A + n_B P_B}{n_A + n_B} \tag{10.26}$$

Q10.14   A student compares the germination of similar seeds under two different conditions. In condition A, 85 out of 100 seeds germinate and, in condition B, 60 out of 80 seeds germinate.

Test, at 0.05 significance, whether the proportion of seeds germinating in condition A is significantly different to the proportion in condition B.

11

Chi-squared Tests

Overview

The family of chi-squared, χ^2, tests is based on counting the occurrences of events within different categories. The tests assess the probability that an observed distribution of events could have occurred purely by chance or whether an underlying factor had caused a deviation from random behaviour.

It is assumed that the reader is familiar with the general principles of hypothesis testing as given in Unit 9.3.

The first Unit, 11.1, gives the basic operation of the test applied to simple frequencies, and outlines the main issues associated with using the χ^2 statistic. The next Unit, 11.2, applies the χ^2-test to assessing whether there is an association (or contingency) between two factors that affect the frequency of events. Finally, in Unit 11.3, the χ^2-test is used to assess whether observed distributions may, or may not, follow certain predicted distributions (e.g. Poisson).

We include below a summary of the main points for reference when using any χ^2-test. The theory underlying these points is developed through the chapter.

Chi-squared statistic (see [11.1]):

$$\chi^2 = \sum_i \frac{(O_i - E_i)^2}{E_i}$$

The data values in the cells must be *frequencies* (i.e. counts of the number of occurrences). Ideally, the total of all frequencies should exceed 50. Chi-squared tests cannot be used for percentages or ratios. (See Unit 10.4 for testing for proportions.)

All expected values must be 5 or more. In some cases where cells contain less than 5 counts, it is possible to merge two cells together (see 11.1.6).

Yates' correction (see [11.3]) should be used when **df = 1**, which includes 2×2 contingency tables:

$$\chi^2 = \sum_i \frac{(|O_i - E_i| - 0.5)^2}{E_i} \quad \text{when } df = 1$$

Degrees of freedom:

One factor with n categories [11.4]: $df = n - 1$

Two-factor contingency table with c columns and r rows [11.6]: $df = (c - 1)(r - 1)$

Goodness of fit for a Poisson distribution with n categories [11.7]: $df = n - 2$

Goodness of fit for a normal distribution with n categories [11.7]: $df = n - 3$

Typical chi-squared tests are 1-tailed tests. The alternative hypothesis will be accepted if:

$$\chi^2 \geq \chi^2_{\text{CRIT}}$$

11.1 Test for frequencies

11.1.1 Introduction

This Unit gives the basic operational method for the chi-squared, χ^2, test.

When there is a random variation in the number of events occurring (e.g. in counting 'heads' when tossing a coin), we would be surprised to see numbers that are *very different* from the 'expected'. We would also be surprised, given the randomness of the situation, to find that the numbers were *very close* to those 'expected'. In general, we would expect to see some randomness of behaviour – but not too much! The χ^2-test is a hypothesis test that provides information for deciding how much deviation would be too much or too little.

11.1.2 Observed and expected frequencies

Example E11.1 gives an example of the type of question that a χ^2-test can address.

E11.1 🖥

A normal playing die is thrown 120 times and we record the number occurrences of each of the six possible scores, giving *observed* frequencies, O_i. Each of the six scores represents a different category, i. We might *expect* that, on average, each particular score should occur $120/6 = 20$ times.

The results, for both observed and expected frequencies, are shown in the table. A quick visual examination of the data might suggest that the die is biased because, for the score '6', there appears to be a large difference between *observed* (31) and *expected* (20) frequencies.

Category, i	Score	1	2	3	4	5	6	Sum
Observed frequency	O	15	22	20	14	18	31	120
Expected frequency	E	20	20	20	20	20	20	120

Can we deduce from the data, with a confidence level of 95%, that the die is indeed biased?

This solution to this example is worked out in the text.

The relevant hypotheses (9.3.2) for a χ^2-test, as in E11.1, would be:

Null hypothesis, H_O:	The differences between observed and expected frequencies could have occurred just by chance.
Alternative hypothesis, H_A:	The differences between observed and expected frequencies could not have occurred just by chance, and hence an additional factor must be significant.

The basic χ^2 statistic is calculated using the equation:

$$\chi^2 = \sum_i \frac{(O_i - E_i)^2}{E_i}$$

[11.1]

where O_i is the observed frequency in category, i, and E_i is the expected frequency in category, i.
The calculation for χ^2 for the data in E11.1 is set out in Table 11.1.

Table 11.1. Calculation for Example E11.1.

Category, i	Score	1	2	3	4	5	6	Sum
Observed frequency	O_i	15	22	20	14	18	31	120
Expected frequency	E_i	20	20	20	20	20	20	120
	$O_i - E_i$	−5	2	0	−6	−2	11	
	$(O_i - E_i)^2$	25	4	0	36	4	121	
	$(O_i - E_i)^2/E_i$	1.25	0.2	0	1.8	0.2	6.05	9.5

In Table 11.1, the chi-squared statistic, $\chi^2 = 1.25 + 0.2 + 0 + 1.8 + 0.2 + 6.05 = 9.5$.

In general, a *large* value for the χ^2 statistic would indicate that observed frequencies were *very different* from the expected frequencies.

However, we need to know *how large* the value of the χ^2 statistic must be before we can accept that the difference is due to more than just random chance (i.e. accept the alternative hypothesis, H_A).

Tables of critical values for χ^2 (see Appendix III) and the Excel function CHIINV usually give the 1-tailed value (11.1.4): $\chi^2{}_{CRIT} = \chi^2{}_{1,\alpha,df}$.

The number of *degrees freedom, df*, will depend on the particular problem.

The decision on the hypothesis test can be made:

$\chi^2 \geq \chi^2{}_{CRIT}$: accept the alternative hypothesis

$\chi^2 < \chi^2{}_{CRIT}$: do not reject the null hypothesis

[11.2]

For the data in E11.1, we see in [11.4] that degrees of freedom $df = 6 - 1 = 5$.

Referring to the $\chi^2{}_{CRIT}$ table for $df = 5$ at $\alpha = 0.05$, we find that $\chi^2{}_{CRIT} = 11.07$. In this example, the value of χ^2 is 9.5, giving

$$\chi^2 < \chi^2{}_{CRIT}$$

Hence we conclude that there is *no evidence that the pattern of scores differs from that which could be expected by chance* – we do not reject the null hypothesis.

Q11.1 Look up the critical values, $\chi^2{}_{CRIT}$, for the following conditions:

(i) 4 degrees of freedom at a significance level of 0.05.

(ii) 1 degree of freedom at a significance level of 0.01.

> **Q11.2** A χ^2-test records a value, $\chi^2 = 3.6$. Is it possible to say, with a confidence level of 95% and degrees of freedom, $df = 3$, that the differences between observed and expected frequencies did not occur just by chance?

11.1.3 Yates' correction for $df = 1$

For χ^2 calculations with the minimum number of categories, giving $df = 1$, the normal calculation for the χ^2 statistic [11.1] tends to overestimate the true χ^2-value. This means that the test would sometimes accept the alternative hypothesis in borderline cases when the alternative hypothesis should not be accepted.

The overestimate for $df = 1$ is reduced by using Yates' corrected equation:

$$\chi^2 = \sum_i \frac{(|(O_i - E_i)| - 0.5)^2}{E_i} \qquad [11.3]$$

where $|(O_i - E_i)|$ is the *absolute value* of $(O_i - E_i)$ and means 'take the positive value of $(O_i - E_i)$'.

After the calculation of the χ^2 statistic using the Yates' correction, the rest of the chi-squared test is performed normally.

E11.2

A random sample of 50 frogs is taken from a lake, and it is found that 37 are female and 13 are male. The expected proportion of female to male frogs is 60:40. Are the observed numbers significantly different to those expected?

Use both equations [11.1] and [11.3] to calculate the value for χ^2_{STAT}.

Observed	O_i	37	13	
Expected	E_i	30	20	
	$O_i - E_i$	7	-7	
Normal χ^2	$(O_i - E_i)^2 / E_i$	1.63	2.45	$\chi^2 = 1.63 + 2.45 = 4.08$
Yates' χ^2	$(|O_i - E_i| - 0.5)^2 / E_i$	1.41	2.11	$\chi^2 = 1.41 + 2.11 = 3.52$

The critical value for $df = n - 1 = 2 - 1 = 1$ and $\alpha = 0.05$:

$$\chi^2_{CRIT} = 3.84$$

We find that, using the normal χ^2 value [11.1], $\chi^2_{STAT} > \chi^2_{CRIT}$, suggesting that we should accept the alternative hypothesis.

However, using the Yates' correction [11.3], $\chi^2_{STAT} < \chi^2_{CRIT}$, suggesting that we should not reject the null hypothesis. The Yates' correction gives a more cautious conclusion.

It is interesting to note, nevertheless, that the parametric calculations performed in Q8.23 and Q10.13 both suggest that the alternative hypothesis is just significant for this particular problem.

The Yates' correction is not always used in statistical software. It is necessary to check with the particular package being used.

11.1.4 Tails in a χ^2-test

The typical χ^2-test is a *1-tailed* test (see 9.3.3), i.e. testing that the variation in frequency values is *greater than* would be expected by chance. The critical value, χ^2_{CRIT}, introduced in 11.1.2 is an *upper* critical value.

However, in any real system, natural *random* variations would normally mean that the observed frequencies would differ to a certain extent from the expected frequencies. For example, if a die were rolled 120 times, it is very unlikely that every score (1 to 6) would appear exactly 20 times. It is therefore unlikely that the χ^2 statistic would be zero, or even very small.
A *very low value* of χ^2 would be indicative,

- either that the events are not truly random,

- or that there has been an error in the collection of the experimental results.

A low value would be significant at α if:

$$\chi^2 < \chi^2_{1,(1-\alpha),df}$$

The value for $\chi^2_{1,(1-\alpha),df}$ can be found in Excel by using CHIINV$(1 - \alpha, df)$. For values of χ^2 below the lower critical value, it would be important to check the way in which the data was recorded, and ensure that there was no external effect that was producing data more *uniformly* than would be expected by chance.

11.1.5 Procedure for χ^2 frequency test

The basic χ^2 frequency test will compare the observed frequencies, O_i, with the expected frequencies, E_i, for each of n categories.

For the basic χ^2-test to compare frequencies, the degrees of freedom,

$$df = n - 1 \hspace{3cm} [11.4]$$

where n is the number of different categories in the problem.
(Note the Yates' correction (11.1.3) when $df = 1$.)

The use of the χ^2 statistic has been introduced in 11.1.2 by solving the problem presented in E11.1. The null and alternative hypotheses for the test are given in 11.1.2. The basic steps for carrying out the hypothesis test are then:

1 Calculate the frequencies, E_i, that would be expected in each category, i, on the basis of the null hypothesis. Add the frequencies in all categories to get the total frequency, T, and then:
   (i) Distribute the total frequency, T, amongst the n categories according to a known ratio – see E11.3.
   (ii) If there is no bias between the categories, the expected frequency in every category will be equal to the total frequency, T, divided by the number of categories, n – as in E11.1.

2 Provided that $df > 1$ (i.e. $n > 2$), use [11.1] to calculate the value of χ^2_{STAT}, as in Table 11.1.

3 If $df = 1$ ($n = 2$), it is necessary to use the Yates' correction in [11.3] to calculate the value of χ^2_{STAT}, as in E11.2.

4   Using degrees of freedom, $df = n - 1$, look up the appropriate critical value for χ^2_{CRIT}.

5   If $\chi^2_{STAT} \geq \chi^2_{CRIT}$ accept the alternative hypothesis, otherwise do not reject the null hypothesis.

Q11.3   In a switchboard for a large company, the calls should be directed randomly to the various operators. In a 5-hour period, four operators receive the following number of calls: 44, 32, 56, 28.

Is there evidence to suggest that the calls are not actually being directed randomly?

Step 1(i) above states that we can use the chi-squared test even if the frequencies are expected to have *different* values. It is only necessary to calculate each of the expected frequencies, E_i *individually*, as illustrated in the following Example E11.3.

E11.3

Measurement of genotypes is expected is expected to show types AB, Ab, aB, ab in the ratio 9:3:3:1. A sample of $T = 200$ observations gave observed frequencies, O_i: 131, 28, 32 and 9 respectively in the four groups. Test whether this sample shows a significant difference from the expected ratio.

The calculation for the expected frequency, E_i, for each group is given in E7.10. We can then use those values to complete the following table, and then perform the calculations for $(O_i - E_i)^2/E_i$ for each of the four groups.

Category i	AB	Ab	aB	ab	Totals
O_i	131	28	32	9	200
Ratio	9	3	3	1	16
E_i	112.5	37.5	37.5	12.5	200
$(O_i - E_i)^2/E_i$	3.042	2.407	0.807	0.980	7.236

$$\chi^2 = 7.24$$
$$\chi^2_{CRIT} = 7.81 \text{ (for 5\% confidence and } df = 4 - 1 = 3)$$

In this case, because $\chi^2 < \chi^2_{CRIT}$ we should not reject the null hypothesis (at 0.05). The apparent differences in numbers for each category could have occurred by chance.

Q11.4   A self-pollinating pink flower is expected to produce red, pink and white progeny with the following relative frequencies 1:2:1. A student grows 50 flowers and finds the following division of colours: red 18, pink 22 and white 10.

Can the student conclude that his observed distribution of colours is significantly different (at 95% confidence) from the expected frequencies?

11.1.6 Expected frequency less than 5?

If the expected frequency in one or more categories has a value of less than 5, then the result of the χ^2 test becomes unreliable. In this case the test should not proceed.

The way of dealing with this problem is to amalgamate some of the categories so that the 'combined' expected frequency is equal to 5 or more. Example E11.4 shows how this can be achieved in a simple case.

If the total number of 'frequencies', or the number of categories, is so small that this cannot be done, then it is probably necessary to repeat an enlarged experiment and record more data values.

E11.4 🖳

A college expects that, in an 'average' group of students, the numbers obtaining examination marks (%) will be in the ratios as in the table below. A group of 60 students obtain marks as given in the row 'Observed 1'. Is there evidence, at 95% confidence, that this group of students has a significantly different spread of abilities compared to an 'average' group?

Marks (%)	< 40	40–49	50–59	60–69	> 69	Totals
Ratios	0.06	0.23	0.42	0.23	0.06	1
Observed 1	10	15	26	8	1	60
Expected 1	3.6	13.8	25.2	13.8	3.6	60
Observed 2		25	26	9		60
Expected 2		17.4	25.2	17.4		60
$(O - E)^2/E$		3.3195	0.0254	4.0552		7.4001

The expected numbers of students in each mark range is calculated in 'Expected 1'.
However, two mark ranges contain less than '5' counts.

Amalgamate the first two cells by adding the values. Similarly, amalgamate the last two cells, giving the values in 'Expected 2' and 'Observed 2'.

Calculate the χ^2_{STAT} using [11.1], giving $\chi^2_{STAT} = 7.40$.

Degrees of freedom, $df = 3 - 1 = 2$.

Critical value, $\chi^2_{CRIT} = 5.99$.

Hence, as $\chi^2_{STAT} > \chi^2_{CRIT}$, we accept the hypothesis that the spread of abilities in this student group is significantly different from the expected.

11.1.7 Chi-squared in Excel

The following functions are available in Excel:

- CHIDIST calculates the p-value for the given value of χ^2 and the relevant number of degrees of freedom.

- CHITEST calculates the p-value, but it is necessary to calculate the expected values first, and set out the observed and expected values in two separate arrays. N.B. CHITEST does not use the Yates' correction when $df = 1$.
- CHIINV calculates the *critical value*, χ^2_{CRIT}, for a given number of degrees of freedom and significance level.

11.2 Contingency tests

11.2.1 Introduction

A *contingency test* is used to investigate whether the effect of one factor in a system depends on the value of a second factor, i.e. is the effect of one factor *contingent on* (dependent on) the value of a second factor? This is equivalent to investigating whether there is an interaction (see also 13.2.9) between two factors. For example, we may find that, if we count the numbers of cars *skidding* in the four categories:

(a)   good tyres and dry road

(b)   good tyres and wet road

(c)   poor tyres and dry road

(d)   poor tyres and wet road

we may find that the frequency for category (d) increases dramatically as the skidding effect of poor tyres increases significantly with (*contingent* on) an increase in the wetness of the road.

11.2.2 Contingency table

The operation of the contingency test, illustrated in the following example E11.5, aims to investigate whether patient improvement is *contingent on* (i.e. depends on) the use of a drug rather than a placebo.

E11.5 ⌨

In a double-blind drug trial, 150 randomly selected patients who are all suffering from a specific disease are given either the drug or a placebo, and the numbers of each group who show no improvement, some improvement, or much improvement, are recorded and reproduced in the table below:

	Drug	Placebo
Much improvement	26	7
Some improvement	42	18
No improvement	32	25

Is there evidence, at a confidence level of 95%, that more patients improve with the drug than with just the placebo?

See text for calculation.

In E11.5, it is necessary to derive the χ^2 statistic to compare the *observed frequencies* with the *expected frequencies* that would occur *if improvement did NOT depend on the drug*.

If the χ^2-test reports a significant difference, then we will conclude that improvement IS affected by taking the drug.

In order to calculate the *expected frequencies*, the first step is to calculate (see Table 11.2):

- totals, R_1, R_2 and R_3, for each row,

- totals, C_1 and C_2, for each column, and

- total, T, of all rows (= total of all columns).

Table 11.2. Column and row totals for Example E11.5.

	Drug	Placebo	Totals
Much improvement	26	7	$R_1 = 33$
Some improvement	42	18	$R_2 = 60$
No improvement	32	25	$R_3 = 57$
Totals	$C_1 = 100$	$C_2 = 50$	$T = 150$

The 'column totals' show that a total of 100 (= C_1) people took the drug and a total of 50 (= C_2) took the placebo, giving an overall total of 150 (= T) in the trial.

From the 'row totals' column on the right we can see that, *overall*, 33 out of 150 people showed *much improvement*. If there were NO difference between drug and placebo, we would expect that the *same proportions* of people ($33/150 = R_1/T$) would show *much* improvement under *both* drug *and* placebo. As 100 (= C_1) people took the drug we would expect that:

$$\text{Expected number with } \textit{much} \text{ improvement with } \textit{drug} = 100 \times \frac{33}{150} = \frac{C_1 \times R_1}{T} = 22$$

As 50 (= C_2) people took the placebo we would expect that:

$$\text{Expected number with } \textit{much} \text{ improvement with } \textit{placebo} = 50 \times \frac{33}{150} = \frac{C_2 \times R_1}{T} = 11$$

Similarly the *proportions*, for both drug and placebo, for *no improvement* and *some improvement* would be $60/150 (= R_2/T)$ and $57/150 (= R_3/T)$ respectively. We can then perform calculations for the numbers in the other cells:

$$\text{Expected number with } \textit{some} \text{ improvement with drug} = 100 \times \frac{60}{150} = \frac{C_1 \times R_2}{T} = 40$$

$$\text{Expected number with } \textit{some} \text{ improvement with placebo} = 50 \times \frac{60}{150} = \frac{C_2 \times R_2}{T} = 20$$

$$\text{Expected number with } \textit{no} \text{ improvement with drug} = 100 \times \frac{57}{150} = \frac{C_1 \times R_3}{T} = 38$$

$$\text{Expected number with } \textit{no} \text{ improvement with placebo} = 50 \times \frac{57}{150} = \frac{C_2 \times R_3}{T} = 19$$

In general, we can write a formula for calculating the expected frequency, E_{xy}, for the cell in column x and row y:

$$E_{xy} = \frac{C_x \times R_y}{T} \qquad [11.5]$$

where C_x is the column total for column x and R_y is the row total for row y.

We can now write the values (see Table 11.3) for both the observed and the expected frequencies:

Table 11.3. Observed and expected frequencies for E11.5.

	Observed frequencies			Expected frequencies	
	Drug	Placebo		Drug	Placebo
Much improvement	26	7		22	11
Some improvement	42	18		40	20
No improvement	32	25		38	19

We calculate the value of the χ^2 statistic

$$\chi^2 = \frac{(26-22)^2}{22} + \frac{(42-40)^2}{40} + \frac{(32-38)^2}{38} + \frac{(7-11)^2}{11} + \frac{(18-20)^2}{20} + \frac{(25-19)^2}{19} = 5.32$$

Degrees of freedom for a contingency table with r rows and c columns:

$$df = (r-1) \times (c-1) \qquad\qquad [11.6]$$

(Note the Yates' correction (11.1.3) when $df = 1$.)

For a 3×2 table, $df = (3-1) \times (2-1) = 2$.
Referring to the χ^2 table for $df = 2$ at 5%, we find that $\chi^2_{CRIT} = 5.99$. In this example, the value of χ^2 is 5.32, giving

$$\chi^2 < \chi^2_{CRIT}$$

Hence we conclude that there is *no evidence for a significant difference* between the occurrences of the scores – we do not reject the null hypothesis.

The apparent improvement due to the drug compared to a placebo could have occurred by chance.

E11.6 🖳

Perform the analysis for E11.5, using CHIDIST, CHITEST and CHINV in Excel or Statistical Software.

The analyses give $p = 0.070$. Since $p > \alpha (= 0.05)$, we do not reject the null hypothesis in agreement with the calculation using the χ^2 statistic.

Q11.5 Two students, A and B, are performing similar projects to investigate whether there is an association between the leaf colour (green, yellow/green, yellow) of seedlings and the soil in which they grow. Of those seeds that germinate in each of three soil types, the students count the number of seedlings that fall into each of the categories for leaf colour. The results are given in the table below.

Do both students arrive at the same conclusion concerning the suggestion that there is an association (or interaction) between leaf colour and soil type?

	Student set A			Student set B		
	Soil 1	Soil 2	Soil 3	Soil 1	Soil 2	Soil 3
Green	63	79	60	56	79	86
Yellow/green	20	25	19	11	25	19
Yellow	4	11	19	6	10	8

11.2.3 2 × 2 contingency table

A contingency table with two rows ($r = 2$) and two columns ($c = 2$) is a special case of a general $r \times c$ table. The degrees of freedom for the 2 × 2 contingency table, $df = (2 - 1) \times (2 - 1) = 1$.

As $df = 1$, it is necessary to use Yates' correction [11.3]:

$$\chi^2 = \sum_i \frac{(|(O_i - E_i)| - 0.5)^2}{E_i}$$

After the calculation of the χ^2 statistic, the rest of the chi-squared test is performed normally.

Q11.6 A student on an education course decided to investigate whether mature female students had greater success in learning mathematics than mature male students. The student recorded the number of students, male and female, mature and not mature, who scored more that 50% in the mathematics assessment of a foundation science course. The numbers are given in the table below.

	Male	Female
Mature	32	28
Not mature	87	42

(i) Calculate the numbers that would be expected if mature male and mature female students were equally successful in learning mathematics.

(ii) Perform a contingency test, using the Yates' correction, to assess whether there is a difference between male and female students when learning mathematics as a mature student.

(iii) Repeat the test in (ii) without using the Yates' correction. Does this give a different result? If yes, which is the *correct* answer?

11.3 Goodness of fit

11.3.1 Introduction

We saw in 11.1.5 how it is possible to test whether the distributions of observed frequencies 'fit' a calculated distribution of expected frequencies. In this Unit we develop the concept of 'goodness of fit' when applied to standard distributions.

The chi-squared test can be useful in testing whether the distribution of some experimentally observed frequencies might follow a known distribution pattern, e.g. a normal distribution or a Poisson distribution.

There are also other methods (e.g. Anderson–Darling) available in statistics software for testing for 'goodness of fit', that are particularly useful for testing whether a given set of sample data is likely to have been drawn from a normal distribution.

11.3.2 Basic steps

If it is believed that the observed frequencies might follow a particular distribution or curve, then the first step is to derive the best estimates for the parameter values that are needed to define that distribution or curve. These best estimates must be calculated from the values of the observed frequencies themselves.

For a Poisson distribution we need a best estimate for the mean, μ.

For a normal distribution, we need values for *both* the mean, μ, and the standard deviation, σ.

The next step is to use the parameters for the distribution to calculate the expected frequencies within each of the observed categories. The rest of the procedure then follows the normal chi-squared process.

In a 'goodness of fit' test with n data values, the number of degrees of freedom is given by:

$$df = n - 1 - d \tag{11.7}$$

where d is the number of parameters that must be *calculated* to define the expected distribution.

For a Poisson distribution it is necessary to calculate the *mean*, μ, of the expected distribution – i.e. just one parameter, $d = 1$. Hence degrees of freedom when testing for a Poisson distribution: $df = n - 1 - 1 = n - 2$.

For a normal distribution it is necessary to calculate the *mean*, μ, and *standard deviation*, σ, of the expected distribution – i.e. two parameters, $d = 2$. Hence degrees of freedom when testing for a normal distribution: $df = n - 1 - 2 = n - 3$.

11.3.3 Poisson distribution

We can illustrate a goodness of fit analysis through the solution to E11.7.

E11.7 🖳

It is believed that a particular rare birth defect in children is a totally random occurrence. The country was divided into regions with approximately similar populations. The table shows the number (frequency, f) of regions that record specific numbers of defects, e.g. 18 regions recorded two defects per 10 000:

No. of defects per 10 000 births, r	0	1	2	3	4	5	6	7	Total
Observed frequency, O_r	12	25	18	14	12	10	5	2	98

If the birth defects occur randomly with the same low probability in all areas, we would expect a Poisson distribution of frequencies. If, however, the defects occur because of some form of infection or local pollution, then clusters of cases may occur, resulting in a deviation from the Poisson distribution.

Use a chi-squared test to investigate whether the observed values might deviate from the Poisson distribution.

Calculations in following text.

Stating the hypotheses for E11.7:

Null Hypothesis, H_O: The observed distribution is random and could be explained using a Poisson distribution.

Alternative Hypothesis, H_A: The observed distribution is not random and could not be explained using a Poisson distribution.

In E11.7, we need to compare the distribution of the 98 observed values with the way that a true Poisson distribution would distribute 98 values amongst the different categories.

The first step, before we can calculate the probability values, $p(r)$, for a Poisson distribution, is to calculate a *best estimate* for its *mean value*, μ. This *best estimate* will be given by the mean value, $\bar{r}$, of our *observed* data.

Calculation of the mean value, $\bar{r}$, for the number of births:

(see also 7.2.6)

$$\bar{r} = \frac{12 \times 0 + 25 \times 1 + 18 \times 2 + 14 \times 3 + 12 \times 4 + 10 \times 5 + 5 \times 6 + 7 \times 2}{98} = 2.5$$

The expected frequency, E_r, for each category, r, is calculated using

$$E_r = n \times p(r)$$

where $n = 98$ is the total number of births, and $p(r)$ is the Poisson probability (8.5.2) for recording each particular number, r, of births. Using [8.38], with $\bar{r}$ as the best estimate for μ:

$$E_r = 98 \times p(r) = 98 \times \frac{e^{-2.5} \times 2.5^r}{r!}$$

Table 11.4.

No. of defects per 10 000 births, r	0	1	2	3	4	5	6	7	Total
Observed frequency, O_r	12	25	18	14	12	10	5	2	98
Expected frequency, E_r	8.0	20.1	25.1	21.0	13.1	6.6	2.7	1.0	97.6

The actual calculation can be performed easily using the POISSON function in Excel, giving the results in Table 11.4. There are two points to note:

- The expected frequencies for the categories, $r = 6$ and $r = 7$, are less than '5', and this falls below the limit for using chi-squared (see 11.1.6).

- Categories for $r > 7$ are missing, giving an error in the total for the expected frequencies.

We can correct for these two problems by amalgamating all the categories for $r = 5$ and above. Repeating the calculation of expected frequencies now gives the values as in Table 11.5.

Table 11.5.

No. of defects per 10 000 births, r	0	1	2	3	4	$r \geq 5$	Total
Observed frequency, O_r	12	25	18	14	12	17	98
Expected frequency, E_r	8.0	20.1	25.1	21.0	13.1	10.7	98
$(O_r - E_r)^2/E_r$	1.95	1.19	2.03	2.30	0.09	3.76	11.32

The chi-squared statistic, $\chi^2 = 1.95 + 1.19 + 2.03 + 2.30 + 0.09 + 3.76 = 11.32$.

From [11.7], degrees of freedom for this problem, $df = 6 - 1 - 1 = 4$.

Referring to the critical values table (Appendix III):

For $df = 4$ at 0.05, we find that $\chi^2_{\text{CRIT}} = 9.49$.

For $df = 4$ at 0.01, we find that $\chi^2_{\text{CRIT}} = 13.28$.

In this example, the value of χ^2 is 11.32, which falls between the two critical values. Hence, at a confidence level of 0.05 we would be able to reject the null hypothesis and claim that there was a significant difference between the observed distribution and a Poisson distribution. We would then conclude that there was a factor, other than pure chance, which influenced the occurrence of these birth defects. However, we would be unable to make the same claim if a confidence level of 99% were required.

11.3.4 Normal distribution

To calculate the expected values for a normal distribution, it is necessary to calculate both the mean and standard deviation, as illustrated in E11.8.

E11.8 🖥

The number of students obtaining marks in different ranges is given in the table below. Is there evidence, at a confidence of 95% (significance level, 0.05), which suggests that the marks do not follow a normal distribution.

The calculations for mean and standard deviation of the observed values:

Mean value, $\bar{x} = 53.50$

Standard deviation, $s = 11.47$

are given in E7.7.

The calculations for the expected values, using the Excel function NORMDIST, are given in E8.7. These values (to 2 dp) have been included in the table below.

Range	0–9	10–19	20–29	30–39	40–49	50–59	60–69	70–79	80–89	90+
Observed	0	1	4	18	39	80	47	9	2	0
Expected	0.01	0.29	3.34	18.58	50.50	67.18	43.79	13.96	2.17	0.16

It can be seen directly that there are categories where the expected values are less than 5. We must amalgamate some cells, as shown in the revised table below:

Range		0–39	40–49	50–59	60–69	70+
Observed		23	39	80	47	11
Expected		22.22	50.50	67.18	43.79	16.29
$(O - E)^2/E$		0.027	2.619	2.446	0.235	1.718

Chi-squared statistic, $\chi^2 = 0.027 + 2.619 + 2.446 + 0.235 + 1.718 = 7.05$.

From [11.7], degrees of freedom for this problem, $df = 5 - 1 - 2 = 2$.

Referring to the critical values table (Appendix III):

For $df = 2$ at 0.05, we find that $\chi^2_{\text{CRIT}} = 5.99$.

Since $\chi^2_{\text{STAT}} > \chi^2_{\text{CRIT}}$ we accept the alternative hypothesis.

The observed distribution of marks follows a distribution that is significantly different from a normal distribution.

12

Non-parametric Tests

Overview

Chapter 9 introduced the concept of a 'hypothesis test', and Chapter 10 developed a range of *parametric* tests for performing hypothesis tests. The term 'parametric' refers to the fact that the data values are used in the calculations for the various test statistics. This can be contrasted with 'non-parametric' tests where the numerical values are only significant in establishing the 'ranking', or order, of the data values.

In this chapter, we introduce a range of non-parametric tests: the Wilcoxon, Mann–Whitney, Kruskal–Wallis, Friedman and Sign tests for differences, and the Spearman rank test for correlation.

The Wilcoxon tests are the non-parametric equivalents of one-sample *t*-tests (10.2.7) and paired *t*-tests (10.2.9), and the Mann–Whitney test is equivalent to a two-sample *t*-test (10.2.8). The Spearman rank correlation coefficient is the non-parametric equivalent to the Pearson's correlation coefficient (10.3.2) for testing for correlation between two variables.

The Kruskal–Wallis and Friedman tests analyse experimental data with more than two samples and are the non-parametric equivalents of the one-way and two-way ANOVAs that are to be introduced in Chapter 13.

Non-parametric statistics *must* be used for *ordinal data*, because ordinal data (see Chapter 2, 'Overview') does not have an inherent numerical value, but does have a sense of progression. For example, 'opinion' scores in a questionnaire may be ranked: 1 = excellent; 2 = good; 3 = satisfactory; 4 = poor; 5 = bad. However these values cannot be used directly in calculations of parametric variables such as *mean* and *standard deviation*. Non-parametric statistics use the *median* value and *interquartile range* (7.1.7) as the equivalent measures of location and spread.

Non-parametric tests can sometimes be used in place of their parametric equivalents for *quantitative* data. For example, the *parametric* data values, 21.2 m, 23.7 m and 22.3 m, can be *ranked* in order as 1, 3 and 2 respectively, and the data could be used in *either* parametric or non-parametric tests. However, due to the loss of information in transforming parametric data to non-parametric data, the non-parametric tests are usually *less powerful* (9.3.7) than their parametric equivalents.

The only data type (see Chapter 2, 'Overview') that *cannot* be ranked is nominal data, because it has no sense of progression from one category to the next, e.g., categorizing people on the basis of their preferences for different sports.

Before proceeding with the following tests, it is recommended that the student confirm that he/she has a good understanding of the principles of hypothesis testing developed in Unit 9.3.

It is not appropriate in this book to include full tables of critical values for non-parametric tests. Some limited values are given in the appendixes to illustrate the use of the 'test-statistic' method, but, for a more complete set of values, the reader is referred to published sets of statistical tables.

Current versions of Excel do not have functions or tools to perform non-parametric tests directly. However, dedicated statistics software can perform non-parametric tests very easily, albeit with a variety of approaches between the different commercial packages.

The answers (on the web site), to the Examples given in the chapter, demonstrate the indirect use of Excel in performing non-parametric calculations, in addition to the direct use of different statistics software.

⌨ Answers & Examples > Examples

12.1 Wilcoxon tests

12.1.1 Introduction

The Wilcoxon test is basically a **one-sample** test – the non-parametric equivalent of the one-sample t-test (10.2.7). The Wilcoxon test can also be used within a *paired* test in the same way that a one-sample t-test can be used within a paired t-test (10.2.9).

Due to the strong parallels between the Wilcoxon test and the one-sample t-test, it would be useful for the reader to review the use of t-tests in Unit 10.2.

12.1.2 One-sample Wilcoxon test

The one-sample Wilcoxon test aims to test whether the observed data sample has been drawn from a population that has a median value, m, that is significantly different from (or greater/less than) a *specific value*, m_O. The tests may be 1-tailed or 2-tailed as for the one-sample t-test.

The first steps are to state the null and alternative hypotheses and to choose the significance level, α, of the test.

Hypotheses for a **2-tailed Wilcoxon test**:

Null hypothesis, H_O: $m = m_O$

Alternative hypothesis, H_A: $m \neq m_O$

Hypotheses for the alternative 'directions' of a **1-tailed Wilcoxon test**:

Null hypothesis, H_O: $m \leq m_O$ **or** $m \geq m_O$

Alternative hypothesis, H_A: $m > m_O$ $m < m_O$

The procedure for this test is illustrated using example E12.1.

E12.1 ⌨

The generation times (5.2.5), t, of ten cultures of the same micro-organisms were recorded.

Time, t(hr)	6.3	4.8	7.2	5.0	6.3	4.2	8.9	4.4	5.6	9.3

The microbiologist wishes to test whether the generation time for this micro-organism is significantly greater than a *specific value* of 5.0 hours.

See following text for calculations.
(The use of statistics software is given in Examples on the web site.)

The test required in E12.1 is 1-tailed, i.e. testing for 'greater than 5.0'. The hypotheses for the Wilcoxon test are therefore:

Null hypothesis, H_O: $m \leq 5.0$

Alternative Hypothesis, h_A: $m > 5.0$

We choose the significance level, $\alpha = 0.05$.

The next step is to take the *differences*, $(t_i - m_O)$, between each data value, t_i, and the target median, $m_O = 5.0$. The differences are then *ranked in order of their absolute values* (i.e. ignoring the $+/-$ signs) giving the results in Table 12.1.

Table 12.1. Ranking of data for E12.1.

Time, t (hr)	6.3	4.8	7.2	5.0	6.3	4.2	8.9	4.4	5.6	9.3
Differences $(t - m_O)$	1.3	−0.2	2.2	0.0	1.3	−0.8	3.9	−0.6	0.6	4.3
Ranks (ignoring sign)	5.5	1	7	—	5.5	4	8	2.5	2.5	9
Sign	+	−	+		+	−	+	−	+	+

Three main **rules** should be observed when calculating rank values:

- Data items that have a **zero difference** should be excluded from the ranking process and ignored in subsequent calculations. For example, in Table 12.1, one item has a zero difference and is then ignored, leaving only nine data values for subsequent calculations.

- Data items are ranked in order of their **absolute** values (i.e. ignoring their signs).

- If two (or more) data items have the **same absolute values (ties)**, then the rank value is *shared* between the values. For example in Table 12.1, the differences −0.6 and +0.6 are the joint second and third values and receive the *shared* ranking of 2.5. Similarly the difference '1.3' occurs twice and the two values are given equal ranking of 5.5.

The **test statistic,** $W(+)$, is the *sum* of the rankings for all *positive* differences. Similarly, the **test statistic,** $W(-)$, is the *sum* of the rankings for all *negative* differences.

The test statistics can be calculated In Table 12.1:

$$W(+) = 5.5 + 7 + 5.5 + 8 + 2.5 + 9 = 37.5$$
$$W(-) = 1 + 4 + 2.5 = 7.5$$

It should be noted that the sum of all rank values is given by:

$$W(+) + W(-) = 0.5 \times n \times (n + 1) \qquad [12.1]$$

where n is the number of values included in the calculation, *after any with zero differences have been excluded*.

In Example E12.1, $n = 9$, giving from [12.1],

$$W(+) + W(-) = 0.5 \times n \times (n + 1) = 45$$

which is in agreement with the calculated values.

If the sample median, m, is close to the specific value, m_O (null hypothesis), we would expect the values of $W(+)$ and $W(-)$ to be very similar. For the alternative hypothesis to be true, we would expect there to be a significant difference between $W(+)$ and $W(-)$, with one of them having a significantly *low value*.

The critical values normally given for the Wilcoxon test are the 'lower critical values', W_L.
Appendix V gives critical values at $\alpha = 0.5$ for sample sizes, n.
The conditions for accepting the alternative hypotheses, H_A, are:

2-tailed Wilcoxon test:

$$\text{Accept } H_A \text{ if } \quad \textit{either } W(+) \leq W_L \textit{ or } W(-) \leq W_L \qquad\qquad [12.2]$$

1-tailed Wilcoxon test:

$$\text{For} \quad m > m_0 \quad \text{accept } H_A \text{ if} \quad W(-) \leq W_L \qquad\qquad [12.3]$$

$$\text{For} \quad m < m_0 \quad \text{accept } H_A \text{ if} \quad W(+) \leq W_L \qquad\qquad [12.4]$$

In E12.1, the lower critical value, W_L, for a 1-tailed test with $n = 9$ and $\alpha = 0.5$ is $W_L = 8$.
Since $W(-) < W_L$ we accept the alternative hypothesis.

Q12.1 In a survey to assess whether 18 trainees found a particular exercise regime useful, they were asked to reply on a scale from -5 (not at all useful) to $+5$ (very useful). Their scores are listed below:

3	5	−3	5	4	−2	2	4	−1
0	5	0	−4	−2	3	0	1	3

(i) Use a non-parametric test to assess whether the results show that the regime was considered to be useful (i.e. the median score is greater than '0').

(ii) What result would have been obtained for a test to assess whether the results show a median value which is not equal to '0', i.e. either greater than or less than '0' ?

12.1.3 Paired Wilcoxon test

The one-sample Wilcoxon test also forms the basis of a *paired* test, in a similar way to the one-sample t-test which forms a basis for a paired t-test (10.2.9).

The procedure for this test is illustrated using E12.2.

E12.2 🖳

The reaction times of nine subjects, before and after being given a particular drug, are recorded below. The differences between the reaction times have also been calculated, together with the ranking of these differences.

Subject	1	2	3	4	5	6	7	8	9
Before	15.5	16.6	24	11.8	14.5	19.4	14.7	23.1	22.7
After	17.6	17.5	25.3	10.8	17.6	19.4	17.2	26.1	20.6
Differences	2.1	0.9	1.3	−1.0	3.1	0	2.5	3.0	−2.1
Ranks (ignoring sign)	4.5	1	3	2	8	—	6	7	4.5
Sign	+	+	+	−	+		+	+	−

Is there sufficient evidence to show that the drug has affected reaction time?

See following text for calculations.
(The use of statistics software is given in Examples on the web site.)

The hypotheses for the paired Wilcoxon test are therefore:

Null hypothesis, H_O: $m_{\text{AFTER}} = m_{\text{BEFORE}}$

Alternative hypothesis, H_A: $m_{\text{AFTER}} \neq m_{\text{BEFORE}}$

The differences between each pair of the data values in E12.2 are ranked using the conditions that:

- Pairs with *zero difference* should be excluded from further calculations.

- Pairs with *equal differences* should share the ranking values.

The test statistics can be calculated for E12.2:

$$W(+) = 4.5 + 1 + 3 + 8 + 6 + 7 = 29.5$$
$$W(-) = 2 + 4.5 = 6.5$$

In E12.2 the hypothesis test is 2-tailed, and if we assume that $\alpha = 0.5$, the critical value for $n = 8$ is $W_L = 3$. Since $W(-) > W_L$ and $W(+) > W_L$ we do not reject the null hypothesis.

Q12.2 This is the same as Q10.9, but assume that the data may not necessarily be normally distributed.

Seven 'experts' in the taste of real ale have been asked to give a 'taste' score to each of two brands of beer, Old Whallop and Rough Deal. The results, a_i and b_i for each expert are as follows, together with the differences, d_i, between the scores for each expert:

Experts	A	B	C	D	E	F	G
Old Whallop, a_i	60	59	65	53	86	78	56
Rough Deal, b_i	45	62	53	47	65	80	46
Difference, $d_i = a_i - b_i$	15	−3	12	6	21	−2	10

Perform a paired Wilcoxon test on the two data sets to test whether their medians are significantly different.

12.2 Mann–Whitney test

12.2.1 Introduction

The Mann–Whitney test is a **two-sample** test, the non-parametric equivalent of the two-sample t-test (10.2.8).

Because of the strong parallels between the Mann–Whitney test and the two-sample t-test, it would be useful for the reader to review the use of t-tests in Unit 10.2.

12.2.2 Mann–Whitney test

The Mann–Whitney test compares the median values of two samples.
Hypotheses for a **2-tailed Mann–Whitney test**:

Null hypothesis, H_O: $m_A = m_B$

Alternative hypothesis, H_A: $m_A \neq m_B$

Hypotheses for the alternative 'directions' of a **1-tailed Mann–Whitney test**:

Null hypothesis, H_O: $m_A \leq m_B$ **or** $m_A \geq m_B$

Alternative hypothesis, H_A: $m_A > m_B$ $m_A < m_B$

The procedure for this test is illustrated using E12.3.

E12.3 🖥

Is there a significant difference between the median values of the populations, from which the following data sets, A and B, were drawn?

Set A	21	23	24	18	20	25
Set B	18	16	20	18	22	

See following text for calculations.
(The use of statistics software is given in Examples on the Web Site.)

The hypotheses for the 2-tailed Mann–Whitney test in E12.3 are:

Null hypothesis, H_O: $m_A = m_B$

Alternative hypothesis, H_A: $m_A \neq m_B$

We choose the significance level, $\alpha = 0.05$.

The first step is to rank the data values in order, keeping a record of the data set to which each value belongs – see Table 12.2. Where two or more data items have the same value, then they are given the same average rank, e.g., the three data items that have the value '18', share ranks '2', '3' and '4', giving the *average* rank '3' for all of them.

The sums of ranks are calculated for each set separately:

$$W_A = 3 + 5.5 + 7 + 9 + 10 + 11 = 45.5$$

$$W_B = 1 + 3 + 3 + 5.5 + 8 = 20.5$$

Table 12.2. Ranking data for E12.3.

Set	B	A	B	B	A	B	A	B	A	A	A
Value	16	18	18	18	20	20	21	22	23	24	25
Rank	1	3	3	3	5.5	5.5	7	8	9	10	11

The test statistic is the ***U*-statistic** that is calculated for each set, x:

$$U_x = W_x - n_x(n_x + 1)/2 \qquad\qquad [12.5]$$

where n_x is the number of data values in set 'x'.

In E12.3 we get

$$U_A = 45.5 - 6 \times (6 + 1)/2 = 24.5$$
$$U_B = 20.5 - 5 \times (5 + 1)/2 = 5.5$$

If the sample medians, m_A and m_B, are similar (null hypothesis), we would also expect the values of U_A and U_B to be similar. For the alternative hypothesis to be true, we would expect there to be a significant difference between U_A and U_B, with one of them having a significantly *low value*.

The critical values normally given for the Mann–Whitney are the lower critical values, U_L.

Appendix VI gives critical values at $\alpha = 0.5$ for sample sizes, n_A and n_B.

The conditions for accepting the alternative hypotheses, H_A, are:

2-tailed Mann–Whitney test:

$$\text{Accept } H_A \text{ if} \quad \textit{either} \quad U_A \leq U_L \quad \textit{or} \quad U_B \leq U_L \qquad\qquad [12.6]$$

1-tailed Mann–Whitney test:

$$\text{For} \quad m_A > m_B \quad \text{accept } H_A \text{ if} \quad U_B \leq U_L \qquad\qquad [12.7]$$
$$\text{For} \quad m_A < m_B \quad \text{accept } H_A \text{ if} \quad U_A \leq U_L \qquad\qquad [12.8]$$

In E12.3, the lower critical value, U_L, for a 2-tailed test with $n_A = 6$, $n_B = 5$, and $\alpha = 0.5$ is $U_L = 3$.

Since $U_A > U_L$ and $U_B > U_L$ and using [12.6], we do not reject the null hypothesis.

Q12.3   The generation times (5.2.6), t, of cultures of the same micro-organisms were recorded under two conditions A and B and reproduced in the table below:

Condition A	6.7	5.8	6.9	9.6	8.9	8.2	6.1	4.8	9.2
Condition B	3.9	6.3	4.4	5.6	6.3	4.2	7.2		

Is there evidence that the generation times are significantly different under the two conditions?

12.3   Kruskal–Wallis and Friedman tests

12.3.1   Introduction

The Mann–Whitney test (Unit 12.2) looks for differences in median values between *two* samples. Both the Kruskal–Wallis and Friedman tests look for differences in median values between *more than two* samples.

The parametric equivalents for the Kruskal–Wallis and Friedman tests are not introduced until the following chapter, but it is logical to group them here with the other non-parametric tests.

The Kruskal–Wallis test is used to analyse the effects of more than two levels of just *one factor* on the experimental result. It is the non-parametric equivalent of the one-way ANOVA (13.2.7).

The Friedman test analyses the effect of two factors, and is the non-parametric equivalent of the two-way ANOVA (13.2.8).

12.3.2 Kruskal–Wallis test

E12.4 🖳

The data in the table below, gives the efficiency of a chemical process using three different catalysts (A, B and C) on each of four days:

Catalyst	Day 1	Day 2	Day 3	Day 4
A	84.5	82.8	79.1	80.2
B	78.4	79.1	78.0	76.0
C	83.1	79.9	77.8	77.9

Is there evidence that the different catalysts result in different efficiencies?
Assume in this example, that the data may not be normally distributed and that it is necessary to use non-parametric statistics. (Example E13.13 answers the same problem using the parametric one-way ANOVA.)

See following text for calculations.
(The use of statistics software is given in Examples on the web site.)

In E12.4, the *factor* involved is the choice of catalyst, which occurs at three *levels*: catalyst A, catalyst B and catalyst C. Each level of the factor is tested with a number of replicate measurements, giving three sample sets overall.
The hypotheses for the test are:

Null hypothesis, H_O: Catalysts have no differential effect on efficiency

Alternative hypothesis, H_A: Catalysts do have a differential effect on efficiency

The first step is to assign ranks to *all* data values. Set out the data values in order and rank them sequentially, as in Table 12.3. Data items that have equal values are given the *average rank* of those items (see also 12.2.2).

Table 12.3. Ranking of data for E12.4.

Data	76	77.8	77.9	78	78.4	79.1	79.1	79.9	80.2	82.8	83.1	84.5
Rank	1	2	3	4	5	6.5	6.5	8	9	10	11	12

Each data value is then replaced, in the original data table, by its ranking. The next step is to calculate the sum of the ranks, W_i, for each level, i, of the factor – as shown in Table 12.4.

Table 12.4. Calculation of rank sums for each sample.

Catalyst	Day 1	Day 2	Day 3	Day 4	Rank sum, W_i
A	12	10	6.5	9	37.5
B	5	6.5	4	1	16.5
C	11	8	2	3	24

The general test statistic, H, for 'r' levels of a factor is calculated from the sums, W_i, of each of the r samples, i.e. for $i = 1$ to $i = r$:

$$H = \frac{12}{N(N+1)} \left\{ \frac{W_1^2}{n_1} + \frac{W_2^2}{n_2} + \frac{W_3^2}{n_3} + \cdots + \frac{W_r^2}{n_r} \right\} - 3(N+1) \qquad [12.9]$$

where n_i is the number of data items in sample, i, and N is the total number of all data items:

$$N = n_1 + n_2 + n_3 + \cdots + n_r = \sum_i n_i \qquad [12.10]$$

If there is a significant difference between the medians of different samples, then values for W_i will also be significantly different. It can be shown that such a variation will result in a larger value for the test statistic, H.

The alternative hypothesis is accepted if $H \geq H_{CRIT}$.

For small data samples, the critical value, H_{CRIT}, for the Kruskal–Wallis statistic is dependent on the individual sample sizes, n_i, and the specific values can be found in published tables.

For larger data samples, and as an *approximation* for smaller samples, the critical value, H_{CRIT}, is equal to the chi-squared critical value, χ^2_{CRIT}, with degrees of freedom given by:

$$df = r - 1 \qquad [12.11]$$

The value of H is calculated for E12.4, using [12.9]:

$$H = \frac{12}{12(12+1)} \left\{ \frac{37.5^2}{4} + \frac{16.5^2}{4} + \frac{24^2}{4} \right\} - 3(12+1) = 4.356$$

Degrees of freedom, $df = 3 - 1 = 2$, giving an approximate critical value:

$$H_{CRIT} \approx \chi^2_{CRIT} = 5.99$$

Since $H < H_{CRIT}$ we do not reject the null hypothesis. We are unable to detect a significant effect due to the different catalysts.

N.B. For the data in E12.4, the *more accurate* critical value $H_{CRIT} = 5.69$, which, in this case, leads to the same conclusion.

Q12.4 The fungi species richness was measured on a 10-point scale on four different species of trees over a period of 4 days, and the results tabulated as below:

	Day 1	Day 2	Day 3	Day 4
Tree 1	6	4	3	3
Tree 2	4	3	3	2
Tree 3	4	2	1	1
Tree 4	2	1	2	1

Perform a Kruskal–Wallis test to investigate whether there is a significant difference in fungi species richness between the trees. Take the measurements on different days as being replicate measurements.

12.3.3 Friedman test

In the Kruskal–Wallis test, it is assumed that all data items, *within each sample*, are *replicates* recorded under exactly the same conditions.

However, it is possible that a *second* factor is also influencing the results. Example E12.5 considers the same problem as E12.4, but with the possibility that, as well as the catalyst, the *choice* of day of measurement may also be having some effect. The calculation takes this into account.

E12.5 🖥

Using the same data as in E12.4:

Catalyst	Day 1	Day 2	Day 3	Day 4
A	84.5	82.8	79.1	80.2
B	78.4	79.1	78.0	76.0
C	83.1	79.9	77.8	77.9

Investigate whether the 'day' as well as the 'catalyst' may be a factor in affecting the efficiency of the chemical process.

The data is already **blocked** (13.3.5) so that values appropriate to specific days have been grouped (blocked) within specific columns.
(Example E13.14, answers the same problem using the parametric two-way ANOVA.)

See following text for calculations.
(The use of statistics software is given in Examples on the web site.)

The hypotheses are the same as E12.4, but, in E12.5, we are now looking for an effect on the efficiency of the process **by** the catalyst **blocked by** the day of measurement (13.3.5).

In this case, the procedure is to rank the data variables **separately** within each of the (vertical) blocks – see Table 12.5. The next step is to calculate the sum of the ranks, W_i, for each level, i, of the factor:

Table 12.5. Calculation of rank sums for each sample for E12.5.

Catalyst	Day 1	Day 2	Day 3	Day 4	Rank Sum, W_i
A	3	3	3	3	12
B	1	1	2	1	5
C	2	2	1	2	7

The general test statistic, S, for 'r' levels of the factor (rows) and 'c' blocks (columns) is calculated from the sums, W_i, of each of the r levels, i.e. for $i = 1$ to $i = r$:

$$S = \frac{12}{cr(r+1)}\left\{W_1^2 + W_2^2 + W_3^2 + \cdots + W_r^2\right\} - 3c(r+1) \qquad [12.12]$$

If there is a significant difference between the medians of different samples, then values for W_i will also be significantly different. It can be shown that this variation will result in a large value for the test statistic, S.

The alternative hypothesis is accepted if $S \geq S_{CRIT}$.

For small data samples, the critical value, S_{CRIT}, for the Friedman statistic is dependent on the number of factor levels, r, and the number of blocks, c, and the specific values can be found in published tables.

For larger data samples, and as an *approximation* for smaller samples, the critical value, S_{CRIT}, is equal to the chi-squared critical value, $\chi^2{}_{CRIT}$, with degrees of freedom given by:

$$df = r - 1 \qquad [12.13]$$

The value of S is calculated for E12.5, using [12.12]:

$$S = \frac{12}{4 \times 3 \times (3+1)} \{12^2 + 5^2 + 7^2\} - 3 \times 4 \times (3+1) = 6.5$$

Degrees of freedom, $df = 3 - 1 = 2$, giving an approximate critical value:

$$S_{CRIT} \approx \chi^2{}_{CRIT} = 5.99$$

Since $S \geq S_{CRIT}$ we accept the alternative hypothesis.

Blocking the data to take account of the effect of the 'day' has made the test more sensitive to the factor effect of the different catalysts. Compare the use of the Friedman test with the use of the two-way ANOVA for the same problem in 13.2.8.

N.B. For the data in E12.5, the *more accurate* critical value $S_{CRIT} = 6.50$, and we would still accept the alternative hypothesis since $S \geq S_{CRIT}$.

Q12.5 Use the same data as in Q12.4 for fungi species richness:

	Day 1	Day 2	Day 3	Day 4
Tree 1	6	4	3	3
Tree 2	4	3	3	2
Tree 3	4	2	1	1
Tree 4	2	1	2	1

Perform a Friedman test to investigate whether there is a significant difference in fungi species richness between the trees (treatment), while blocking the data by 'day'.

12.3.4 Friedman test – swapping factors

In E12.5, there are, effectively, *two* factors – the catalyst and the day. We may choose to re-analyse the data to check principally for the effect of the day, while taking the effect of the catalyst into account – see E12.6:

E12.6 🖥️

Using the same data as in E12.5, we can choose to look for an effect on the efficiency of the process *by* the day of measurement **blocked by** the catalyst. To do this, we first *transpose* the data so that the *day* becomes the *factor* being tested – the rows of the table now represent the different factor levels (days):

Catalyst	A	B	C
Day 1	84.5	78.4	83.1
Day 2	82.8	79.1	79.9
Day 3	79.1	78	77.8
Day 4	80.2	76	77.9

See following text for calculations.
(*The use of statistics software is given in Examples on the web site.*)

The *transposed* problem in E12.6 is analysed using the same process as in E12.5, but noting the new values:

$$c = 3, r = 4 \text{ and } df = 4 - 1 = 3$$

The calculation of rank sum values is given in Table 12.6.

Table 12.6. Calculation of rank sums for each sample for E12.6.

Catalyst	A	B	C	Rank Sum, W_i
Day 1	4	3	4	11
Day 2	3	4	3	10
Day 3	1	2	1	4
Day 4	2	1	2	5

Using [12.12] the value of the test statistic in E12.6 is calculated:

$$S = \frac{12}{3 \times 4 \times (4+1)} \{11^2 + 10^2 + 4^2 + 5^2\} - 3 \times 3 \times (4+1) = 7.4$$

Degrees of freedom: $df = 4 - 1 = 3$, giving an approximate critical value:

$$S_{\text{CRIT}} \approx \chi^2_{\text{CRIT}} = 7.81$$

Since $S < S_{\text{CRIT}}$ we do not reject the null hypothesis.

The Friedman test has been unable to show that the effect of the 'day' is significant at 0.05. Compare this with the more *powerful* two-way ANOVA, which gives a *significant* p-value $= 0.022$ for the effect of the 'day'.

Q12.6 Use the same data as in Q12.4 and Q12.5 for fungi species richness.

Perform a Friedman test to investigate whether there is a significant difference in fungi species richness between the days (treatment), while blocking the data by 'tree'.

12.4 Sign test

12.4.1 Introduction

The sign test is a **one-sample** test that compares the median of a data set with a specific target value. The sign test performs the same function as the one-sample Wilcoxon test, but it *does not rank* data values. Without the information provided by ranking, the sign test is less powerful (9.3.7) than the Wilcoxon test. However, the sign test is useful for problems involving 'binomial' data, where observed data values are either above or below a target value.

12.4.2 Sign test

The sign test aims to test whether the observed data sample has been drawn from a population that has a median value, m, that is significantly different from (or greater/less than) a specific value, m_O. The hypotheses are the same as in 12.1.2.

We will illustrate the use of the sign test in E12.7 by using the same data as in E12.1.

E12.7 🖥️

The generation times, t (5.2.6), of 10 cultures of the same micro-organisms were recorded.

Time, t (hr)	6.3	4.8	7.2	5.0	6.3	4.2	8.9	4.4	5.6	9.3

The microbiologist wishes to test whether the generation time for this micro-organism is significantly greater than a specific value of 5.0 hours.

See following text for calculations.
(The use of statistics software is given in Examples on the web site.)

In performing the sign test, each data value, t_i, is compared with the target value, m_O, using the following conditions:

$$t_i < m_O \quad \text{replace the data value with '}-\text{'}$$
$$t_i = m_O \quad \text{exclude the data value}$$
$$t_i > m_O \quad \text{replace the data value with '}+\text{'}$$

giving the results in Table 12.7.

Table 12.7. Signs of differences in E12.7.

Difference	+	−	+		+	−	+	−	+	+

We now calculate the values of

r^+	= number of '+' values:	$r^+ = 6$
n	= total number of data values *not excluded*:	$n = 9$

Decision making in this test is based on the probability of observing particular values of r^+ for a given value of n. For example, when choosing n data values at random from the population, it is more likely that they would be evenly spread on either side of the true median value, rather than predominately either above or below the median.

We are therefore concerned with the *cumulative* binomial probability distribution, $p(r^+)$, of observing r^+ specific outcomes from n random selections.

The p-value for the test can be calculated from the binomial distribution, using the Excel function BINOMDIST with a probability $p = 0.5$ (see 8.4.4), and using the entry 'TRUE' to return the 'cumulative' value:

1-tailed sign test:

$$\text{For } r^+ > n/2 \quad p\text{-value} = p(r \geq r^+) = 1 - \text{BINOMDIST}(r^+ - 1, n, 0.5, \text{TRUE}) \qquad [12.14]$$

$$\text{For } r^+ < n/2 \quad p\text{-value} = p(r \leq r^+) = \text{BINOMDIST}(r^+, n, 0.5, \text{TRUE}) \qquad [12.15]$$

(If $r^+ = n/2$, then there is no significant difference!)

2-tailed sign test:

$$p\text{-value} = 2 \times p\text{-value for 1-tailed test} \qquad [12.16]$$

Example E12.7 is a 1-tailed problem with $n = 9$ and $r^+ = 6$, giving $r^+ > n/2$, and we calculate the p-value using [12.14]:

$$p\text{-value} = p(r \geq 6) = 1 - \text{BINOMDIST}(6 - 1, 9, 0.5, \text{TRUE}) = 0.2539$$

Since p-value > 0.05 — we do not reject the null hypothesis.

The sign test is not sufficiently powerful to detect a significant difference in this data. However, the more powerful one-sample Wilcoxon test, which ranks data values, was capable of detecting a significant difference with the same data as in E12.1.

Q12.7 In monitoring animal response to stimuli, an index on a scale -10 to $+10$ was used to record the response of each animal to the stimulus. The values of the indices for 20 trials is given in the table:

0	7	−2	8	3	6	1	9	1	9	2
7	6	6	3	5	0	3	−3	−2	−2	−1

Use the sign test to test whether the animals showed a significant response (positive or negative) to the stimuli.

12.5 Spearman's rank correlation

12.5.1 Introduction

Spearman's rank correlation coefficient, r_S (sometimes written as ρ, 'rho'), is the non-parametric equivalent of Pearson's product moment coefficient, r (10.3.2), and is used as a measure of association between two variables.

When using the *parametric* correlation coefficient, r, it is required that the data values for the two variables are derived from populations with *normal* distributions. There is no such requirement for the Spearman's rank correlation, and it can be used with any data whose values can be ranked in order.

12.5.2 Spearman's rank correlation coefficient

The first step is to rank the x-values and the y-values **separately**. The correlation coefficient, r_S, can then be calculated in one of two ways:

1   The value of r_S equals the value of the *parametric* correlation coefficient, r, between the *rank values* of the 'x' and 'y' data, and can be calculated in Excel using the functions CORREL or PEARSON.

2   Alternatively, the differences, d_i, in rank can be calculated between each pair of data values:

$$d_i = \text{Rank of } x_i - \text{Rank of } y_i \qquad [12.17]$$

and then using the formula:

$$r_S = 1 - \frac{6 \sum d_i^2}{n(n^2 - 1)} \qquad [12.18]$$

where n is the number of data pairs.

The two above methods for calculating, r_S, show slight differences if there are values of equal rank (ties) in either of the data sets.

A selected set of critical values, $r_{S(\text{CRIT})}$, for the Spearman's correlation coefficient are given in Appendix V. If $r_S \geq r_{S(\text{CRIT})}$ we would accept that there is a significant correlation.

E12.8 🖥

Referring to the same data as in E10.8, use a *non-parametric* method to investigate whether the data shows a *1-tailed* correlation between performance in the number of 'press-ups' and 'sit-ups' achieved by seven children (subjects).

Subject	1	2	3	4	5	6	7
Press-ups	10	8	2	6	7	3	5
Sit-ups	24	27	12	14	21	16	22

See following text for calculations.

(The use of statistics software is given in Examples on the web site.)

The ranks are calculated *separately* for the data values for 'press-ups' and 'sit-ups' in E12.8 – see Table 12.8.

Using method 1 above, the correlation coefficient between the *ranks* for the two sets of data in Excel is found to be $r_S = 0.7857$.

Table 12.8. Ranking data for Spearman's correlation in E12.8.

Subject	1	2	3	4	5	6	7	Total
Number of 'press-ups'	10	8	2	6	7	3	5	
Ranks for 'press-ups'	7	6	1	4	5	2	3	
Number of 'sit-ups'	24	27	12	14	21	16	22	
Ranks for 'sit-ups'	6	7	1	2	4	3	5	
Difference in ranks, d	1	−1	0	2	1	−1	−2	
d^2	1	1	0	4	1	1	4	12

Using method 2, taking the differences between each of the ranks, squaring the differences, and then taking the sum of the differences gives $\Sigma d^2 = 12$. Substituting this value, together with $n = 7$ into [12.18] gives $r_S = 0.7857$.

The critical 1-tailed value for $\alpha = 0.05$ from Appendix V is $r_{S(CRIT)} = 0.714$.

Since $r_S > r_{S(CRIT)}$ we accept the alternative hypothesis that the number of 'press-ups' shows a positive correlation with the number of 'sit-ups'.

Note that the results obtained for E12.8 using the non-parametric method are consistent with those obtained for E10.8 using the parametric method.

12.5.3 Using the p-value in Excel

The 2-tailed p-value can be calculated in Excel by first calculating the separate ranks (in 12.5.2) of the two data sets and then using:

$$\text{Tools} > \text{Data Analysis} > \text{Regression}$$

as explained in 10.3.4.

E12.9 🖥

Use Excel to derive the p-value for example E12.8.

Entering the rank values into two columns in Excel, and using

$$\text{Tools} > \text{Data Analysis} > \text{Regression}$$

for the data in the two columns gives $p(\text{2-tailed}) = 0.036$.

Calculating the 1-tailed value: $p(\text{1-tailed}) = 0.036/2 = 0.018$.

Since $p < 0.05$ we accept the alternative hypothesis (as in E12.8).

Q12.8 Ten trainees attended different lengths of training, and their individual progress was assessed on a five-point scale, from 0 (poor) to 5 (excellent).

Using the data given below, assess (at 0.05) whether their assessment score increased with the number of training days.

Trainee	1	2	3	4	5	6	7	8	9	10
Days	8	3	4	6	7	5	9	10	6	5
Assessment	2	1	4	1	4	2	5	3	5	4

(i)   Calculate Spearman's rank correlation coefficient and compare with the critical value to investigate possible correlation.

(ii) Use the p-value method to investigate possible correlation.

13

Experimental Design and Analysis

Overview

In planning any quantitative investigation, the scientist must include the *analysis* of data as an integral part of the overall design of any experiment. This chapter develops the close links between the principles of good experiment design and the availability of suitable statistical tests for subsequent data analysis.

In particular, this chapter develops the influence that *data variance* has in the analysis of experimental results and, consequently, in the design of the original experiment.

We saw in Unit 8.3 that the variances in the separate factors in an experimental system combine together in a process of *synthesis* to give a single value for the variance in the overall result.

We now introduce some new statistical terms in Unit 13.1 that are used in the analysis of variance. This then opens up a wider understanding of the role of variance across a range of topics – correlation, regression, uncertainty – leading to the analysis of complex experiments.

Unit 13.2 then changes the emphasis from the *synthesis* of variances to the *analysis* of variances. It looks at the process of analysing the *observed* variance in experimental results to provide information about the separate contributions from various possible factors. It is this *reverse* process of *analysis of variance* that provides a range of powerful statistical techniques, called ANOVAs, that integrate so effectively with the analysis of complex experimental measurements.

The processes of experimental design and the use of ANOVAs have developed *separately* in different experimental disciplines. The unfortunate consequence is the emergence of different terminologies used for similar aspects of the analysis. However, we will continue to use a generic terminology in which an experiment typically involves treating 'subjects' in different ways to investigate how different 'levels' of a 'factor' result in different 'outcomes' (see Unit 9.1). In this context a 'subject' may be any animate or inanimate object, or even a process, and the 'factor' could be generated specifically within the experiment or be merely the observation of subjects under different natural conditions. Examples of different types of experiments include:

- Exposing different Petri dishes of bacteria (subjects) to different 'levels' of light intensity (factor) to observe the effect on the bacteria growth (outcome).

- Monitoring the health (outcome) of different groups of people (subjects) who are smokers and non-smokers respectively (two levels of the 'smoking' factor).

Experimental design includes the planning of the experimental process, the choice of data to be collected, and the methods by which the data will be analysed. However, there is a wide variety of possible experiments, and it is important that each particular design is tailored to fit the particular experiment.

It is impossible to give a step-by-step guide that fits all experimental situations. The ability to produce a good experiment design requires a knowledge of the basic techniques available (replication, randomization, blocking, repeated measures, allocation of subjects), plus the experience of knowing how these techniques can be applied in a variety to different experiment types.

In some experiments, it is possible to exercise total control over the subjects that are being measured, e.g., in an experiment in plant physiology, plants may be grown under controlled conditions and then destructively analysed at the conclusion of the test. However, such control is not physically possible in many systems, e.g. in experiments in meteorology or astronomy. It is also impossible to exercise total control in experimentation involving human or animal subjects, as there are clear ethical limitations to the ways in which the subjects can be selected and treated.

At a minimum level, the scientist should be able to control, within the ethical and practical limits of the experiment, exactly what data will be collected. It should then be possible to design the experiment in such a way that it reduces the effects of bias and uncertainty in the experimental process.

There are two broad categories of investigation:

• **Hypothesis testing in an unknown system** – postulates, and then tests for, *new relationships* within a system, for example, tests to see if a new drug may have an effect as an anti-clotting agent in blood.

• **Measurements within a known system** – occurs after a hypothesis has been confirmed, and the purpose of the investigation is to measure the *strengths* of the factors and their interactions, for example, measures the *effectiveness* of an agent in blood that is *known* to have some anti-clotting characteristics.

In addition, there is a variety in the form of experiments between scientific disciplines. At one extreme a physicist may be concerned with a single measurement to a very high accuracy, while at the other, an environmental scientist is trying to determine the impact of global warming within a complex ecosystem.

The mathematical and statistical analyses required in the different forms of experimentation may appear to be different, but, nevertheless, the underlying concepts are very similar. This final chapter develops the detailed analysis of variance that is needed to quantify and manage experimental uncertainty.

13.1 Data variance

13.1.1 Introduction

The concept of variance was introduced in 7.1.6:

$$\text{Variance} = (\text{standard deviation})^2 \qquad [13.1]$$

$$\text{Population variance:} \; \sigma^2 = \sigma_n^2 = \frac{\sum_1^n (x_i - \mu)^2}{n} \qquad [13.2]$$

$$\text{Sample variance:} \; s^2 = \sigma_{n-1}^2 = \frac{\sum_1^n (x_i - \bar{x})^2}{(n-1)} \qquad [13.3]$$

The variance, s^2, calculated for a sample is the 'best estimate' for the variance, σ^2, of the population from which the data sample was drawn.

Excel uses the function VARP to calculate a population variance and VAR to calculate a sample variance. The coefficient of dispersion, CD, for a sample was introduced in [7.14].

$$\text{Coefficient of dispersion:} \; CD = \frac{s^2}{\bar{x}} \qquad [13.4]$$

In this Unit we introduce the use of variance in the complex analysis of subject and measurement variations. We start by introducing two new terms, related to variance, that are used extensively in this form of analysis – sum of squares, *SS*, and mean square, *MS*.

13.1.2 Sum of squares and mean squares

For a set of n values of x_i, the mean value

$$\bar{x} = \frac{\sum_i (x_i)}{n}$$ [13.5]

Sum of squares, SS_x, for the x-data values is *defined by*:

$$SS_x = \sum_i (x_i - \bar{x})^2$$ [13.6]

The sum of squares, SS_x, is the same as the *sum of squares of the deviations* [7.5].

E13.1 🖳

Calculate the 'sum of squares', SS_x, for the data set, x, in the table below, by first calculating the mean value, $\bar{x}$, followed by $\sum_i (x_i - \bar{x})^2$

Answer:

x	10	15	20	25	$\bar{x} = 17.5$	
$(x_i - \bar{x})$	−7.5	−2.5	2.5	7.5		
$(x_i - \bar{x})^2$	56.25	6.25	6.25	56.25	$SS_x = \sum_i (x_i - \bar{x})^2 = 125$	

Mean Square, MS_x, for the x-data values is *defined by*:

$$\text{Mean square} = \frac{\text{Sum of squares}}{\text{Degrees of freedom}}$$

giving

$$MS_x = \frac{SS_x}{df}$$ [13.7]

where df is the number of *degrees of freedom* for the particular calculation.

13.1.3 Mean square value for a sample of n data values

The mean square, MS, for a sample of n data values is given by:

$$MS_x = \frac{SS_x}{df} = \frac{\sum_i (x_i - \bar{x})^2}{n - 1} = s^2$$ [13.8]

where s is the sample standard deviation, and degrees of freedom, $df = n - 1$.

The *mean square* of a set of data values equals the sample *variance* of the set.

Consequently,

$$SS_x = s^2 \times df \qquad\qquad [13.9]$$

E13.2 💻

For the data sets, x and y, calculate (i) to (iii) and answer (iv):

x	10	15	20	25
y	0.37	0.48	0.7	0.81

(i)   variances, s_x and s_y,

(ii) degrees of freedom, df_x and df_y,

(iii) 'sums of squares', SS_x and SS_y.

(iv) Compare the result for SS_x with the result in E13.1.

Answer:

(i) Using Excel, VAR gives the sample variances,

$$s_x^2 = 41.667$$
$$s_y^2 = 0.0403$$

(ii) Degrees of freedom for both samples, $df_x = df_y = (n - 1) = 3$.

(iii) Using equation [13.9]:

$$SS_x = s_x^2 \times df_x = 41.667 \times 3 = 125$$
$$SS_y = s_y^2 \times df_y = 0.0403 \times 3 = 0.121$$

(iv) The methods used in E13.1 and E13.2 to calculate the value for SS_x give the same result.

Q13.1 Complete the missing values in the table below. Each row in the table refers to a different sample data set, where n is the size of each data set.

n	Variance, s^2	Standard deviation, s	SS	df	MS
5		8.0			
6					0.35
	25			8	
			8.8		2.2

Use the following relationships:

Variance = (standard deviation)^2

MS = variance

$SS = MS \times df$

$df = n - 1$

13.1.4 Root mean square value for a population of *n* values

$$\text{Root mean square value} = \sqrt{(\text{mean square})} = \sqrt{(\text{variance})}$$

Hence:

$$\text{root mean square value} = \text{population standard deviation} \qquad [13.10]$$

The root mean square (rms) value of the voltage is often used in electricity to give a measure of an 'average' voltage in an AC signal. It is equal to the voltage of a DC signal that would give the same *average power* as the AC signal.

Q13.2 The following voltage values occur at 2 millisecond intervals in one cycle of an AC mains voltage signal in the UK:

Time, ms	2	4	6	8	10	12	14	16	18	20
Voltage, V	+200	+323	+323	+200	0	−200	−323	−323	−200	0

Calculate the root mean square (rms) of the data values.

13.1.5 Statistics of linear regression of 'y on x'

The process of linear regression calculates the coefficients of the 'best-fit' straight line to a set of 'linear' data, and was introduced in Unit 4.2.

The 'best-fit' line of regression of *y* on *x*, can be described by:

$$y = mx + c$$

where the regression coefficients are:

$$m = slope \text{ (also called the } coefficient \text{ of 'x')}$$

$$c = intercept$$

For the 'regression of *y* on *x*', it is assumed that the only errors in the data are in the measurement of the *y*-values. Hence it is important that the measured variable with the greatest uncertainty is placed on the *y*-axis; this is normally the *dependent* variable. Reversing the data would give slightly *different* results.

The **Residual, R**, for a given data point, (x, y), is the difference in the *y*-values between the data point (y) and the point on the 'best-fit' straight line (y') at the same *x*-value – see Figure 13.1.

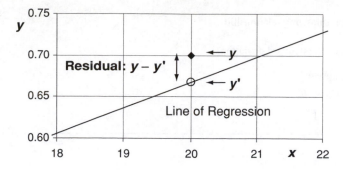

Figure 13.1. Residual on a best-fit line (using data from E13.2).

In general, for a point (x_i, y_i), the residual is given by:

$$R_i = y_i - y_i'$$

[13.11]

E13.3

A data point with co-ordinates (20, 0.7) in Figure 13.1 is part of a data set (see E13.2) that has a 'best-fit' line of linear regression given by the equation:

$$y = 0.0308x + 0.051$$

Calculate:

(i)   the y-value, y', of the point on the line of regression at the same value of $x = 20$.

(ii) the 'vertical' difference between the data point and the line of regression.

Answer:

(i)   Substituting $x = 20$ into the equation of the line gives:

$$y' = 0.0308 \times 20 + 0.051 = 0.667$$

(ii) The 'vertical' difference is the Residual, $R = y - y' = 0.700 - 0.667 = +0.033$

The **residual sum of the squares** (SS_{RESID}) is defined by taking the sum of the squares of the residuals for all the data points:

$$SS_{\text{RESID}} = \sum_i R_i^2$$

[13.12]

The calculations for regression work by adjusting the regression coefficients (m and c) of the straight line to obtain a *minimum value* for the 'residual sum of the squares'. The best-fit straight line has the *smallest possible* value for SS_{RESID}.

For the above reason, the process of finding the 'best-fit' straight line is also described as the **'method of least squares'**.

Once the 'best-fit' has been achieved, residual differences in y-values still exist between each data point and the line of regression. The overall residual difference in the 'fit' is quantified by the **standard error of regression, SE_{xy}**:

$$SE_{xy} = \sqrt{\frac{SS_{\text{RESID}}}{n-2}} \qquad\qquad [13.13]$$

The standard error of regression is a 'best estimate' for the *standard deviation of the experimental measurement*, σ_E.

The standard error of regression can be calculated directly in Excel by using the STEYX function.

E13.4 🖳

The data in the table below (same values as E13.2) gives the result of a spectrophotometric measurement, where the y-variable is the absorbance, A, and the x-variable is the concentration, C.

The slope and intercept of the 'best-fit' straight are calculated using the SLOPE and INTERCEPT function in Excel, giving $m = 0.0308$ and $c = 0.051$ respectively. For each value of C, calculate the values of

(i) A' (the y-value for the point on the calibration line), and

(ii) $A - A'$ (the residual for that point).

Calculate the values of

(iii) SS_{RESID}

(iv) SE_{XY}

C	(x)	10	15	20	25
A	(y)	0.37	0.48	0.7	0.81

Answer:

(i) Each value of A' is calculated using the equation: $A' = m \times C + c$:

	A'	(y')	0.359	0.513	0.667	0.821
(ii)	$A - A'$	$(y - y')$	0.011	-0.033	0.033	-0.011
	$(A - A')^2$	$(y - y')^2$	0.121×10^{-3}	1.089×10^{-3}	1.089×10^{-3}	0.121×10^{-3}

(iii) $SS_{\text{RESID}} = \sum (A - A')^2 = 0.00242$

(iv) $SE_{xy} = \sqrt{\dfrac{SS_{\text{RESID}}}{n-2}} = \sqrt{\dfrac{0.00242}{4-2}} = 0.0348$

13.1.6 Statistics of linear correlation

The *hypothesis test* for linear correlation is given in Unit 10.3, which also introduces the Pearson's product moment correlation coefficient, r.

The amount of linear correlation between two data sets, x and y, is defined as the extent to which the 'spread' of the x-values *will predict* the 'spread' of the y-values.

It can be shown that the regression line drawn through a set of 'linear' data points will pass through the **centroid** of the data, whose co-ordinates are given by:

$$\bar{x} = \text{mean value of the } x\text{-values for all data points}$$

and

$$\bar{y} = \text{mean value of the } y\text{-values for all data points}$$

Calculating the sum of squares of the deviations for both the x-data and the y-data:

$$SS_y = \sum_i (y_i - \bar{y})^2 \qquad\qquad [13.14]$$

and

$$SS_x = \sum_i (x_i - \bar{x})^2 \qquad\qquad [13.15]$$

If there is **perfect linear correlation**, then all the data points will lie exactly on the line of regression. In this case, the 'spread' of the y-values are predicted *exactly* by the 'spread' of x-values.

$$\text{Perfect linear correlation, } r^2 = 1: \qquad r = +1 \quad \text{or} \quad r = -1 \qquad\qquad [13.16]$$

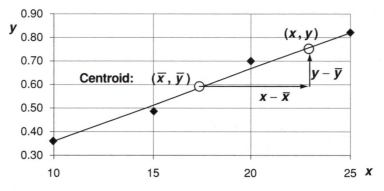

Figure 13.2. Deviations about the centroid.

For any point that *lies on the line of regression* (see Figure 13.2), equation [4.8] shows that the y-deviation, $(y - \bar{y})$, is related to the x-deviation, $(x - \bar{x})$, by the equation:

$$(y - \bar{y}) = m \times (x - \bar{x}) \qquad\qquad [13.17]$$

Hence, for *perfectly correlated x–y data*, where all data points lie on the line of 'best–fit', combining [13.14], [13.15] and [13.17] gives,

$$SS_y = m^2 \times SS_x \text{ (for perfect correlation)} \qquad\qquad [13.18]$$

However, if there is **less than perfect correlation**, the actual data points will be spread above and below the line or regression (as in Figure 13.2), giving an *additional* 'spread' in the y-values.

$$\text{Less than perfect linear correlation, } r^2 < 1: \qquad -1 < r < +1 \qquad [13.19]$$

It can be shown that the total variance in the y-values will be the contribution, $m^2 \times SS_x$, predicted by equation [13.18], plus the additional variance, SS_{RESID}, due to the spread of the data points away from the line:

$$SS_y = m^2 \times SS_x + SS_{RESID} \qquad [13.20]$$

E13.5

Using the same set of data values as in E13.2 and E13.4, with $SS_y = 0.121$, $SS_x = 125$, $SS_{RESID} = 0.002\,42$, and $m = 0.0308$:

(i)   Calculate the value of $m^2 \times SS_x + SS_{RESID}$.

(ii) Does $SS_y = m^2 \times SS_x + SS_{RESID}$?

Answer:

(i) $m^2 \times SS_x + SS_{RESID} = (0.0308)^2 \times 125 + 0.002\,42 = 0.121$.

(ii) Yes. See (iii) in E13.2.

The **square of the correlation coefficient, r^2**, is now defined as the *proportion of the variation in y predicted by the variation in x*.

$$r^2 = \frac{m^2 \times SS_x}{SS_y} = \frac{SS_y - SS_{RESID}}{SS_y} = 1 - \frac{SS_{RESID}}{SS_y} \qquad [13.21]$$

For perfect correlation, $SS_{RESID} = 0$, and $r^2 = 1$.

Excel also defines the following in Regression calculations:

$$SS_{REG} - \text{sum of squares predicted by regression} = m^2 \times SS_x$$
$$SS_{TOTAL} - \text{total sum of squares in ordinate variable} = SS_y \qquad [13.22]$$

Note that there is also an 'adjusted' r-value, r', which is defined using the mean square values (MS), rather than the sum of the squares (SS).

$$(r')^2 = 1 - \frac{MS_{RESID}}{MS_y} \qquad [13.23]$$

In practice, there is often very little difference between r and r', but the adjusted value (r') gives a more accurate indication of the degree of correlation when n is small and when comparing different linear models in multiple regression calculations.

E13.6 🖳

For the data set introduced in E13.2 and E13.4:

(i)   Use [13.21] and data from E13.2 and E13.4 to calculate the value of r^2.

(ii) Use the Excel function CORREL to calculate r.

(iii) Compare the results obtained in (i) and (ii).

Answer:

(i) $r^2 = 1 - \dfrac{SS_{RESID}}{SS_y} = 1 - \dfrac{0.002\,42}{0.121} = 1 - 0.02 = 0.98$

(ii) The function CORREL returns the value $r = 0.9899$.

(iii) Squaring the value from (ii), gives $r^2 = 0.98$ which agrees with (i).

Q13.3 Using the same data as in Q10.11, with the jump distances as the dependent 'y-variable' and the heights as the 'x-variable':

Student		1	2	3	4	5	6	7	8	9	10
Height (m)	x	1.59	1.63	1.67	1.70	1.72	1.80	1.74	1.75	1.82	1.84
Jump (m)	y	1.49	1.45	1.78	1.60	1.90	1.55	1.69	1.81	1.69	1.93

Performing a linear regression analysis on the data gives:

Slope of 'best-fit' line, $m = 1.11$

Sample variance of height data, $s_x{}^2 = 0.006\,63$

Sample variance of jump data, $s_y{}^2 = 0.0277$

(i)   Calculate the following:

$$SS_x,\ SS_y,\ SS_{RESID}$$

(ii) Using the data from (i) calculate the correlation coefficient, r.

13.1.7 Uncertainty in linear calibration

Calibration lines (4.2.6) are often used in science. Typically the line will be drawn with n *calibration data points*, and a 'best estimate' y-value, y_O, of the unknown sample is calculated by taking the mean value of k *replicate sample measurements*. The 'best estimate' x-value, x_O, of the unknown sample is then calculated by using the equation:

$$x_O = \frac{y_O - c}{m} \tag{13.24}$$

There is a statistical uncertainty in the position of the 'true' calibration line, which is illustrated by the 95% confidence limits in Figures 13.3(a) and 13.3(b). Note that the confidence limits 'allow' for possible errors in the data points, based on the calculated variation in data values.

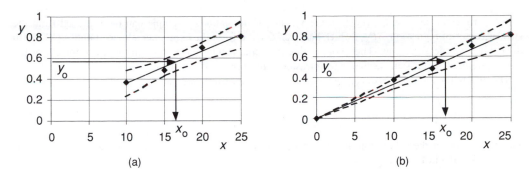

Figure 13.3. 95% confidence limits of best-fit straight lines.

Figure 13.3 shows data from E13.4. The dashed lines in the two diagrams show the ranges within which it is possible to be 95% confident of quoting the position of the 'true' calibration line.

- Figure 13.3 (a) shows the limits for the line for a 'free fit' of both slope *and* intercept.

- Figure 13.3 (b) shows the line when it is known to go through the origin (i.e. intercept $= 0$), and it is only necessary to choose a 'best-fit' value for the slope.

Defining some data values:

(x, y) – co-ordinates of an experimental data point,

(x, y') – co-ordinates of the point on the calibration line at the same value of x,

$u(x_O, y_O)$ – uncertainty in the y-value of the calibration line at the point (x_O, y_O),

y_O – average of k measurements of the y-values of the unknown sample,

$u(y_O)$ – uncertainty in the true y-value of the unknown sample,

x_O – 'best estimate' for the x-value of the unknown sample,

$u(x_O)$ – standard uncertainty in the true x-value,

μ – 'true' x-value of the unknown sample.

The overall uncertainty, $u(x_O)$, in the value of x_O is due to a combination of three factors:

- slope, m, of the calibration line – for example, a 'flatter' slope (small m) will increase the range of uncertainty in x-value for a given uncertainty, $u(y_O)$, in the y-value,

- uncertainty, $u(y_O)$, in the true y-value of the unknown sample, and

- uncertainty, $u(x_O, y_O)$, in the y-value of the calibration line.

The three factors are combined together, by combining the variances (8.3.3) of the uncertainties and dividing by the slope of the line:

$$u(x_O) = \frac{\sqrt{u(y_O)^2 + u(x_O, y_O)^2}}{m} \quad [13.25]$$

The experimental uncertainty, $u(y_O)$, in the y-value of the unknown sample can be estimated from the standard error of regression, SE_{XY}, which has been derived from the experimental uncertainties [13.13] in the calibration measurements. This assumes that the experimental uncertainty in the measurement of the unknown sample is the same as that in the measurement of the calibration data.

The uncertainty, $u(y_O)$, is reduced by taking k replicate measurements of the unknown:

$$u(y_O) = SE_{xy} \times \sqrt{\frac{1}{k}} \quad [13.26]$$

The calculation of the calibration line uncertainty, $u(x_O, y_O)$, is different in the two cases illustrated in Figures 13.3(a) and 13.3(b):

With a 'free fit' of both slope and intercept – Figure 13.3(a):

$$u(x_O, y_O) = SE_{xy} \times \sqrt{\left\{ \frac{1}{n} + \frac{(y_O - \bar{y})^2}{m^2 \times \sum_i (x_i - \bar{x})^2} \right\}} \quad [13.27]$$

where $\bar{x}$ and $\bar{y}$ are the mean values of all x-values and y-values respectively, and x_i is the x-value of the ith data point.

When the calibration line is *known* to pass through the origin, $u(x_O, y_O)$, equation [13.27] becomes:

$$u(x_O, y_O) = SE_{xy} \times \sqrt{\left\{ \frac{y_O^2}{m^2 \times \sum_i (x_i)^2} \right\}} \quad [13.28]$$

We can now combine equations [13.25], [13.26] and [13.27] or [13.28], to give the overall *standard uncertainty*, $u(x_O)$, in the value of x_O:

For the 'free' fit calibration line:

$$u(x_O) = \frac{SE_{xy}}{m} \times \sqrt{\left\{ \frac{1}{k} + \frac{1}{n} + \frac{(y_O - \bar{y})^2}{m^2 \times \sum_i (x_i - \bar{x})^2} \right\}} \quad [13.29]$$

For the line that is *known* to pass through the origin:

$$u(x_O) = \frac{SE_{xy}}{m} \times \sqrt{\left\{ \frac{1}{k} + \frac{y_O^2}{m^2 \times \sum_i (x_i)^2} \right\}} \quad [13.30]$$

See E13.7 for using Excel to perform the calculations involved in the above equations.

Equation [13.30] should be used when the origin is a *more accurate* data point than the other calibration points. If this is not the case, but the origin is still believed to be a valid data 'point', then the point (0,0) should be added as additional data, and a 'free fit' regression used.

In the case of the 'free-fit' line in Figure 13.3(a), the last term in the square root of [13.29] is the factor that is responsible for the 'opening-up' of the uncertainty range at both ends of the calibration line. This term becomes *zero* (and disappears) when y_O equals the mean y-value ($\bar{y}$) of the points used to generate the best-fit line, i.e. when the measured value falls in the 'middle' of the calibration values.

For *well-designed* experiments where the measured value of y_O falls within the *middle one half of the calibration line*, we can make the approximation that the last term in the square root will then be small, and can usually be omitted. This gives the approximate equation for the free fit calibration line:

$$u(x_O) \approx \frac{SE_{xy}}{m} \times \sqrt{\left\{\frac{1}{k} + \frac{1}{n}\right\}} \qquad\qquad [13.31]$$

The standard uncertainties, $u(x_O)$, calculated in [13.29], [13.30] and [13.31] give a 68% confidence range for the unknown true value, μ.

The confidence interval, $CI(\mu, X\%)$, for $X\%$ confidence (see 8.2.6) is given by:

$$CI = x_O \pm t_{2,\alpha,n-2} \times u(x_O) \qquad\qquad [13.32]$$

where the t-value is calculated for degrees of freedom, $df = n - 2$ and $\alpha = 1 - X\%/100$.

E13.7 ⌨

The data in the table below (same values as E13.4) give the result of a spectrophotometric measurement, where the y-variable is the absorbance, A, and the x-variable is the concentration, C.

C	(x)	10	15	20	25
A	(y)	0.37	0.48	0.7	0.81

Three replicate measurements of an unknown solution give absorbance values: 0.57, 0.58 and 0.61.

Calculate the 95% confidence interval for the concentration, C_O, of the unknown solution using the three following methods:

(i)   Using a 'free-fit' calibration line.

(ii) Assuming that the origin of the graph is an additional data point.

(iii) Forcing the calibration line through the origin.

Answer:

Performing the necessary calculations in Excel gives:

The mean of the ($k =$) 3 absorbance values gives: $A_O = y_O = 0.587$

Standard error, $SE_{XY} = 0.0348$

Number of calibration points, $n = 4$

Appropriate t-value, $t_{2,0.05,2} = 4.30$

(i)   slope, $m = 0.0308$; intercept, $c = 0.051$

using [13.24], $C_O = x_O = 17.39$; using [13.31], $u(x_O) = 0.863$

$CI(C_O, 95\%) = 17.39 \pm 4.3 \times 0.863 = 17.4 \pm 3.7$

(ii) slope, $m = 0.0327$; intercept, $c = 0.014$

using [13.24], $C_O = x_O = 17.50$; using [13.31], $u(x_O) = 0.812$

$CI(C_O, 95\%) = 17.50 \pm 4.3 \times 0.812 = 17.5 \pm 3.5$

(iii) slope, $m = 0.0334$; intercept, $c = 0$

using [13.24], $C_O = x_O = 17.54$; using [13.30] (see web site), $u(x_O) = 0.779$

$CI(C_O, 95\%) = 17.54 \pm 4.3 \times 0.779 = 17.5 \pm 3.3$

Q13.4 Using the same data as Q4.22, different masses w(in grams) of a sugar were dissolved in 100 mL of water and the angle of rotation θ (in degrees) of polarized light in the solution was measured in a polarimeter. The following results were obtained.

w (g)	1.5	3	4.5	6	7.5
θ (degrees)	1.3	2.7	4	5.2	6.6

Three replicate solutions of an unknown sugar solution gave an average rotation of $4.9°$.

Calculate the 95% confidence interval for the concentration of the sugar solution in grams per 100 mL.

13.1.8 Linear Regression data in Excel

The following functions are useful in Excel:

SLOPE – to calculate the slope, m, in linear regression,

INTERCEPT – to calculate the intercept, c, in linear regression,

LINEST – to calculate the slope, m, in when forcing a zero intercept,

STEYX – to calculate the standard error of regression, SE_{xy},

CORREL or PEARSON – to calculate the linear correlation coefficient, r.

The use of **Tools > Data Analysis > Regression** can be used for linear regression and produces an output of the form illustrated in E13.8.

E13.8

A spectrophotometric calibration line between absorbance, A, and concentration, C (mg L^{-1}), gives the following values (see E13.7):

C	10	15	20	25
A	0.37	0.48	0.7	0.81

An extract from the results produced by Tools > Data Analysis > Linear Regression in Excel is reproduced below:

	A	B	C	D	E	F
12						
13	*Regression Statistics*					
14	Multiple R	0.98995				
15	R Square	0.98000				
16	Adjusted R Square	0.97000				
17	Standard Error	0.03479				
18	Observations	4				
19						
20	ANOVA					
21		*df*	*SS*	*MS*	*F*	*Significance F*
22	Regression	1	0.11858	0.11858	98	0.01005
23	Residual	2	0.00242	0.00121		
24	Total	3	0.121			
25						
26		*Coefficients*	*Standard Error*	*t Stat*	*P-value*	
27	Intercept	0.051	0.05716	0.89227	0.4664	
28	X Variable 1	0.0308	0.00311	9.89950	0.0100	

The meanings of many of the values derived in E13.8 are outlined in Table 13.1.

Table 13.1. Data available using Excel regression.

Cell	Name	Explanation
B27	Intercept	Intercept, c
B28	X Variable 1	Slope, m
B14	Multiple R	Correlation coefficient, r
B15	R Square	Coefficient of determination, r^2 [13.21]
B16	Adjusted R Square	Adjusted r-value squared, r'^2 [13.23]

(Continued)

Table 13.1. (*Continued*)

Cell	Name	Explanation
B17	Standard Error	Standard error of regression, SE_{xy} [13.13]
B18	Observations	Number of data pairs, n
A20	ANOVA	See Unit 13.2 for the meaning of 'ANOVA'
C22	*SS* Regression	Sum of squares for regression, SS_{REG} [13.22]
C23	*SS* Residual	Sum of squares for residual, SS_{RESID} [13.12]
E22	*F*	$F = MS(\text{regression})/MS(\text{residual})$ [13.42]
F22	*Significance F*	p-value for the F-test in E22
E28	*P-value*	p-value for the test for linear correlation (10.3.4) This is the same value as appears in F22

13.2 Analysis of variance (ANOVA)

13.2.1 Introduction

The two-sample t-test (10.2.8) allows us to test whether the difference in the mean values of *two* data samples is significantly greater than could be expected from random variations in the experimental results. It allows us to test whether *two levels* of a *factor* (e.g. different temperatures) give different experimental outcomes.

In practice, however, we often want to test *simultaneously* for *more than two* levels of a factor in the one experiment. For examples, we may test for variations in

- reaction rates at four different temperatures,

- plant growth rates between five different soil nutrients,

- muscle performance after three different types of warm-up regime.

In these, more complicated experiments, we cannot use the *two-sample t*-test. We now develop a new analytical approach by investigating the sources of variation that can contribute to the final variations in the experimental data.

We see that an ANOVA (analysis of variance) calculation can test whether there may be some difference between the effects of the three or more levels of a factor. If the ANOVA *does* indicate a significant difference, then the next step is to investigate where the difference(s) may lie. This subsequent step is beyond the scope of the book, but the web site introduces the use of other tests (e.g. Tukey) to identify which factor levels may have effects that are significantly different to the others.

13.2.2 Calculation of variances

Example E13.9 sets out a typical type of problem in which there are TWO sources of variation associated with plant growth:

- the possible effects of different nutrients, and

- the inherent growth variability between plants (subjects).

The aim of an experiment may be to identify any differences in the effectiveness of various nutrients.

E13.9 ⌨

We wish to test whether four different nutrients (A, B, C, D) may have different effects on plant growth. We grow three plants in each of the four nutrients, giving us four *sample sets* with three *replicates* in each. The 12 measurements of plant growth are recorded in the Table 13.2.

The *mean values* and *sample variances* are calculated in Table 13.2 for each of the four sample sets of three replicate measurements.

Table 13.2.

Nutrient	Growth rates				Means	Variance
A	32.1	21.9	24.9		26.30	27.48
B	12.6	16.9	14.6		14.70	4.63
C	22.9	22.5	30.3		25.23	19.29
D	16.6	23.7	18.5		19.60	13.51
	Mean of sample variances, MSV:					16.23
	Variance of sample means, VSM:				28.94	

The **variance of the sample means, *VSM***, and the **mean of the sample variances, *MSV***, are also calculated:

28.94 is the sample variance of 26.30, 14.70, 25.23, and 19.60

16.23 is the mean value of 27.48, 4.63, 19.29 and 13.51

We now investigate, *in the text below*, how we can separate the effects of the different nutrients from the natural variations in plant growth rates.

Q13.5 Using data from Table 13.2 in E13.9, perform a two-sample *t*-test to investigate whether there is any significant difference between the growth rates for nutrients A and B.

Use the pooled standard deviation (see [10.12]):

$$s' = \sqrt{\{[(3 - 1) \times 27.48 + (3 - 1) \times 4.63]/(3 + 3 - 2)\}} = 4.01$$

Referring to E13.9, we could possibly use *t*-tests (as in Q13.5) to look for differences between every data sample (A, B, C and D) and every other data sample. However, using multiple *t*-tests is not appropriate for two reasons:

1 The number of *t*-tests required would increase rapidly with an increasing number of samples. For example, if we only have four samples, this would require *six* *t*-tests to test all the possible pairs of samples.

2 Every t-test has a probability of giving a false result. Even though this probability is low for a single test (typically set at less than 5%), the use of *multiple* tests means that the overall probability of *at least one* false positive result becomes unacceptably high.

The **analysis of variance** technique analyses the variations (or *variances*) in the data values, and aims to separate the variation that may be due to the different levels of the factor from the inherent variability in the experimental process itself.

This analysis is developed in the following section.

13.2.3 Adding-up variances

In a simple set of experimental results, as in E13.9 above, we can expect to find TWO main sources of variation:

- **Experimental variation**: this variation is inherent in the experimental process (Unit 1.2), and will normally be due to:
  - *measurement* uncertainty (e.g. variations in instrumental results), or
  - *subject* uncertainty due to the natural variation in the subjects being measured (e.g. different growth rates from similar biological systems).

  Experimental variations are described by the *experimental variance*: σ_E^2.

- **Factor variation**: this variation may be due to the different factor levels between the samples. In E13.9 the 'factor' is the nutrient and the 'level' is the choice of particular nutrient.

  Factor variations are described by the *factor variance*: σ_F^2.

The total variance in the observed results, σ_{TOT}^2, is equal to the addition of the separate variances, σ_E^2 and σ_F^2:

$$\sigma_{TOT}^2 = \sigma_E^2 + \sigma_F^2 \qquad\qquad [13.33]$$

Q13.6 In measuring soil pH, the standard deviation of pH for different samples is given by $\sigma_S = 0.08$. If the pH meter reading also has a standard deviation uncertainty of $\sigma_M = 0.05$, estimate the standard deviation that might be expected in a large number of repeated measurements from different soil samples.

13.2.4 Separation of variances

Adding up the variances (as in 13.2.3) is a simple addition process.

However, it is more complicated to perform the *reverse process*, and analyse (or separate) a total variance into experimental and factor variances.

The reverse process for the *analysis (or separation) of variances* can be illustrated as follows:
Using the data in E13.9 as an example, assume that

- the *response variable* is the growth rate of the plants, with

- k different samples sets, each representing a different '*level*' of the external factor, e.g. different types of nutrients applied to plants, and

- n *replicate* plants within each of the sample sets.

The different 'levels' of the factor are often called different **'treatments'**.

This data can be given in the form of a simple $k \times n$ table of measured growth values, as shown in E13.9 with $k = 4$ and $n = 3$.

Step 1 – Estimating the experimental uncertainty, σ_E^2   The variance *within* each sample, is *only* dependent on the *experimental uncertainty*, because the factor 'level' is the same for every subject within the sample. The variance *within* each sample is therefore an *estimate* for the experimental uncertainty, σ_E^2.

Taking the **mean of all the sample variances, MSV** (see E13.9), gives the *best overall estimate* for σ_E^2.

$$\sigma_E^2 = MSV \qquad\qquad [13.34]$$

Step 2 – Estimating the factor variance, σ_F^2   Assuming first that there are *no factor variations* between the samples, there will still be a variation in the *mean* values due to the experimental uncertainty, i.e. the sample mean values will still show a variation. The standard deviation in sample means will be given by the standard error (see 8.2.3):

Standard deviation in *mean values* due to experimental variation $= \sigma_E/\sqrt{n}$.

Variance in *mean values* due to experimental variation alone $= \sigma_E^2/n$ $\qquad$ [13.35]

If we now include the effect of *factor level variations*, the **variance of the sample mean, VSM** (see E13.9), will be a combination of the variances due to both the factor and the experimental process:

$$VSM = \sigma_F^2 + (\sigma_E^2/n) = \sigma_F^2 + (MSV/n) \qquad\qquad [13.36]$$

This allows us to derive an expression for the factor variance:

$$\sigma_F^2 = VSM - (MSV/n) \qquad\qquad [13.37]$$

E13.10 🖳

Using data from E13.9, estimate:

(i)   Experimental variance, σ_E^2

(ii) Factor variance, σ_F^2

(i)   We can estimate the experimental variance directly from the value of *MSV*:

$$\sigma_E^2 = MSV = 16.23$$

(ii) Apparent factor variance, $\sigma_F^2 = VSM - MSV/n = 28.94 - 16.23/3 = 23.53$

13.2.5 Mean squares

We now define two 'mean square' values (see 13.1.2), which are produced in Excel (and other statistics software) in analysis of variance calculations:

- Mean square (within) gives a measure of the variance 'within' one level of the factor, and will be dependent *only* on the experimental variation.

- Mean square (between) gives a measure of the variance including *both* experimental and factor effects.

$$\text{Mean square (within)} = MS(\text{W}) = MSV = \sigma_E{}^2 \qquad [13.38]$$

$$\text{Mean square (between)} = MS(\text{B}) = n \times \text{VSM} = (n \times \sigma_F{}^2) + \sigma_E{}^2 \qquad [13.39]$$

giving best estimates for the experimental and factor variances:

$$\sigma_E{}^2 = MS(W) \qquad [13.40]$$

$$\sigma_F{}^2 = \{MS(B) - MS(\text{W})\}/n \qquad [13.41]$$

E13.11 🖳

Using data from Examples E13.9 and E13.10, calculate:

(i) Mean square (within), $MS(\text{W})$

(ii) Mean square (between), $MS(\text{B})$

(i) $MS(\text{W}) =$ Mean square (within) $= MSV = 16.23$

(ii) $MS(\text{B}) =$ Mean square (between) $= n \times \text{VSM} = 3 \times 28.94 = 86.82$

Q13.7 Three different samples (A, B, C) of a batch of plant material are taken, and five replicate measurements are made on each sample for the determination of potassium content, giving the following results (μg mL^{-1}).

A	2.76	2.75	2.81	2.81	2.72
B	2.69	2.63	2.65	2.66	2.69
C	2.69	2.7	2.65	2.79	2.68

Calculate the following:

(i) Experimental variance, $\sigma_E{}^2$

(ii) Variance between different samples, $\sigma_F{}^2$

(iii) Mean square (within), $MS(\text{W})$

(iv) Mean square (between), $MS(\text{B})$

13.2.6 Significance of the factor effect

We now test whether we can be confident that the estimated factor variance, $\sigma_F{}^2$, is a significant effect, i.e. it is not a value arising purely through statistical fluctuations in the experimental data.

We can use an F-test with the following hypotheses:

Null hypothesis, H_O:	The combined effect of variances σ_F^2 and σ_E^2 are NOT greater than σ_E^2 alone.
Alternative Hypothesis, H_A:	The combined effect of variances σ_F^2 and σ_E^2 ARE greater than σ_E^2 alone.

The appropriate 1-tailed F-statistic (Unit 10.1) is:

$$F = \frac{MS(B)}{MS(W)} \qquad [13.42]$$

Degrees of freedom Numerator $df(B) = k - 1$ and Denominator, $df(W) = N - k = k(n - 1)$
where N = total number of data values = $n \times k$.   [13.43]

E13.12 🖥

Using data from Examples E13.9, E13.10 and E13.11

(i)   State the hypotheses appropriate for testing whether there is a significant factor effect.

(ii) Derive the appropriate value of F_{CRIT}.

(iii) Calculate the F-statistic: $F = \dfrac{MS(B)}{MS(W)}$.

(iv) Decide whether there is a significant factor effect.

Answer:

(i) Null hypothesis, H_O:	There is NO difference between the effects of the different nutrients on growth rate.
Alternative Hypothesis, H_A:	There IS a difference between the effects of the different nutrients on growth rate.
(ii) Significance level, α:	0.05 (default level).
Tails:	Uses a 1-tailed F-test to test for *increased* variance due to the different nutrients.
Degrees of freedom:	Numerator, $df = k - 1 = 3$.
	Denominator, $df = N - k = 12 - 4 = 8$.
	$F_{CRIT} = 4.07$.

(iii) $F = MS(B)/MS(W) = 86.82/16.23 = 5.35$.

(iv) Since $F > F_{CRIT}$, we accept the alternative hypothesis that there is a significant factor effect. The nutrients do have significantly different effects.

Q13.8 Use data from Q13.7.

    (i) State the hypotheses appropriate for testing whether there is a significant variation between the different samples, A, B and C.

    (ii) Derive the appropriate value of F_{CRIT}.

    (iii) Calculate the value of F_{STAT}.

    (iv) Decide whether there is a significant variation.

13.2.7 One-way ANOVA

The one-way ANOVA is a *statistical test* that is based on the theory developed in section 13.2.6.

The one-way ANOVA tests for the possible effect of one factor (the treatment) in the presence of experimental (including measurement and subject) variance. The data is arranged as k levels of treatment, with each level having n replicate measurements.

It compares (using the F-test):

* the variance due to the factor plus the experimental process – *between* levels (treatments), with

* the variance inherent within the experimental process – *within* levels (treatments).

The organization of the experiment is called a **completely randomized design**, with subjects being randomly allocated to each particular level (see 13.3.4).

Most statistics software packages perform ANOVA tests with ease. An important advantage is that a software analysis usually gives the result of the F-test with both the F-statistic *and* the equivalent p-value (9.3.6). The data produced by a one-way ANOVA is illustrated in E13.13, which uses Excel Data Analysis Tools.

E13.13 🖥

An example is given in the table below, showing the efficiency of a chemical process using three different catalysts (A, B and C) on each of 4 days:

Catalyst	Day 1	Day 2	Day 3	Day 4
A	84	83	79	80
B	78	79	78	76
C	83	80	78	78

A one-way ANOVA performed in Excel (with the data organized in rows) gives the following results:

ANOVA						
Source of Variation	*SS*	*df*	*MS*	*F*	*P-value*	*F crit*
Between Groups	28.16667	2	14.08333	3.292208	0.084525	4.256492
Within Groups	38.5	9	4.277778			
Total	66.66667	11				

In E13.13, the 'Groups' refer to the different catalysts. The 'Within Groups' source of variation is dependent only on the experimental variance, but the 'Between Groups' source of variation depends on variances due to the catalyst and the day, in addition to the experimental variance. From the results, we can conclude that the one-way analysis of this data is *unable* to detect any significant 'catalyst' effect, because $p > 0.05$ and $F < F_{CRIT}$.

In the ANOVA results, the total 'sum of squares' equals the total of the 'sums of squares' from between and within the groups:

$$SS_{TOT} = SS(B) + SS(W) \qquad\qquad [13.44]$$

Q13.9 Conduct a one-way ANOVA in Excel on the data in E13.9, and compare the results with the results produced in E13.12.

Q13.10 In an ecological study of the lifespan of a certain species of protozoan, 50 individuals were divided into 5 groups of 10. Each group was fed on a different feed. The results of a one-way ANOVA are shown below with some of the data removed.

Calculate the missing data in the table below.
Use equation [13.44], [13.7], [13.42], [13.43]
(Use Excel to calculate the p-value.)

ANOVA						
Source of Variation	*SS*	*df*	*MS*	*F*	*P-value*	*F crit*
Between Groups	67.7072					
Within Groups						
Total	304.1872					

13.2.8 Two-way ANOVA (without replication)

Visual examination of the data in E13.13, suggests that the values tend to decrease from left to right in the table. Is this likely to be an actual outcome of randomized data, or could there be some effect due to the 'day' in our experiment that was not taken into account?

In the one-way ANOVA, it was assumed that there is no difference between the conditions, from day to day, under which the *replicate* measurements were made. It was assumed that within each catalyst sample the only variation was the experimental variation.

However, in many experimental set-ups, we may suspect that there is actually a real difference in experimental conditions between the replicates. For example, in the catalyst experiment (E13.13) we may suspect that one of the reagents may be deteriorating over the four days and affecting the efficiency of the process.

We could then introduce the 'day' as a new factor in the process, and each column would become a different 'level' in that second factor. In terms of experiment design (13.3.5) we now *block* these 'replicates' in an organized way, with each block representing different days.

We now have *two* factors affecting the response – the *catalyst variation* itself and the *day variation*. Example E13.14 introduces the use of the *two-way* ANOVA to test for both factors.

E13.14 💻

This example uses the same data as E13.13, showing the efficiency of a chemical process using three different catalysts (A, B and C) on each of four days:

Catalyst	Day 1	Day 2	Day 3	Day 4
A	84	83	79	80
B	78	79	78	76
C	83	80	78	78

A two-way ANOVA (without replication) is performed in Excel giving the following results:

ANOVA						
Source of Variation	*SS*	*df*	*MS*	*F*	*P-value*	*F crit*
Rows	28.16667	2	14.08333	8.59322	0.017328	5.143249
Columns	28.66667	3	9.555556	5.830508	0.032756	4.757055
Error	9.833333	6	1.638889			
Total	66.66667	11				

The **two-way** ANOVA in E13.14 is able to analyse the two-way (or 2-factor) set of data producing a similar format of results to the one-way ANOVA. The statistics involved in this calculation will not be covered in this book, but, in practice, it is easy to use one of the software packages to produce the results.

The results now indicate that there is a significant catalyst (*rows*) effect, because $p = 0.017$ giving $p < 0.05$.

The results also show a significant variation between the days (*columns*), because $p = 0.033$, giving $p < 0.05$.

The one-way ANOVA in E13.13 was unable to detect a significant catalyst effect owing to the experimental uncertainty introduced by the variation between days. The two-way ANOVA in E13.14 is now able to identify the contribution by the 'blocks' (or days) to the total variance, and can therefore perform a more sensitive test for the significance of the variance due to the catalyst.

In the Excel two-way ANOVA, the 'Error' is the variation that remains after the 'Row' and 'Column' variations have been subtracted from the total variations.

$$SS_{ERROR} = SS_{TOTAL} - SS_{ROW} - SS_{COLUMN} \qquad [13.45]$$

The 'Error' variation will now include contributions from:

- experimental (system and measurement) variance, and

- variance due to an *interaction* between 'Row' and 'Column' factors (see 13.2.9).

The experimental design that *blocks* the *replicate* samples is called a **randomized block design** – see 13.3.5.

Q13.11 As part of a project in forensic science, a student measures the times to emergence for larvae collected from a recently deceased pig. The times given in the table below were recorded for larvae collected on different days (Day1, Day2 and Day3) and grown on different media (M1, M2, M3 and M4).

	Day1	Day2	Day3
M1	27	19	33
M2	25	24	32
M3	20	19	24
M4	21	20	28

Analyse the data to investigate whether day of collection and/or media have any significant effect (at 0.05) on the emergence times of the larvae.

13.2.9 Two-way ANOVA with replication

In the context of an ANOVA, a *replicate* measurement requires repeating a measurement under exactly the same combinations of conditions. For example in E13.14, this would require at least two measurements for *each* combination of catalyst and day.

In a system that is affected by two (or more factors), there is the possibility that an *interaction* between the two factors may also affect the outcome. For example, the detrimental side effects of two drugs, A and B, may be quite low when each drug is given independently, but an *interaction* between the drugs may greatly increase the side effects when the drugs are given together – see also 11.2.1.

When performing a two-way ANOVA *without* using replicates, there is not enough information to separate the effect of an interaction between the 'Row' factor and 'Column' factor from the effects of the inherent experimental variations. The 'Error' variation includes any interaction effects mixed together with the experimental variation.

However, when *replicate* measurements are introduced, the ANOVA is able to estimate the magnitude of the experimental (system and measurement) variance. Having calculated a value for the experimental variance, the ANOVA can then estimate a value for the variation due to a possible *interaction* between 'Row' and 'Column' factors

---

E13.15 🖳

The table below gives the relative efficiency of two different motor engines when run on three different varieties of fuel, A, B, C. There are 3 *replicate* measurements for each combination of Engine and Fuel.

	Fuel mixture		
	A	B	C
Engine 1	66.2	67.2	73.1
	65.3	68.4	71.8
	68.0	69.6	70.9
Engine 2	78.8	78.0	76.6
	76.1	76.0	77.6
	79.8	77.0	75.4

Analyse the data to investigate for significant effects on efficiency due to:

(i)   type of engine used,

(ii) variety of fuel used,

(iii) interaction between the variety of fuel and the type of engine.

In E13.15, the replicate measurements give additional information (compared to the basic two-way ANOVA), that enables the *experimental* variation to be calculated directly. The ANOVA can then calculate the effect of any possible *interaction* between the treatment and the blocking factors.

A two-way ANOVA with replication in Excel for E13.15 gives the results shown in Table 13.3.

Table 13.3. ANOVA results for E13.15.

Source of Variation	SS	df	MS	F	P-value	F crit
Sample	310.8356	1	310.8356	178.8124	1.43E-08	4.747221
Columns	11.89333	2	5.946667	3.420901	0.066735	3.88529
Interaction	38.35111	2	19.17556	11.031	0.001912	3.88529
Within	20.86	12	1.738333			
Total	381.94	17				

In Table 13.3, 'Sample' now refers to the row factor (i.e. the choice of engine), and 'Columns' refers to the choice of fuel.

The conclusions that can be drawn from this example are that there is:

- significant difference between the performance of the two engines, since $p = 1.43 \times 10^{-8} < 0.05$. Excel also gives the overall averages (given in E13.16) for the engine efficiencies, which show that engine 2 is more efficient than engine 1.

- significant *interaction* between the engine and the fuel, since $p = 0.0019 < 0.05$. The calculated averages in E13.16 show that fuel C *increases* the efficiency of engine 1 but *decreases* the efficiency of engine 2 – this is an interaction between engine and fuel.

- no significant *overall* difference between the effects of the fuels, since $p = 0.0667 > 0.05$.

Q13.12 The percentage yields from a chemical reaction are recorded as functions of both temperature and pressure. Three replicate measurements were made at each of three different pressures, *P1*, *P2* and *P3*, and two different temperatures, *T1* and *T2*, as recorded in the table below:

Temperature	Pressures		
	P1	*P2*	*P3*
T1	79	68	62
T1	76	71	65
T1	77	69	65
T2	76	69	68
T2	72	70	71
T2	73	73	69

Analyse the above data, using a two-way ANOVA with replication, to assess whether temperature, pressure and/or an interaction between the two, have significant effects on the efficiency of the process.

13.2.10 Interaction plot

It is possible to represent the interaction between two factors graphically using an interaction plot. In this plot, the mean outcome for each level of one factor is plotted against the levels of the other factor.

E13.16 🖳

Taking the data from example E13.15, the mean values are given in the table

	A	B	C
Engine 1	66.5	68.4	71.93
Engine 2	78.23	77	76.53

For each engine a 'line' is drawn on an Excel chart to give efficiency as a function of fuel. The two lines represent the two different engines.

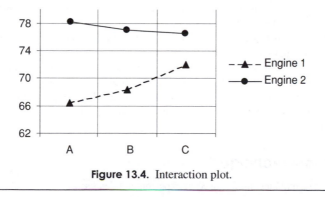

Figure 13.4. Interaction plot.

From the graph in Figure 13.4 it can be seen that:

- Engine 2 is more efficient than engine 1 for all fuels – agrees with $p < 0.05$ for the 'Sample' row in Table 13.3.

- There is an interaction between the fuels and engine – changing from fuel A to fuel C improves the performance for engine 1, but degrades the performance for engine 2 – agrees with $p < 0.05$ for the 'Interaction' row in Table 13.3.

- The effect of changing the fuel has no 'overall' effect as the effects are opposite for the two engines – agrees with $p > 0.05$ for the Columns row in Table 13.3.

The interaction plot can be interpreted approximately:

- A significant effect due to the variable that labels each 'line' will give differences between the *vertical positions* of the lines – this is the case with the significant difference in performance between the two engines.

- A significant interaction effect will cause the *slopes* of the lines to be different – this is the case with the interaction between engine and fuel.

- A significant effect due to the abscissa ('*x*') variable will give an *overall slope up or down* – in this case, one line goes up and the other down with no significant overall fuel effect.

The interaction plot could also be redrawn with the other factor as the abscissa.

Q13.13   Re-plot the interaction plot in E13.16 using the 'engine' as the abscissa variable.

13.3 Experiment design

13.3.1 Introduction

There is an enormous variety of possible designs, and it is important that each design is tailored to fit the particular experiment. Consequently, it is impossible to give a step-by-step guide to producing the correct design in any particular situation. The only valid approach is to develop each new experiment on the basis of a knowledge and experience of the range of fundamental design principles and data analysis techniques that may be appropriate. This Unit aims to provide an overview of key aspects of experimental design and their links with suitable statistical tests.

The objective of good experiment design is to reduce the errors in any measurement and improve the confidence in the final experimental result. In particular, a good experiment design will *counteract* the effects of possible systematic errors, and *take account of* the magnitude of possible random errors.

The process of experiment design proceeds hand-in-hand with the availability of statistical techniques for analysing data. In particular, the reader is recommended to read Chapters 9, 10, 11 and 12 in conjunction with this chapter, and follow the development of statistical tests in parallel with their use in supporting good experimental design.

13.3.2 Principal techniques

In general there are five main techniques used in experimental design:

1 **Replication** – counteracts the effect of random experimental uncertainties, and makes their effect more explicit in the analysis.

2 **Randomization** – counteracts hidden systematic errors (bias) and makes the effects of random subject uncertainties more explicit in the analysis.

3 **Repeated measures or pairing** – counteracts for systematic differences between subjects by including each subject in all factor levels,

4 **Blocking of factors** – used to separate the effects of known factors, including possible systematic (bias) effects.

5 **Allocation of subjects** – used to counter the effects of remaining factors and bias.

13.3.3 Replication

Replication is the process of repeating the experimental measurement under exactly the same conditions. With replicated measurements, the only variations are those due to the experimental process itself – all other factors remain the same.

If the replication is performed with the same subject – *within subjects* – then any variation will only be due to *measurement* uncertainty, for example, repeating the measurement of blood alcohol level on the *same sample* of blood will demonstrate the variability of the measurement itself.

If the replicate experiments involve different subjects – *between subjects* – then any variation will be due to *subject* uncertainty in addition to *measurement* uncertainty, for example, the growth of several bacteria colonies exposed to the *same* experimental conditions will demonstrate the variability of bacterial growth.

The process of replication provides information that can be used to directly calculate the magnitude of the experimental uncertainty (measurement or subject).

E13.17 🖳

An experimental measurement of the sodium content of a water sample is repeated, giving five replicate results:

Sodium content (mg L^{-1}):	31.2	32.3	31.9	31.7	31.5

Give the best-estimate for the true sodium content, together with an estimate of measurement uncertainty and the 95% confidence interval.

Mean value, $\bar{x} = 31.72$

Sample standard deviation, $s = 0.4147$

Standard uncertainty [8.9], $u(\bar{x}) = s/\sqrt{n} = 0.415/\sqrt{5} = 0.185$

t-value (8.2.7) for $df (= n - 1 = 4)$, default significance level, $\alpha(= 0.05)$, 2-tails, $t = 2.78$

Confidence deviation [8.11], $Cd(\mu, \alpha) = t \times u(\bar{x}) = 2.78 \times 0.185 = 0.515$

Confidence interval [8.13] of the mean value:

$$CI(\mu, \alpha) = \bar{x} \pm Cd(\mu, \alpha) = 31.72 \pm 0.52$$

The process of *linear regression* (13.1.5) is also based on a form of replication. In linear regression:

- the dependent variable, recorded on the ordinate (*y*-axis), is the experimental 'outcome'.
- the independent variable, recorded on the abscissa (*x*-axis), represents the different 'levels' of the experimental 'factor'.

It only takes two points to define a straight line, so that additional points provide 'replicate' information for the calculation of the 'best-fit' straight line.

The calculation of experimental uncertainty is based on the spread of points (*residuals*) around the 'best-fit' straight line, and is expressed using the standard error of regression [13.13].

> **Q13.14** Using the same data as Q13.4, calculate the uncertainty in the measurement of the rotation of polarized light.

Replication is the only way of clearly separating experimental variance from other factor variances that may be present in the system. For example, in a two-sample *t*-test, the data variation *within* each sample depends only on the experimental variation. The best-estimate of the experimental standard deviation is then given by the pooled standard deviation [10.12].

> **Q13.15** Using the data from E10.5 given in the table below, calculate:
>
> (i)   the pooled standard deviation [10.12] for the sample sets before and after,
>
> (ii) the standard deviation of the differences.
>
> | Subject | 1 | 2 | 3 | 4 | 5 | 6 | 7 |
> |------------|-----|-----|-----|-----|-----|-----|-----|
> | BP before | 115 | 122 | 145 | 133 | 147 | 118 | 130 |
> | BP after | 125 | 127 | 155 | 136 | 142 | 124 | 139 |
> | Difference | 10 | 5 | 10 | 3 | −5 | 6 | 9 |

Within more complex experimental designs, replicate measurements continue to be essential for separating experimental variance from other possible factors.

13.3.4 Randomization

Randomization is the process of making a truly random choice in the allocation of the subjects to the different factor levels.

In an experiment where the major uncertainty may be due to a subject variability or bias, it is necessary to make a random selection of subjects between samples or factor levels. Although we may believe that we could pick subjects randomly (e.g. seedlings for a growth experiment, patients for a drug trial), experience shows that we would almost always be biased subconsciously in some way.

The process of random allocation can be performed simply by giving each subject a number, and then allocating each subject to a factor level (treatment) by reference to a set of random numbers.

E13.18

As a simple example of random allocation, use the following set of random numbers:

$$7, 3, 7, 8, 2, 6, 6, 7, 9, 1, 2, 2, 9, 4, 6, 6, 2, 6, 5, 3, 8, 2, 4$$

to produce a random allocation of the eight letters ABCDEFGH to two samples, X and Y, of four letters each.

Allocate a number to each of the letters:

$$A = 1, B = 2, C = 3, D = 4, E = 5, F = 6, G = 7, H = 8$$

Work through the list of random numbers and allocate the letter that corresponds to the first number to sample X, and the letter that corresponds to the next number to the sample Y. Continue with allocations to alternative samples. If a number appears which does not correspond to an available letter just pass on to the next number. Resulting allocation:

Sample X: G, H, F, D

Sample Y: C, B, A, E

Drugs trials are conducted as 'double-blind' trials, in which the patients are allocated randomly to the drug or placebo option in such a way that:

- the medical staff do not know which patients are taken the drug, so that their treatment does not become biased, and

- the patients do not know whether they are taking the drug, so that their response to the treatment is not affected subconsciously.

An experiment where subjects are randomly allocated to the different levels of a factor is often called a **completely randomized design**. The experiment described in E13.9 is an example of such a design.

13.3.5 Blocking of factors

Blocking is a process in which groups (blocks) of subjects are identified with particular combinations of levels for different factors. The process of 'blocking' is aimed at separating the effects of known, and possibly unknown, factors that may affect the outcome of the experiment.

The analysis performed in E13.13 assumed that it was a *completely randomized design* experiment where the effects of three different catalysts were each measured with four replicates. A one-way ANOVA (E13.13) gave a p-value, $p = 0.0845$, which suggested that, at a significance level, $\alpha = 0.05$, there was NOT sufficient evidence to identify any difference between the catalysts.

Example E13.14 used the same data as E13.13 but treated the 'days' as a possible *second factor*, and 'blocked' the repeated measurements into columns for each of the 4 days. Example E13.14 then analysed the two-factor data using a two-way ANOVA, giving p-values for both catalyst (row) and day (column) effects:

$$p \text{ (row)} = 0.0173$$
$$p \text{ (column)} = 0.0328$$

which suggested that, at a significance level $\alpha = 0.05$, there was now sufficient evidence for a difference between the catalysts, and also between the days.

The effect of *blocking* in E13.14 enabled the analytical software to allow for systematic variations between the days and provide a more powerful (9.3.7) analysis of the possible effects of the catalysts.

The experiment design used in E13.14 is a **randomized block design**.

It is possible to extract still more information from the basic two-factor *randomized block design* if we also add *replication* – i.e. for every 'cell' in the table, we make at least two replicate measurements. In this case, the replicate information allows an ANOVA calculation to estimate the magnitude of the experimental variation, and then any remaining variation in the data can be explained by an **interaction** between the two factors – see 13.2.9 and 13.2.10.

13.3.6 Repeated measures and pairing

An effective way of counteracting *subject* variation is to ensure that each subject is 'tested' at all levels of each factor, and that the results are compared individually for each subject. The simplest example is the paired t-test (10.2.9) where the differences of response, d_i, are calculated between the responses for the two factor levels, and then a one-sample test applied to the set of difference values. See the example in Q13.15 and E10.5, which uses a *pairing* of data values to reduce the uncertainty due to variations between subjects.

The procedure of testing the same subject at all levels of a test factor is also used in more complex experiments, and it then referred to as a **repeated measures** experiment.

Within 'repeated measures' or 'paired' designs it is necessary to remember that a subject might gain 'experience' from one test which can then affect a subsequent test, e.g., in a sporting test there may be an element of training in having already performed the test under one set of conditions. Example E13.19 shows that this problem can be counteracted by using a *cross-over* design, where different groups of subjects are allocated to do the repeated measures in a reverse order.

E13.19 💻

As part of a project to investigate the factors that affect reaction times, 10 subjects perform standard reaction tests whose timing is measured through a computer. Five subjects (1 to 5) perform the test, first with a high level of background noise, and then repeat the test with no background noise. Five other subjects (6 to 10) perform the same tests, but first with no noise, followed by the test with the high level of background noise.

The experiment is a 'cross-over' design to ensure that the *order of the testing* is opposite for the two groups. This counteracts the possibility that performing one test may affect performance in the second test, e.g. through 'training'.

We perform a paired t-test to investigate whether there is any significant difference (at 0.05) between the reaction times under the two conditions. The results are given in the table below, where 'Difference' = 'Noise' − 'No noise'

Subject	1	2	3	4	5	6	7	8	9	10
Noise	6.5	8.2	4	7.5	4.8					
No noise	5.6	8.7	3.4	6.2	4.4	6.9	4.2	5.9	3.3	7.6
Noise						7.6	4	6.3	4.3	8
Difference	0.9	−0.5	0.6	1.3	0.4	0.7	−0.2	0.4	1	0.4

Each subject links unique 'pairs' of data in the 'no noise' and 'noise' data sets.

Take the differences, d_i, between the values in each data pair.

Sample standard deviation of the differences, $s = 0.5395$.

Mean value of differences $\bar{x} = 0.50$.

Using the one sample t_{STAT} in [10.9], compare $\bar{x}$ with $\mu_O = 0$

$$t_{STAT} = \frac{(\bar{x} - \mu_O)}{s/\sqrt{n}} = \frac{0.50}{0.5395/\sqrt{10}} = 2.93$$

Degrees of freedom, $df = n - 1 = 9$.

$t_{CRIT} = 2.26$.

Since $t_{STAT} > t_{CRIT}$, we accept the alternative hypothesis, that there is a difference in reaction times between the different conditions.

13.3.7 Allocation of subjects

We saw in 13.3.4 that, where there are *no systematic differences* between the subjects, they should be allocated *randomly* to the various factor levels. However, in situations where there may be bias between subjects or between the measurement of the subjects, it is appropriate to allocate the subjects in such a way that is likely to counteract the possible effects of the bias.

In particular, hidden bias between subjects might give the appearance of a significant factor effect when, in fact, no such effect actually exists. For example, in testing for difference in athletic performance, an allocation of *young* athletes to one sample, and *older* athletes to the other, might produce a biased effect due to age that masks the factor being tested. Undetected bias can lead directly to a type I error (9.3.5).

We can illustrate this situation with the following examples that investigate the effect of subject allocation.

E13.20 🖳

An analytical laboratory decided to test whether there was any difference between the effects of three different catalysts (*C1*, *C2*, *C3*) in an industrial process. The process was repeated using three different temperatures (*T1*, *T2*, *T3*) for each catalyst. The yields from each experiment, recorded in the table below, are those that would be recorded assuming that *no systematic measurement errors* are made in recording the data (see also the following example, E13.21).

	T1	T2	T3
C1	39	33	32
C2	36	36	31
C3	32	30	29

Use a two-way ANOVA to test whether there is a significant difference between the effects of the three catalysts.

See text for calculation of results.

Table 13.4 gives the results of a two-way ANOVA for example E13.20.

Table 13.4. ANOVA results for E13.20.

Variation	SS	df	MS	F	P-value	F crit
Rows	34.88889	2	17.44444	5.607143	0.069122	6.944276
Columns	37.55556	2	18.77778	6.035714	0.061946	6.944276
Error	12.44444	4	3.111111			
Total	84.88889	8				

The results in Table 13.4:

$$p\text{-value for catalyst} = 0.069$$
$$p\text{-value for temperature} = 0.062$$

show that there is not sufficient evidence (at 0.05) to suggest that either the catalyst or temperature have a significant effect on the experiment yield.

The next example considers the effect of possible bias due to a specific allocation of 'subjects'. In this case each 'subject' represents an experiment carried out by a particular scientist.

E13.21 🖥

Consider the situation where the experiments given in the previous example E13.20 are actually carried out by three scientists, A, B and C, two of whom produce systematic errors:

> Scientist A produces results that are too high by 1.0
>
> Scientist B produces accurate results
>
> Scientist C produces results that are too low by 1.0

Scientist A measures the performance of catalyst 1 at each of the three temperatures, and scientists B and C measure the performances of catalysts 2 and 3 respectively – the allocation of 'subjects' to scientists is shown in Table 13.5(a).

The actual data values recorded by the scientists can then be calculated from the data in E13.20, and these values are given in the Table 13.5(b).

Table 13.5.

	T1	T2	T3
C1	A	A	A
C2	B	B	B
C3	C	C	C

(a)
Allocation of scientists

lead to data values:

	T1	T2	T3
C1	40	34	33
C2	36	36	31
C3	31	29	28

(b)
Experimental results

Use a two-way ANOVA to decide what conclusions would be drawn from these results.

See text for calculation of results.

The results of a two-way ANOVA for example E13.21 give the results,

$$p\text{-value for catalyst} = 0.025, \text{ and}$$
$$p\text{-value for temperature} = 0.062$$

which *appears to show* that the catalyst *does* have a significant effect.

We know from the analysis in E13.20 that this is a *false* result (type I error). It has arisen because of the *bias between the different scientists*.

Example E13.22 now investigates the possible effects of different subject allocation patterns.

E13.22

Repeat the calculations performed in the previous example, E13.21, but for the different allocations for the three scientists as given below:

	T1	T2	T3
C1	A	B	C
C2	A	B	C
C3	A	B	C

(1)

	T1	T2	T3
C1	B	A	A
C2	B	C	A
C3	C	C	B

(2)

	T1	T2	T3
C1	A	B	C
C2	C	A	B
C3	B	C	A

(3)

See text for calculation of results.

The results of a two-way ANOVA carried out for each of the allocations in E13.22 give the results,

p-value for catalyst =	0.069	0.013	0.217
p-value for temperature =	0.021	0.049	0.200
Subject allocation:	(1)	(2)	(3)

Compared to the 'true' results of Example E13.20, it can be seen that subject allocation (1) gives a false result for the effect of temperature, and allocation (2) gives false results for both the effects of the catalyst and temperature.

The results for subject allocation (3) show *increased p*-values (compared to E13.20), which indicates a *reduced confidence* for the existence of significant effects. This is consistent with the fact that greater errors have been introduced into the measurement process because of the bias of two scientists.

In subject allocations (2) and (3) the systematic errors in measurement have created the *appearance* of significant effects in the results, leading to type I errors.

The significant aspect of allocation (3) is that *each of the values of A, B and C appears once only in each row and column*. By ensuring that every level of every factor has at least one occurrence of each subject type (A, B or C), the differences between subject types are more likely to be reduced.

Any arrangement of items, where each item appears just once in every row and column, is called a **Latin square**. Tables of Latin squares are available for each size of square, and the particular one to be used should be selected at random.

In the above example, the allocation of scientists to the labels A, B and C should be made at random.

As a final note, it is important to understand that any particular allocation of subjects may *coincidentally* work with other variations in measurement to produce *increased* errors. However, provided that the proper procedures are followed, then it is *more likely* that an appropriate allocation of subjects will *reduce* the probability of errors.

13.3.8 Planning a research project

Carrying out a science project as an undergraduate is likely to be the first major piece of original work carried out by the student.

Many student projects are marred by the fact that the student did not anticipate how the data would be analysed and failed to collect data that was essential for the process of analysis. Too often a student reports 'no measurable effect was discovered' because the design of the experiment did not ensure that the data collected would match the methods used to analyse it.

If the process of data collection is not planned to match the requirements of the analytical process, then information will be lost. A loss of information will then lead to

- increased uncertainty in the experimental results, and

- a reduction in the 'power' of a hypothesis test (see 9.3.7).

Before starting the project, the student should:

- be clear about the type of investigation – hypothesis testing or direct measurement,

- focus clearly on the types of quantitative conclusions that could be drawn from the results of the experiments,

- identify the possible sources of uncertainty (random and systematic) in the measurements,

- decide what mathematical or statistical process could be used for analysing the data,

- create an experiment design,

- plan the complete experimental process from measurement to analysis,

- carry out a short 'pilot' experiment to assess the methods to be used, and to obtain some example data,

- check all stages of the analytical process using 'test' data derived, either from the pilot experiment, or by taking an intelligent guess at the sort of values that might be produced in the experiments.

A set of short 'pilot' measurements at the very beginning of the project can be extremely useful in picking up any unforeseen problems at an early stage, and can often suggest useful modifications to the main experimental plan. In addition, 'pilot' measurements will produce 'test' data with appropriate precision and variance suitable for checking the data analysis procedures.

This procedure of working through the whole process of data collection and analysis allows the student to check that nothing has been left out, and prevents the embarrassing scenario of preparing to write up the project only to realize that an essential element of data was never recorded!

Appendixes

Appendix I: Microsoft™ Excel

Tutorials on the use of Excel for science are available on the web site:

💻 Computer Techniques > Excel Tutorials

Functions

Excel uses many different **functions** to perform specific calculations. Useful functions for use in science are given at the end of this Appendix.

A typical example is the TTEST function for a two-sample t-test:

$$= \text{TTEST}(array1, array2, tails, type)$$

The expression is placed in an Excel cell. The four components inside the brackets are called the **arguments**, and identify the values that the TTEST function will use in its calculation. The values for the arguments can be entered:

- directly as values, e.g., in the above example:1 or 2 for *tails*; 1, 2 or 3 for *type*,

- as a calculation using cell references, e.g., 2*A2/B2, or

- as blocks, or **arrays**, of data, for example entering B2:E4 for *array1* in TTEST would include all data values shown shaded below:

Array functions

There are some special *array functions* in Excel, which are sometimes difficult to use. In the book, we only use the LINEST function (4.2.5), to calculate the slope of a line with zero intercept. In this particular case it acts just like an ordinary function.

On the website we also introduce the function FREQUENCY which can be used to 'count' the number of data values that fall within defined classes (or bins).

Data Analysis Tools

The **Data Analysis Tools** facility contains a range of specific statistical calculations, e.g. *ANOVA tests, F-test, t-tests* and *regression*. These are available from the Menu bar using <u>T</u>ools > <u>D</u>ata Analysis.

If this option does not appear, go to <u>T</u>ools > Add-<u>I</u>ns, select the Analysis ToolPak and then click OK. This should enable the installation of Data Analysis Tools.

Note that the results of the calculations using Data Analysis Tools are *not dynamic*; if the original data is changed it is necessary to *re-calculate* the results. This is in contrast to all Excel functions, which are *dynamic* and automatically change immediately the original data is changed.

Index of Excel Function

Computer Techniques > Excel Functions

ACOS	32	NORMDIST	178
AND	145	NORMINV	179
ASIN	32	NORMSDIST	173
ATAN	32	NORMSINV	174
AVERAGE	124	NOT	147
BINOMDIST	198, 202, 274	OR	146
CHIDIST	251	PEARSON	238, 275, 290
CHIINV	249, 252	PERMUT	163
CHITEST	252	PI	32
COMBIN	164	POISSON	205
CORREL	238, 275, 290	RADIANS	27, 32
COS	32	SIN	32
DEGREES	27, 32	SLOPE	77, 290
EXP	90	SQRT	48
FACT	160	STDEV	128
FDIST	223	STDEVP	127
FINV	221	STEYX	283, 290
FREQUENCY	314	TAN	32
FTEST	223	TDIST	234
INTERCEPT	77, 290	TINV	188
LINEST	79, 290, 313	TTEST	234, 313
LN	91	VAR	221, 278
LOG	91	VARP	278

Appendix II: Cumulative z-areas for standard normal distribution

Area shown is the cumulative probability
between $z = -$ Infinity and the given value of z

Initial example for $z = 0.445$ gives
Probability Area $= 0.6718$

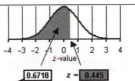

z	0.00	0.01	0.02	0.03	0.04	0.05	0.06	0.07	0.08	0.09
0.0	0.5000	0.5040	0.5080	0.5120	0.5160	0.5199	0.5239	0.5279	0.5319	0.5359
0.1	0.5398	0.5438	0.5478	0.5517	0.5557	0.5596	0.5636	0.5675	0.5714	0.5753
0.2	0.5793	0.5832	0.5871	0.5910	0.5948	0.5987	0.6026	0.6064	0.6103	0.6141
0.3	0.6179	0.6217	0.6255	0.6293	0.6331	0.6368	0.6406	0.6443	0.6480	0.6517
0.4	0.6554	0.6591	0.6628	0.6664	0.6700	0.6736	0.6772	0.6808	0.6844	0.6879
0.5	0.6915	0.6950	0.6985	0.7019	0.7054	0.7088	0.7123	0.7157	0.7190	0.7224
0.6	0.7257	0.7291	0.7324	0.7357	0.7389	0.7422	0.7454	0.7486	0.7517	0.7549
0.7	0.7580	0.7611	0.7642	0.7673	0.7704	0.7734	0.7764	0.7794	0.7823	0.7852
0.8	0.7881	0.7910	0.7939	0.7967	0.7995	0.8023	0.8051	0.8078	0.8106	0.8133
0.9	0.8159	0.8186	0.8212	0.8238	0.8264	0.8289	0.8315	0.8340	0.8365	0.8389
1.0	0.8413	0.8438	0.8461	0.8485	0.8508	0.8531	0.8554	0.8577	0.8599	0.8621
1.1	0.8643	0.8665	0.8686	0.8708	0.8729	0.8749	0.8770	0.8790	0.8810	0.8830
1.2	0.8849	0.8869	0.8888	0.8907	0.8925	0.8944	0.8962	0.8980	0.8997	0.9015
1.3	0.9032	0.9049	0.9066	0.9082	0.9099	0.9115	0.9131	0.9147	0.9162	0.9177
1.4	0.9192	0.9207	0.9222	0.9236	0.9251	0.9265	0.9279	0.9292	0.9306	0.9319
1.5	0.9332	0.9345	0.9357	0.9370	0.9382	0.9394	0.9406	0.9418	0.9429	0.9441
1.6	0.9452	0.9463	0.9474	0.9484	0.9495	0.9505	0.9515	0.9525	0.9535	0.9545
1.7	0.9554	0.9564	0.9573	0.9582	0.9591	0.9599	0.9608	0.9616	0.9625	0.9633
1.8	0.9641	0.9649	0.9656	0.9664	0.9671	0.9678	0.9686	0.9693	0.9699	0.9706
1.9	0.9713	0.9719	0.9726	0.9732	0.9738	0.9744	0.9750	0.9756	0.9761	0.9767
2.0	0.9772	0.9778	0.9783	0.9788	0.9793	0.9798	0.9803	0.9808	0.9812	0.9817
2.1	0.9821	0.9826	0.9830	0.9834	0.9838	0.9842	0.9846	0.9850	0.9854	0.9857
2.2	0.9861	0.9864	0.9868	0.9871	0.9875	0.9878	0.9881	0.9884	0.9887	0.9890
2.3	0.9893	0.9896	0.9898	0.9901	0.9904	0.9906	0.9909	0.9911	0.9913	0.9916
2.4	0.9918	0.9920	0.9922	0.9925	0.9927	0.9929	0.9931	0.9932	0.9934	0.9936
2.5	0.9938	0.9940	0.9941	0.9943	0.9945	0.9946	0.9948	0.9949	0.9951	0.9952
2.6	0.9953	0.9955	0.9956	0.9957	0.9959	0.9960	0.9961	0.9962	0.9963	0.9964
2.7	0.9965	0.9966	0.9967	0.9968	0.9969	0.9970	0.9971	0.9972	0.9973	0.9974
2.8	0.9974	0.9975	0.9976	0.9977	0.9977	0.9978	0.9979	0.9979	0.9980	0.9981
2.9	0.9981	0.9982	0.9982	0.9983	0.9984	0.9984	0.9985	0.9985	0.9986	0.9986
3.0	0.9987	0.9987	0.9987	0.9988	0.9988	0.9989	0.9989	0.9989	0.9990	0.9990
3.1	0.9990	0.9991	0.9991	0.9991	0.9992	0.9992	0.9992	0.9992	0.9993	0.9993
3.2	0.9993	0.9993	0.9994	0.9994	0.9994	0.9994	0.9994	0.9995	0.9995	0.9995
3.3	0.9995	0.9995	0.9995	0.9996	0.9996	0.9996	0.9996	0.9996	0.9996	0.9997
3.4	0.9997	0.9997	0.9997	0.9997	0.9997	0.9997	0.9997	0.9997	0.9997	0.9998
3.5	0.9998	0.9998	0.9998	0.9998	0.9998	0.9998	0.9998	0.9998	0.9998	0.9998
3.6	0.9998	0.9998	0.9999	0.9999	0.9999	0.9999	0.9999	0.9999	0.9999	0.9999
3.7	0.9999	0.9999	0.9999	0.9999	0.9999	0.9999	0.9999	0.9999	0.9999	0.9999
3.8	0.9999	0.9999	0.9999	0.9999	0.9999	0.9999	0.9999	0.9999	0.9999	0.9999
3.9	1.0000	1.0000	1.0000	1.0000	1.0000	1.0000	1.0000	1.0000	1.0000	1.0000

Appendix III: Critical values – t-statistic and chi-squared, χ^2

Tails	Degrees	t-statistic		Chi-squared	
		Significance		Significance	
	of freedom	95%	99%	95%	99%
1 or 2	df	0.05	0.01	0.05	0.01
2	1	12.71	63.66		
1		6.31	31.82	3.84	6.63
2	2	4.30	9.92		
1		2.92	6.96	5.99	9.21
2	3	3.18	5.84		
1		2.35	4.54	7.81	11.34
2	4	2.78	4.60		
1		2.13	3.75	9.49	13.28
2	5	2.57	4.03		
1		2.02	3.36	11.07	15.09
2	6	2.45	3.71		
1		1.94	3.14	12.59	16.81
2	7	2.36	3.50		
1		1.89	3.00	14.07	18.48
2	8	2.31	3.36		
1		1.86	2.90	15.51	20.09
2	9	2.26	3.25		
1		1.83	2.82	16.92	21.67
2	10	2.23	3.17		
1		1.81	2.76	18.31	23.21
2	11	2.20	3.11		
1		1.80	2.72	19.68	24.73
2	12	2.18	3.05		
1		1.78	2.68	21.03	26.22
2	13	2.16	3.01		
1		1.77	2.65	22.36	27.69
2	14	2.14	2.98		
1		1.76	2.62	23.68	29.14
2	15	2.13	2.95		
1		1.75	2.60	25.00	30.58
2	16	2.12	2.92		
1		1.75	2.58	26.30	32.00
2	17	2.11	2.90		
1		1.74	2.57	27.59	33.41
2	18	2.10	2.88		
1		1.73	2.55	28.87	34.81
2	19	2.09	2.86		
1		1.73	2.54	30.14	36.19
2	20	2.09	2.85		
1		1.72	2.53	31.41	37.57
2	25	2.06	2.79		
1		1.71	2.49	37.65	44.31
2	50	2.01	2.68		
1		1.68	2.40	67.50	76.15
2	Infinity	1.96	2.58		
1		1.64	2.33		

Appendix IV: Critical *F* values at 0.05 (95%) significance

Tails	$df_N=$	1	2	3	4	5	6	7	8	9	10	15	20	30	50	100
		df_N = degrees of freedom for numerator														
		df_D = degrees of freedom for denominator														
1 or 2	df_D															
2	1	648	799	864	900	922	937	948	957	963	969	985	993	1001	1008	1013
1		161	199	216	225	230	234	237	239	241	242	246	248	250	252	253
2	2	38.5	39.0	39.2	39.2	39.3	39.3	39.4	39.4	39.4	39.4	39.4	39.4	39.5	39.5	39.5
1		18.5	19.0	19.2	19.2	19.3	19.3	19.4	19.4	19.4	19.4	19.4	19.4	19.5	19.5	19.5
2	3	17.4	16.0	15.4	15.1	14.9	14.7	14.6	14.5	14.5	14.4	14.3	14.2	14.1	14.0	14.0
1		10	10	9.28	9.12	9.01	8.94	8.89	8.85	8.81	8.79	8.70	8.66	8.62	8.58	8.55
2	4	12.2	10.6	9.98	9.60	9.36	9.20	9.07	8.98	8.90	8.84	8.66	8.56	8.46	8.38	8.32
1		7.71	6.94	6.59	6.39	6.26	6.16	6.09	6.04	6.00	5.96	5.86	5.80	5.75	5.70	5.66
2	5	10.0	8.43	7.76	7.39	7.15	6.98	6.85	6.76	6.68	6.62	6.43	6.33	6.23	6.14	6.08
1		6.61	5.79	5.41	5.19	5.05	4.95	4.88	4.82	4.77	4.74	4.62	4.56	4.50	4.44	4.41
2	6	8.81	7.26	6.60	6.23	5.99	5.82	5.70	5.60	5.52	5.46	5.27	5.17	5.07	4.98	4.92
1		5.99	5.14	4.76	4.53	4.39	4.28	4.21	4.15	4.10	4.06	3.94	3.87	3.81	3.75	3.71
2	7	8.07	6.54	5.89	5.52	5.29	5.12	4.99	4.90	4.82	4.76	4.57	4.47	4.36	4.28	4.21
1		5.59	4.74	4.35	4.12	3.97	3.87	3.79	3.73	3.68	3.64	3.51	3.44	3.38	3.32	3.27
2	8	7.57	6.06	5.42	5.05	4.82	4.65	4.53	4.43	4.36	4.30	4.10	4.00	3.89	3.81	3.74
1		5.32	4.46	4.07	3.84	3.69	3.58	3.50	3.44	3.39	3.35	3.22	3.15	3.08	3.02	2.97
2	9	7.21	5.71	5.08	4.72	4.48	4.32	4.20	4.10	4.03	3.96	3.77	3.67	3.56	3.47	3.40
1		5.12	4.26	3.86	3.63	3.48	3.37	3.29	3.23	3.18	3.14	3.01	2.94	2.86	2.80	2.76
2	10	6.94	5.46	4.83	4.47	4.24	4.07	3.95	3.85	3.78	3.72	3.52	3.42	3.31	3.22	3.15
1		4.96	4.10	3.71	3.48	3.33	3.22	3.14	3.07	3.02	2.98	2.85	2.77	2.70	2.64	2.59
2	15	6.20	4.77	4.15	3.80	3.58	3.41	3.29	3.20	3.12	3.06	2.86	2.76	2.64	2.55	2.47
1		4.54	3.68	3.29	3.06	2.90	2.79	2.71	2.64	2.59	2.54	2.40	2.33	2.25	2.18	2.12
2	20	5.87	4.46	3.86	3.51	3.29	3.13	3.01	2.91	2.84	2.77	2.57	2.46	2.35	2.25	2.17
1		4.35	3.49	3.10	2.87	2.71	2.60	2.51	2.45	2.39	2.35	2.20	2.12	2.04	1.97	1.91
2	30	5.57	4.18	3.59	3.25	3.03	2.87	2.75	2.65	2.57	2.51	2.31	2.20	2.07	1.97	1.88
1		4.17	3.32	2.92	2.69	2.53	2.42	2.33	2.27	2.21	2.16	2.01	1.93	1.84	1.76	1.70
2	50	5.34	3.97	3.39	3.05	2.83	2.67	2.55	2.46	2.38	2.32	2.11	1.99	1.87	1.75	1.66
1		4.03	3.18	2.79	2.56	2.40	2.29	2.20	2.13	2.07	2.03	1.87	1.78	1.69	1.60	1.52
2	100	5.18	3.83	3.25	2.92	2.70	2.54	2.42	2.32	2.24	2.18	1.97	1.85	1.71	1.59	1.48
1		3.94	3.09	2.70	2.46	2.31	2.19	2.10	2.03	1.97	1.93	1.77	1.68	1.57	1.48	1.39

Example:

Critical *F*-value for 1-tailed test with $df_N = 15$ and $df_D = 8$ at 0.05 significance:

$$F_{1,0.05,15,8} = 3.22$$

Appendix V: Critical values at 0.05 (95%) significance for

- Pearson's correlation coefficient, r
- Spearman's Rank correlation coefficient, r_S
- Wilcoxon lower limit, W_L

Tails	Sample size	Pearson's	Spearman's	Wilcoxon
1 or 2	n	r	r_s	W_L
2	4	0.950		
1		0.900		
2	5	0.878	–	
1		0.805	0.900	
2	6	0.811	0.886	–
1		0.729	0.829	2
2	7	0.754	0.786	2
1		0.669	0.714	3
2	8	0.707	0.738	3
1		0.622	0.643	5
2	9	0.666	0.683	5
1		0.582	0.600	8
2	10	0.632	0.648	8
1		0.549	0.564	10
2	11	0.602	0.623	10
1		0.521	0.523	13
2	12	0.576	0.591	13
1		0.497	0.497	17
2	13	0.553	0.566	17
1		0.476	0.475	21
2	14	0.532	0.545	21
1		0.456	0.457	25
2	15	0.514	0.525	25
1		0.441	0.441	30
2	16	0.497	0.507	29
1		0.426	0.425	35
2	17	0.482	0.490	34
1		0.412	0.412	41
2	18	0.468	0.476	40
1		0.400	0.399	47
2	19	0.456	0.462	46
1		0.389	0.388	53
2	20	0.444	0.450	52
1		0.378	0.377	60
2	25	0.396	0.400	89
1		0.337	0.336	100
2	30	0.361	0.364	137
1		0.306	0.305	151

Appendix VI: Mann–Whitney lower limit, U_L, at 0.05 (95%) significance

Tails 1 or 2	n_2	$n_1=$ 2	3	4	5	6	7	8	9	10	11	12	13	14	15	16	17	18	19	20
2	2										0	1	1	1	1	1	2	2	2	2
1							0	1	1	1	1	1	2	2	3	3	3	3	4	4
2	3				0	1	1	2	2	3	3	4	4	5	5	6	6	7	7	8
1				0	1	2	2	3	4	4	5	5	6	7	7	8	9	9	10	11
2	4			0	1	2	3	4	4	5	6	7	8	9	10	11	11	12	13	14
1			0	1	2	3	4	5	6	7	8	9	10	11	12	14	15	16	17	18
2	5		0	1	2	3	5	6	7	8	9	11	12	13	14	15	17	18	19	20
1			1	2	4	5	6	8	9	11	12	13	15	16	18	19	20	22	23	25
2	6		1	2	3	5	6	8	10	11	13	14	16	17	19	21	22	24	25	27
1			2	3	5	7	8	10	12	14	16	17	19	21	23	25	26	28	30	32
2	7		1	3	5	6	8	10	12	14	16	18	20	22	24	26	28	30	32	34
1		0	2	4	6	8	11	13	15	17	19	21	24	26	28	30	33	35	37	39
2	8	0	2	4	6	8	10	13	15	17	19	22	24	26	29	31	34	36	38	41
1		1	3	5	8	10	13	15	18	20	23	26	28	31	33	36	39	41	44	47
2	9	0	2	4	7	10	12	15	17	20	23	26	28	31	34	37	39	42	45	48
1		1	4	6	9	12	15	18	21	24	27	30	33	36	39	42	45	48	51	54
2	10	0	3	5	8	11	14	17	20	23	26	29	33	36	39	42	45	48	52	55
1		1	4	7	11	14	17	20	24	27	31	34	37	41	44	48	51	55	58	62
2	11	0	3	6	9	13	16	19	23	26	30	33	37	40	44	47	51	55	58	62
1		1	5	8	12	16	19	23	27	31	34	38	42	46	50	54	57	61	65	69
2	12	1	4	7	11	14	18	22	26	29	33	37	41	45	49	53	57	61	65	69
1		2	5	9	13	17	21	26	30	34	38	42	47	51	55	60	64	68	72	77
2	13	1	4	8	12	16	20	24	28	33	37	41	45	50	54	59	63	67	72	76
1		2	6	10	15	19	24	28	33	37	42	47	51	56	61	65	70	75	80	84
2	14	1	5	9	13	17	22	26	31	36	40	45	50	55	59	64	69	74	78	83
1		3	7	11	16	21	26	31	36	41	46	51	56	61	66	71	77	82	87	92
2	15	1	5	10	14	19	24	29	34	39	44	49	54	59	64	70	75	80	85	90
1		3	7	12	18	23	28	33	39	44	50	55	61	66	72	77	83	88	94	100
2	16	1	6	11	15	21	26	31	37	42	47	53	59	64	70	75	81	86	92	98
1		3	8	14	19	25	30	36	42	48	54	60	65	71	77	83	89	95	101	107
2	17	2	6	11	17	22	28	34	39	45	51	57	63	69	75	81	87	93	99	105
1		3	9	15	20	26	33	39	45	51	57	64	70	77	83	89	96	102	109	115
2	18	2	7	12	18	24	30	36	42	48	55	61	67	74	80	86	93	99	106	112
1		4	9	16	22	28	35	41	48	55	61	68	75	82	88	95	102	109	116	123
2	19	2	7	13	19	25	32	38	45	52	58	65	72	78	85	92	99	106	113	119
1		4	10	17	23	30	37	44	51	58	65	72	80	87	94	101	109	116	123	130
2	20	2	8	14	20	27	34	41	48	55	62	69	76	83	90	98	105	119	119	127
1		4	11	18	25	32	39	47	54	62	69	77	84	92	100	107	115	123	138	138

Review Questions

Chapter 2

1 Write the following

   (i)   0.03696 to three significant figures

   (ii)   0.03652 to three decimal places

   (iii)   0.00354 in scientific notation

   (iv)   567.23 in scientific notation

2 Evaluate the following *without* using a calculator

   (i) $3.64 \times 10^{-3} + 2.365 \times 10^{-2}$

   (ii) $\dfrac{2.44 \times 10^3}{8.0 \times 10^{-2}}$

3 Use a calculator to calculate the value of the expression (to 3 sf):

$$\frac{3.098 \times 10^{-2} - 4.98 \times 10^{-3}}{(42.7 - 16.8) \times 0.03}$$

4 How would the number 1.6×10^{-19} be entered into a cell in Excel?

5 In an Excel spreadsheet, certain cells contain numbers as follows:

A1 contains the *number* 3, A2 contains the *number* 14, A3 contains the *number* 2.
The cell B1 contains the *formula*: $= A1^2 + A2/A3$
What value would be seen in cell B1 as a result of the calculation by the formula?

6 Perform the following conversions:

   (i) A volume of 267 mm^3 into units of m^3.

   (ii) A 'weight' of paper of 0.0080 g cm^{-2} into units of kg m^{-2}.

   (iii) A height of 5 feet 8 inches to metres, given that there are 2.54 centimetres in an inch and 12 inches in a foot.

   (iv) Express a speed of 30 mph in m s^{-1}, given that 1 mile $= 1.609$ km.

   (v) A man is 6 feet 1 inch tall and weighs 14 stone and 6 lbs. Calculate his mass body index (*BMI*) in kg m^{-2} where

$$BMI = \text{mass}/(\text{height})^2$$

(1 stone $= 14$ lbs; 1 kg $= 2.2$ lb; 1 foot $= 12$ inches; 1 inch $= 2.54$ cm).

7 Given that Avogadro's number is 6.022×10^{23}, how many molecules are there in 10.3 mg of the compound whose relative molecular mass is 64.5?

8 1.6 g of a compound whose relative molecular mass is 64.5 is dissolved in water to give a solution with a volume of 100 mL.

   Calculate the molar concentration of the solution.

9 20 mL of a solution is put into a 100-mL graduated flask, and further solvent is added to bring the volume up to exactly 100 mL. If the concentration of the final solution is 0.5 mol L^{-1} what was the concentration of the initial solution?

10 What mass of NaOH when dissolved in water to give 100 mL of solution will give a concentration of 0.3 mol L^{-1}? (RMM for NaOH = 40.)

11 Convert the following (to 2 dp):

   (i) 2.3 radians into degrees   (ii) 46° into radians

12 A road on a hill is inclined at 21° to the horizontal. What will be the increase in vertical height of a car as it travels 200 m uphill along the surface of the road?

Chapter 3

1 Replace x with the number 3 and y with the number 4 in the following expressions and calculate the values:

   (i) $(3x)^2 + 4x^2$ (ii) $3(y - x^2)$

2 Cancel where possible in the following fractions:

   (i) $\dfrac{36}{27}$ (ii) $\dfrac{12a^2}{8ba}$

3 Multiply out the brackets

   (i) $5(x - 2)$ (iii) $(a - b)(a + b)$

   (ii) $-3(x + a)$ (iv) $(2x + 3)(x - 4)$

4 Make x the subject of the equations:

   (i) $a - x = 14$ (iv) $p(x + 2) = x + 2q$

   (ii) $2a + x = 4 - 2x$ (v) $a = \sqrt{(x + 2)}$

   (iii) $\dfrac{x}{4} = 2p - x$ (vi) $a = 3\sqrt{x} + 5$

5 Calculate x if

   (i) $\dfrac{(2x + 6)}{2} = 12$ (iii) $6 = \sqrt{(9 - x)}$

   (ii) $9x^2 = 0.81$ (iv) $2^x = 8$

6 Express the following with common denominators:

   (i) $\dfrac{1}{p} + \dfrac{2}{q}$ (ii) $\dfrac{2}{x} + \dfrac{1}{x + 1}$

7 Make x the subject of the equations:

(i) $\dfrac{x+1}{x} + \dfrac{1}{b} = 4$

(ii) $\dfrac{x-2}{x+2} = p$

8 What, if any, are the possible answers for the value of x in the following?

(i) $x^2 - 16 = 0$

(iii) $2x^2 + 2x = 5$

(ii) $2x^2 - 7x + 5 = 0$

(iv) $2x^2 - 6x + 5 = 0$

9 Solve the following pairs of simultaneous equations:

(i) $y = 2x + 2,$
$y = -x + 5$

(iii) $2y = x + 4,$
$y = 2x - 1$

(ii) $y = x + 3,$
$4y + x = 2$

(iv) $1.6 = 4.2x + 1.8y$
$1.2 = 0.9x + 3.2y$

10 Find the points where the curve $y = x^2$ intersects the line $x + y = 2$.

11 Solve the following equations:

(i) $10^x = 1000$

(iv) $4.6^{(3-2x)} = 14$

(ii) $4^x = 56$

(v) $\log(x) = -1.32$

(iii) $0.5^{2x} = 0.0025$

(vi) $\ln(2x - 3) = 1.38$

Chapter 4

1 At what points do the following lines pass through the y-axis?

(i) $3y = x - 5$

(ii) $y = 4x - 2$

2 At what points do the following lines pass through the x-axis?

(i) $3y = x - 5$

(ii) $y = 4x - 2$

3 Calculate the slopes of the lines that pass through the following pairs of points:

(i) $(4, 3)$ and $(6, 6)$

(ii) $(-1, 2)$ and $(2, -4)$

4 Derive the equations of the straight lines with

(i) slope, $m = 1.5$, that passes through the point $(5, 2)$

(ii) slope, $m = -2.0$, that passes through the point $(4, 4)$

5 Derive the equations of the lines that pass through the following pairs of points:

(i) $(0.8, 0.98)$ and $(0.6, 0.72)$

(ii) $(-14, 32)$ and $(+14, 18)$

6 A straight line on an x–y graph passes through the points $(3, 2)$ and $(5, 6)$.

(i) Calculate the value of y for the line when, $x = 8$

(ii) Calculate the value of x for the line when, $y = -4$

7 A second line is parallel to the line $y = 3x + 2$, and passes through the point $(3, 2)$. What is the equation of this second line?

8   The velocity, v, of an accelerating car is given by the equation $v = 8.4 + a \times t$ where a is the acceleration, and t is the time. If the velocity of the car is 12.9 m s^{-1} when $t = 9$ s, calculate its velocity when $t = 20$ s.

9   Use Excel, or other statistics software, to calculate the slope and intercept of the best-fit straight line for the following calibration data for the absorbance, A, against the concentration, C, of a chemical solution.

C	0.01	0.02	0.03	0.04
A	0.165	0.306	0.407	0.552

Calculate the concentration, C, of an unknown solution that has an absorbance, $A = 0.365$.

10   A vehicle skidding to a halt from an initial speed, u (m s^{-1}), will stop in a distance, s (m), given by the equation: $u^2 = 2\mu g\, s$, where $g = 9.81$ m s^{-1}, and μ is the coefficient of friction between the tyres and the road. The stopping distances, s, are measured experimentally for different initial speeds, u, and recorded in the following table.

Distance, s (m)	1.5	8	14	28
Speed, u (m s^{-1})	5	10	15	20

Use a process of linearization to calculate a best-estimate for the value of the coefficient of friction, μ.

11   The data in the table show the numbers, N, of a bacterial colony as a function of time, t (min). Assume that the growth of the numbers, N, follows the equation: $N = N_O \exp(kt)$.

Time, t (mins)	50	100	150	200	250	300
Count, N	130	300	740	1200	3500	8200

Use a process of linearization to calculate best-estimate values of k and N_O.

Chapter 5

1   Evaluate the following

   (i) $10^4 \times 10^3$ (iii) $(10^3)^{-2}$

   (ii) 10^0 (iv) $10^4 + 10^3$

2   Calculate the value of x in each of the following:

   (i) $1.46 = e^x$ (iii) $\log_{10}(x) = 2.5$

   (ii) $3e^x = 11.2$ (iv) $\ln(2x + 3) = 0.56$

3   Calculate the value of x in each of the following:

   (i) $5^x = 125$ (iii) $23 = x^{2.4}$

   (ii) $5^x = 62$ (iv) $0.6 = 0.87^x$

4   Calculate the hydrogen ion concentration, [H$^+$] for the following pH values:

   (i) $pH = 7.0$ (iii) $pH = 9.2$

   (ii) $pH = 7.3$ (iv) $pH = 1.2$

5 Calculate the spectrophotometric absorbance, A, corresponding to:

   (i) transmittance, $T = 50\%$ (ii) transmittance, $T = 10\%$

6 Calculate the spectrophotometric transmittance, T, corresponding to:

   (i) absorbance, $A = 0.1$ (ii) absorbance, $A = 2.0$

7 A noise increases in power density from $2.0 \times 10^{-7} \, W \, m^{-2}$ to $8.0 \times 10^{-6} \, W \, m^{-2}$. Calculate the increase in loudness in decibels.

8 It is believed that the fraction, F, of a drug remaining in the blood at time t (in hours) after an injection follows the equation:

$$F = e^{-0.13t}$$

What time (in hours) will it take for the fraction, F, to fall to 0.5?

9 A population growth, P, that increases exponentially by 10% every day can be represented by the following equation:

$$P = P_O \times (1.1)^N$$

where N is the number of days and P_O is the population when $N = 0$.

   (i) If $P_O = 600$, when $N = 0$, calculate the population, P, after 5 days.

   (ii) Calculate the value of k in the equation $P = P_O \, e^{kN}$ that will give the same rate of population increase of 10% every day.

10 The intensity of light, I, given off by some bioluminescent bacteria is believed to vary with time, t (s), according to the equation $I = I_0 \, e^{-rt}$ where I_0 and r are constants. I_0 is the value of I when $t = 0$.

   (i) If $r = 0.001 \, s^{-1}$ and it is found that $I = 400$ units after a period, $t = 5$ minutes, calculate the value of the constant, I_0.

   (ii) Using the result from (i) calculate the intensity of light, I, that could be expected after a period, $t = 10$ minutes.

Chapter 6

1 A car accelerates from rest at time, $t = 0$, and then travels at varying speeds. The distances, s, travelled by the car at a few specific times are given in the table:

Time, t (s)	0	10	20	30	40	50	60	70
Distance, s (m)	0	50	200	400	720	1100	1520	1900
Speed, v (m s^{-1})								
Acceleration, a (m s^{-2})								

   (i) Estimate the speed of the car at times from $t = 10$ s to $t = 60$ s.

   (ii) Estimate the acceleration at times from $t = 20$ s to $t = 50$ s.

2 The trunk diameters, d (cm), of a tree, that was grown in the UK, were calculated for the end of each year by measuring the diameters of its 'rings' after the tree was felled:

Trunk diameter:	3.0	3.6	4.2	4.6	5.2	6.0	7.0	7.2
End of year:	1981	1982	1983	1984	1985	1986	1987	1988

Which ONE of the following statements is true:

(A)   The rate of growth in 1986 was double the rate of growth in 1982.

(B)   Over the years 1984 to 1987 the rate of growth was increasing every year.

(C)   The fastest rate of growth was in 1986.

(D)   The slowest rate of growth was in 1983.

3   A growth curve is described by the equation: $N = 100 \times e^{+0.05 \times t}$. Given that the differential coefficient of the equation

$$N = A \times e^{B \times t} \quad \text{is} \quad dN/dt = A \times B \times e^{B \times t}$$

calculate the rate of growth at time, $t = 20$ s.

4   A rocket accelerates from its launch pad such that its height, h(m), at a time, t(s) after lift-off is given by:

$$h = 40 \times t^2$$

(i)   Derive an expression for the differential coefficient, dh/dt, as a function of the time, t.

(ii)   Hence calculate the speed of the rocket after $t = 5$ seconds.

Chapter 7

1   For the following five data values, 16, 18, 21, 16, 14, calculate (without using a calculator):

(i)   mean value   (iii)   sum of square of deviations

(ii)   sum of deviations   (iv)   population variance

2   The following set of data values, 96.2, 94.3, 97.6, 95.6, gives the results of four replicate experimental measurements of alcohol level in a single blood sample in units of mg $(100 \text{ mL})^{-1}$. Use Excel or other statistics software to calculate the standard deviation of the experimental procedure.

3   The heights, h (in cm), of 40 people were measured, and recorded as below:

153	169	159	158	161	168	176	158	172	155
168	171	163	154	162	173	161	155	160	172
178	175	157	166	159	160	177	167	157	174
171	166	148	153	173	163	162	150	161	171

(i)   Draw a stem and leaf diagram to show the distribution of heights.

(ii)   Using the results of (i) plot the data as a histogram.

(iii)   Show how you have calculated the heights of each of the blocks in the histogram.

(iv)   Using the frequency data from above, calculate a best-estimate for the mean height of the 40 people.

4 Use the data in Figure 7.10, to make a rough estimate of the probability that a student, selected at random, will have a course mark between 40% and 50%.

5 Recording graduate performance over several years, it was found that, out of 500 students, 40 students obtained first class honours grades. Using this information, estimate the probability that a particular student selected at random from the current class of 80 will gain the first class honours grade.

6 In a car park there are 25 red cars, 20 black cars, 20 silver cars and 15 of other colours. Calculate the probability that the first of these cars to leave the car park is not black.

7 A trial has four possible outcomes, A, B, C and D, which have the following probabilities: $P(A) = 0.1$, $P(B) = 0.2$, $P(C) = 0.4$, $P(D) = 0.3$. Calculate the probability that a single outcome will result in either B or C.

8 In a prize draw, the probability of winning first prize is 0.001, the probability of winning any one of the second prizes is 0.004 and the probability of winning any one of the third prizes is 0.009. What is the probability of not winning any of the first three prizes?

9 It is known that, in a given region, the numbers of people reading local newspapers – morning paper *only*, evening paper *only*, *both* morning and evening papers – are in the ratios 8, 10, 2, respectively. Estimate the expected number of people who only take a morning paper out of a total of 89 people who buy papers.

10 In rolling two standard dice together, each with scores 1 to 6, calculate the probability of getting a total score of exactly 9.

11 Evaluate the following:

(i) $4! - 3!$ (ii) $150!/148!$

12 Calculate the ratio $_nP_r/_nC_r$

(i) when $n = 10$ and $r = 3$ (ii) when $n = 8$ and $r = 3$

13 Calculate the number of ways of selecting 5 items from 10 items when

(i) the order is important   (ii) the order is not important

14 Students studying a difficult topic have a choice taking one of two modules, A or B, in the first year. If they pass either A or B, they can then study module C in the second year. Past experience suggests that the probability of passing module A is 0.9, and that the probability of passing module B is 0.6. However, it is also known that module B is a better grounding for module C, so that, on average 80% of people who pass module B also pass module C, but only 50% who pass module A also pass module C. Assume that 60% of students opt to take module A and the rest take module B in the first year.

(i) Use a probability tree to set out the data in this question in a diagrammatic form.

(ii) If 100 students enrol to study this topic in year one, calculate how many are likely to pass module C at the end of the second year.

(iii) Which route (A or B) gives the better chance of overall success?

15 It is known that, on average, one in a hundred people (1 in 100) have a particular disease. A diagnostic test is devised to screen for this disease. A positive result is one that suggests that the person has the disease, and a negative result is one that suggests that the person does not have the disease. The possibility of errors in the test give the following result probabilities:

- For a person who has the disease, the probability of a positive result, $p(P|D) = 0.95$.

- For a person who does not have the disease, the probability of a negative result, $p(\bar{P}|\bar{D}) = 0.90$.

The test is applied to a randomly selected sample of 100 000 people:

(i) Calculate the numbers of people (out of the sample of 100 000) who could be expected to get *each* of the following outcomes:

$$\text{true positive, false positive, true negative, false negative}$$

(ii) Calculate the probability that a single person selected at random will give a positive result (either true or false).

(iii) Calculate the probability that a single person selected at random, who then gets a positive result, will actually be suffering from the disease.

(iv) Explain the difference between the probabilities that are described by the following two expressions: $P(P|D)$ and $P(D|P)$.

16 For question 15 calculate

(i) the prior odds that a person selected at random will have the disease,

(ii) the likelihood ratio for the diagnostic test,

(iii) the posterior odds that a person selected at random, who then tests positive for the disease, will actually have the disease.

Chapter 8

1 Estimate by eye the mean and standard deviation of the normal distribution shown below:

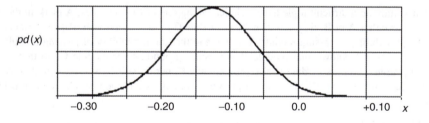

2 For a standard normal distribution, calculate the following:

(i) Cumulative probability area, $p(-\infty < z < +0.67)$.

(ii) Cumulative probability area, $p(-\infty < z < -0.67)$.

(iii) The value, z', such that the probability area, $p(z' < z < \infty) = 0.10$.

(iv) Cumulative probability area between $z = -1.5$ and $z = +0.5$.

3 A normal distribution of x-values has a mean value, $\mu = 18$, and standard deviation, $\sigma = 4$. Calculate the following:

(i) The 'z-value' equivalent to $x = 15$.

(ii) The value of x equivalent to a 'z-value' $= -1.5$.

(iii) The probability area, $p(21 < x < 24)$, between $x = 21$ and $x = 24$.

(iv) The value, x', such that the probability area, $p(x' < x < \infty) = 0.10$.

4 The standard deviation uncertainty in an experimental measurement procedure is 0.36. One hundred students each take the mean of five replicate results using this procedure. Estimate the standard deviation of the mean values recorded by all the students.

5 It is known that the process of measuring the calcium content in a serum sample has a standard deviation of 0.15 mmol L^{-1}. Five replicate measurements from one sample give values:

$$1.95, 2.30, 2.25, 2.20 \text{ and } 2.12$$

Calculate the range of values, within which you would be 95% confident in stating that the true value lies.

6 It is known that the random uncertainty of an analytical method, that is used to measure the concentration (in ppb) of cadmium in river water, is described by a normal distribution with a standard deviation, $\sigma = 3.5$ ppb. The method is used to measure the concentration of a sample whose true concentration 34.5 ppb.

(i) Calculate the probability that, using this method, a *single* experimental result will produce a result greater than 40.0 ppb.

(ii) The same analytical measurement is repeated *eight* times and the mean of the eight measurements is calculated. Calculate the range of values within which you would be 99% confident of recording the *mean value of the eight measurements*, again assuming that the true value of the concentration is 34.5 ppm.

7 What is the 95% confidence interval for an experimental result in which five experimental values give a mean value of 18.8 with a sample standard deviation of 1.5.

8 It is known that the average number of cases of a particular random characteristic is 3 per 1000 people. Calculate the probability that, in a particular group of 500, there will be exactly four people with this condition.

9 In a population, one person in fifty (1 in 50) have a particular characteristic that occurs randomly. For a group of 80 people:

(i) Use the binomial distribution to calculate the probability (to 4 dp) that exactly 3 people out of 80 will have this particular characteristic.

(ii) Calculate the number of people who, on average, would have this characteristic in the group of 80 people.

(iii) Use the Poisson distribution (to 4 dp) to estimate the probability (to 4 dp) that exactly 3 people out of 80 will have this particular characteristic.

(iv) Compare the results in (i) and (iii).

Chapter 9

1   An analyst performs a t-test on a set of data and decides on a significance level, α, and then calculates the p-value for the data. She then decides to accept the alternative hypothesis. Which ONE of the following sets of statements could be true.

   (A) $\alpha = 0.050$ and $p = 0.055$

   (B) $\alpha = 0.010$ and $p = 0.005$

   (C) $\alpha = 0.010$ and $p = 0.035$

   (D) $\alpha = 0.050$ and $p = 0.960$

2   The results of a t-test suggest that the alternative hypothesis should be accepted. Which ONE of the following sets of statements could be true.

   (A) Probability of a type I error is less than the significance level

   (B) $-t_{CRIT} < t_{STAT} < +t_{CRIT}$

   (C) $t_{STAT} < 0.05$

   (D) $\alpha < p$

Chapter 10

1   In an experiment to measure the effect of bio-feedback, 10 students measure their 'at-rest' heart-rate, and then endeavour to see if they can lower their own heart-rate using bio-feedback. They record their heart-rate again while using bio-feedback. What type of test should be used to analyse the results?

2   Two separate analytical methods, X and Y, are used to make replicate measurements of the COD value in the same sample of wastewater. Method X uses six replicates and method Y uses five, giving numerical values:

X	71.3	75.1	72.9	67.9	73.7	70.9
Y	73.7	75.5	74.5	72	74.8	

   (i)   Test for a difference in variance between the two methods.

   (ii) Test for a difference in mean value between the two methods.

   (iii) Using data from (i) and (ii) comment on whether there appears to be any bias and/or difference in measurement uncertainty between the two methods.

3   An analyst wishes to test, with 95% confidence, whether a source of mineral water has a sodium content, μ, that is greater than 40 mg L^{-1}. Five aliquots (physical samples) of the water are analysed. The sodium contents of the five samples are recorded (in mg L^{-1}) as

$$40.7 \quad 41.5 \quad 39.9 \quad 41.6 \quad 40$$

   (i) State the hypotheses that would be appropriate for this test.

   (ii) Perform appropriate calculations to decide what conclusion the analyst should reach concerning the sodium content of the water.

4 Two groups (A and B) of plants were grown under identical conditions except for the pH of the soil. The aim of the experiment was to test whether the difference in pH had a significant effect on the mean growth rate of the plants. The level of confidence was required to be 95%. The growth rates for the members of each group were measured, giving the following results:

Group A (11 plants): mean value $= 14.6$, standard deviation $= 2.6$

Group B (16 plants): mean value $= 9.5$, standard deviation $= 1.8$

(i)   State the hypotheses that would be appropriate for this test.

(ii) Calculate the t-statistic.

(iii) What is the critical t-value appropriate to this problem?

(iv) On the basis of the results in (ii) and (iii), what conclusion can be drawn from this test?

5 An investigation was conducted into whether student's performances in music are related to thier performances in mathematics. The data given in the table show the performance scores of a group of eight students in both mathematics and music.

Student	1	2	3	4	5	6	7	8
Maths mark	65	53	71	63	49	58	54	73
Music mark	59	47	61	65	52	65	51	64

(i)   Investigate whether the mean mathematics mark is significantly different from the mean music mark.

(ii) Investigate whether the data supports the hypothesis (at 95% confidence) that performance in music is related to performance in mathematics.

6 In an experiment to compare effects of different genetic modifications on parasitic resistance, two different variants of plant were grown in two plots. In one plot, 40 out of 200 plants were affected by the parasite, and 52 out of 180 plants were affected in the other plot. Is there evidence of significant difference in resistance between the plants in the two plots?

Chapter 11

1 A student keeps a record of the 'correct' answers in multiple-choice test, and finds that the numbers of times that particular answers (A, B, C or D) were the correct answers were as follows:

A: 25,   B: 14,   C: 21,   D: 10

Perform a chi-squared test (at 95% confidence) to investigate whether the distribution shows a deviation from a random allocation of correct answers.

2 The observed values in a 2×2 contingency test were

19	6
11	14

Calculate the value of the chi-squared statistic using the Yates' correction.

3 A student investigates whether there is any relationship between the colour of the car and the speed at which it is driven. Selecting an open stretch of road, he counts the numbers of cars of each colour that

could be considered to be driving 'fast', 'at a moderate speed', or 'slowly'. The results are given in the
table below.

	Red	Green
Fast	25	25
Moderate	12	18
Slow	3	17

Perform a chi-squared test for association between colour and speed.

4 In an experiment to investigate the preferences of primary school children, 50 boys were asked their
   preferences for colour (red or green) and their preference for sport (football or swimming). The numbers
   choosing particular categories are given in the table below.

	Red	Green
Football	24	11
Swimming	6	9

(i)   Calculate numbers of boys that would be *expected* in each category if the choice of colour and
      sport were independent.

(ii) Complete the calculation of the chi-squared value for a contingency test between colour and sport.

(iii) Identify the critical value for chi-squared in this contingency test, assuming a significance level of
      0.05.

(iv) Explain carefully what can be concluded from the results of the contingency test about the boys'
      preferences for colour and sport.

Chapter 12

1 The pocket money (in £) received by 10 children selected at random from a class is given below:

6	7	15	4	5	3	8	15	12	7

(i) Use a Wilcoxon test to investigate whether there is evidence, at 0.05 significance, that this class of
     students receive more than a median value of £5 pocket money.

(ii) Use a sign test and compare the result with (i).

2 A student aims to test whether exercise increases systolic blood pressure. Seven subjects measure their
   systolic blood pressure (in mm Hg) before and after exercise, with the results given in the table below.

   Perform a suitable *non-parametric* test to investigate whether there is any significant increase (at 0.05)
   in blood pressure:

Subject	1	2	3	4	5	6	7
BP before	115	122	145	133	147	118	130
BP after	125	127	155	136	142	124	139

3 Use a Mann–Whitney test to test whether there is a significant difference between the medians of the
   populations of values from which the following two samples, X and Y, were drawn.

X	184	135	129	95	132	67	152
Y	348	255	244	179	250	126	

4 As part of a project in forensic science, a student measures the times to emergence for larvae collected from a recently deceased pig. The times given in the table below were recorded for larvae collected on different days (Day1, Day2 and Day3) and grown on different media (M1, M2, M3 and M4).

	Day1	Day2	Day3
M1	27	19	33
M2	25	24	32
M3	20	19	24
M4	21	20	28

Use suitable *non-parametric tests* to analyse the data to investigate whether day of collection and/or media have any significant effect (at 0.05) on the emergence times of the larvae.

5 For the following pairs of $x-y$ values:

x	71.3	75.1	72.9	67.9	73.7
y	73.7	74.6	74.5	73.6	74.1

(i) Apply a *non-parametric test* for correlation between x and y,

(ii) Apply a parametric test for correlation between x and y, and compare with the result in (i).

6 Three different samples (A, B, C) of a batch of plant material are taken, and five replicate measurements are made on each sample for the determination of potassium content, giving the following results ($\mu g\ mL^{-1}$).

A	3.66	3.62	3.68	3.64	3.58
B	3.7	3.64	3.76	3.72	3.68
C	3.6	3.68	3.6	3.64	3.58

Use an appropriate *non-parametric test* to test for a difference between the three samples.

Chapter 13

1 The cadmium content of water is measured with an atomic absorption technique, and *six replicate measurements* are made on the *same* sample of river water. Samples were then taken of river water from nine *different places* in the river, and the cadmium content measured, giving the following results in ppb:

Replicates from *same* sample   45   55   52   49   47   50
Nine *different* samples   69   43   40   51   42   46   63   52   50

(i) Calculate the measurement variance in the six replicates.

(ii) Calculate the variance for the nine samples.

(iii) Is the result in (ii) significantly greater than in (i)?

(iv) Do your results suggest that there is any significant variation in cadmium content between the different river locations?

2 Use the data set in the table below, assume that Q is the independent variable:

P	4.4	-10.4	1.8	-1.4	-8.4	2	-6.2	-6.4
Q	10	2	4	8	2	12	6	8

(i) Calculate the 'sum of squares', SS_{RESID}, of the residuals.

(ii) Calculate the 'sum of squares', SS_P, for the P variable.

(iii) Using the data from (i) and (ii), calculate the correlation coefficient, r, for the data.

(iv) Use appropriate software (e.g. REGRESSION in Excel Data Analysis Tools) to calculate values SS_{RESID}, SS_P, and the correlation coefficient, r.

(v) Compare the results from (i) to (iii) with the results in (iv).

3 The following calibration data for a spectrophotometric measurement gives the values of absorbance, A, observed for standard solutions with concentrations, C (in ppm):

C	0	0.5	1	1.5	2	2.5	3
A	0.13	0.24	0.29	0.39	0.46	0.54	0.65

The absorbance of an unknown solution is measured three times giving an average value, $A_O = 0.31$. Calculate the 95% confidence interval for the concentration, C_O, of the unknown solution.

4 Three different samples (A, B, C) of a batch of plant material are taken, and five replicate measurements are made on each sample for the determination of potassium content, giving the results (μg mL^{-1}). The mean value and variance is calculated for each set of five replicate measurements. The mean of sample variances and the variance of sample means are also calculated.

						Mean	Variance
A	3.66	3.62	3.68	3.64	3.58	3.6360	0.001 48
B	3.7	3.64	3.76	3.72	3.68	3.7000	0.002 00
C	3.6	3.68	3.6	3.64	3.58	3.6200	0.001 60
			Column mean:				0.001 69
			Column variance:			0.001 792	

(i) Calculate the values for MS(W) and MS(B).

(ii) Estimate the variance of the measurement procedure.

(iii) Estimate the variance between the potassium content of the samples.

(iv) Test for a significant difference between the potassium content of the different samples.

(v) Use appropriate software to perform an ANOVA test.

(vi) Compare the results in (v) with those in (i) to (iv).

5 The data below gives the values of dissolved oxygen (mL L^{-1}) in standing water at three different depths, measured at four different locations. Analyse the data for possible differences in oxygen content at different depths and/or locations.

	Surface	Middle	Bottom
A	4.76	4.57	3.92
B	4.86	4.23	3.33
C	5.83	5.02	3.12
D	4.12	3.52	3.34

Short Answers to In-text Questions

Worked answers to most questions are available on the web site:

⌨ Answers & Examples

Some answers, given as 'web' below, are only available on the web site.

Chapter 2

Q2.1 4.26×10^4, 3.62×10^{-3}, 1.0×10^4, 1.0×10^{-4}, 4.5×10^2, 2.66×10^4, 3.2×10^3, 4.5×10^{-6}

Q2.2 3.6×10^2, 3.6×10^2, 4.0×10^1, 5.0×10^4, 2.24×10^1, 4.0×10^7

Q2.3 -1.2042×10^3, 2.3771×10^9

Q2.4   0.047, 0.046, 14.0, 7.35×10^3, 27000, 11.2, 11.2, 5.64×10^{-3}

Q2.5   0047, 8.00, 426.89, 1.34

Q2.6   3.89 g

Q2.7   500 mph

Q2.8 4.0×10^{-5}, 2083.33, 0.1556, 12.159, ±0.259

Q2.9   12

Q2.10 3150 kJ, 285.7 kcal, 65.9 kg, 152.4 mm by 25.4 mm, 22.72 L, 11.5 g

Q2.11 100, 7.9×10^3 kg m^{-3}, 0.15 kg m^{-2}, 7.06 L per 100 km,

Q2.12 180.17, 180.17 g mol^{-1}, 180.17 g

Q2.13 106 g, 15.9 g, 0.033 mol

Q2.14 122.2 g mol^{-1}, 122.2

Q2.15 0.2 mol L^{-1}

Q2.16 1.867 mol L^{-1}

Q2.17 0.584 g

Q2.18 6.2425 g

Q2.19 0.1 mol L^{-1}

Q2.20 5 mL

Q2.21   20 mL

Q2.22 $2\pi\ (= 6.283)$, $0.5\pi\ (= 1.57)$, 2.967, $57.30°$, $120.32°$, $630°$

Q2.23   6.98 cm

Q2.24   8.66 m

Q2.25   149.67 m

Q2.26   See web

Q2.27   3840 km

Q2.28 $36.87°$

Q2.29   D1*TAN(RADIANS(D3))

Chapter 3

Q3.1 $12, 36, 4, 3/2, -4, 36, 0, 5$

Q3.2 $3/4, 3/4, 4, 1/3, a/b, a/b, c, 1/h$

Q3.3 $9, 8, 7, 10, [2,6,11], 3, 11, [+5\ \text{or}\ -5]$

Q3.4   Web

Q3.5 $6, 20, 1/3, 4.25, 1, 3$

Q3.6 $6 + 3x, 6 + 12x, 6 - 8x, ax + 2a, -6a + 3ax, ax + a^2$

Q3.7 $2a/b, 2a/b, (4a + 1)/2b, (2a + 3)/2b$

Q3.8   Y, Y, N, Y, Y, Y

Q3.9 $x = 1, x = -2, x = 6 - a, x = 1 + a, x = 7 - a, x = 0.6 + b + c$

Q3.10 $x = a - 6, x = 7 - a, x = 3, x = a - 8$

Q3.11 $x = (a - 6)/5, x = (y - c)/m, x = 3p - 12, x = (p + 4)/(2 - p), x = 5(3 - p),$
$x = 2(3 - p)/(2 - p)$

Q3.12 $x = -6, x = (3 + a)/3, x = (8 + 3p)/4, x = 2p/3, x = 3a/2, x = 11/5$

Q3.13 $3, 5p, (m - 3), (3 - p), -2 - p, p + 2k - 1$

Q3.14 $x + 10, 9x + 8, 10x + 2, 2x(b - a), (a + 5)x - 3a, x$

Q3.15 $x = 3/2, x = -(b + a)/2, x = 3/2, x = 5/(3 - p), x = (5 - p)/(3 - p),$
$x = (5 - p)/(p - 2 - q), x = 12a/(6a - 1), x = (2p - 6)/p, x = (2p - 6)/(p - 1),$

Q3.16 $x = \pm 3, x = \pm\sqrt{(2 + a)}, x = \pm\sqrt{2a}, x = a^2 - 3, x = \pm\sqrt{(a^2 - 25)}$

Q3.17 $x = 2, x = (4 - a)/2, x = 4/(2 + a), x = a/(3a - 1), x = (1 - 2a)/(3a - 1),$
$x = (3a - 1)/(1 - 2a), x = (4a + 6)/(3 - 4a), x = (y + 1)/(y - 1)$

Q3.18 $a^2 - 4a + 4, a^2 - b^2, 3a - a^2 b - 4ab + 12, 3ac + ad + 3bc + bd, px - qx - 2py + 2qy,$
$5a - b + a^2 - b^2 + 6$

Q3.19 $x = -2, x = 4/3, x = -2/10, x = ab/(a + b), x = 3a/(5 + a), x = 4a/7,$
$x = 8(1 - a)/(4 - a)$

Q3.20 $5/6, 1/3, (p + q)/pq, (p + q - 1)/pq, 2/\{p(p + 2)\}, (p + 3)/\{p(p + 2)\}$

Q3.21   4 and $(x + y)$, 2 and $(2x+3y)$, 2 and x and $(2+3x)$, p and x and $(q + b)$

Q3.22   15, 2, -2, 122, 1.5

Q3.23 ± 5, ± 0.158,   19.95,   81,   3.27,   19.88,   2.76, -2

Q3.24 ± 4, $[-14$ or $-4]$, 3, -0.0464, 0.809, 0.391

Q3.25   1.0 or 0.5, 1.025 or -0.244, -2.87 or 0.871, 5.24 or 0.764

Q3.26   Equal roots $= 2$, 1 or -1.5, No real solutions, Equal roots $= -1.5$

Q3.27   2.153, 1.759

Q3.28   1285.7 s, 2042.9 m

Q3.29 $C_1 = 0.210$, $C_2 = 0.150$

Q3.30   (1,1) or $(-1.5, 2.25)$

Q3.31   Web

Chapter 4

Q4.1 T, F, F

Q4.2 $[-1, 1, 3, 5, 7]$, Web, 3, N, Y, N, Web

Q4.3 $L = 2.299 \times 10^{-5} \times T + 1.21$, 2.299×10^{-5}, 1.21

Q4.4 -0.75, -0.5

Q4.5 $7/4$

Q4.6 $T = 40W + 20$, 2.5 kg

Q4.7   (c), 7 s, -260 m

Q4.8   2, $2/3$, -4, -0.5, -1, 6.5

Q4.9 $y = 2x + 3$, $y = 0.5x + 2$, $y = -0.5x + 4$

Q4.10 $F = 1.8\,C + 32$, -17.78

Q4.11 $y = 3x - 2$

Q4.12 $y = -(1/3)x + (4/3)$

Q4.13 $x = -1.5$

Q4.14   3.9, 7.2, 2, -4

Q4.15   16.25, 22

Q4.16   0.125, 7.125°

Q4.17 $y = $ heart rate, $x = $ step-up rate

Q4.18   C, R, C, C

Q4.19 $P = 1.235Q + 2.845$, 7.044

Q4.20   0.0308, 0.051

Q4.21   0.03344

Q4.22 $\theta = 0.8788w$, 5.58 g

Q4.23   No, Yes, RT, 0, Yes, 8.33, Slope increases

Q4.24 $y = 1/v$, $x = 1/S$, $v_{\text{max}} = 1/c$, $K_M = m/c$

Q4.25 0.61 h^{-1}, 60.3

Q4.26   Plot $\ln(k)$ against $(1/T)$, $E = -mR$, $A = \exp(c)$

Q4.27   (i) Use 'log' or 'ln':

$\log(W)$ against h, $T = 10^m$, $K = 10^c$, or

$\ln(W)$ against h, $T = \exp(m)$, $K = \exp(c)$

(ii) Only use 'ln':

$\ln(N)$ against t, $k = m$, $N_O = \exp(c)$

Q4.28 $\ln(V)$ against $\ln(A)\, m = x$, $k = \exp(c)$, or $\log(V)$ against $\log(A)\, m = x$, $k = 10^c$, $\ln(E)$ against $\ln(T + 273)\, m = x$, $A = \exp(c)$, or $\log(E)$ against $\log(T + 273)\, m = x$, $A = 10^c$,

Chapter 5

Q5.1 10^5, 10^6, 10, 10^5, e^2, e^5, e^2, e^6,

Q5.2   1000, 2511.9, 0.001, 66.69, 1, 2.718

Q5.3   2.54, 1.54, 0.54, 2, -2, -0.301, 0.301, 1.301, 1, 2.303

Q5.4   3.09, 1.03, 0.799, 1.255, 0.628, 0.588

Q5.5 -0.3, 0.62, 0.3, 1.3, 2.3, -0.3, 6.9, 0.69, 0.93, 1.1

Q5.6 -5.12, 0.119, 2.56, -0.38, 2.52, 0.748

Q5.7   73 dB, 79 dB, 90 dB, 67 dB, 61 dB, 50 dB

Q5.8 $9.99.. \times 10^{-8}$, $0.099 \ldots$, No, 1.0×10^6, 1.0×10^6, Yes, -6, -6, Yes

Q5.9   8.47, $6.31 \times 10^{-5} \text{ mol L}^{-1}$

Q5.10   Web

Q5.11   Web

Q5.12   1 week, 6, 9.4 weeks

Q5.13 6.25×10^4, 0.5, 4.42×10^4

Q5.14   5771, 9514, 15 686

Q5.15 $N_t = 100 \exp(0.0953t)$

Q5.16 $N_O = 450$, $k = 0.032$, 661

Q5.17   28 min

Q5.18   2000, 8000, 1414, 4755

Q5.19   0.5, 0.25, 0.125

Q5.20   7.825 hours

Q5.21   0.0152 s (or 15.2 ms)

Q5.22   Web (ii) 0.65, 0.85 (iii) 0.632, 0.865

Chapter 6

Q6.1   5000, 300, -2000, -10

Q6.2   0, 0.25, 0.75, 1.5, 2, 1.75, 1, 0.25, 0, -0.25, -0.5, -0.5, -0.25, 0

Q6.3   Estimated values: 9350 cells per minute, 19 600 cells per minute

Q6.4 2.24×10^8, 3.03×10^7, 4.10×10^6

Q6.5 6.22×10^{-3} (web)

Chapter 7

Q7.1   P, S, S, P, P, S

Q7.2   24.0

Q7.3   (i) 4.8, 4.8 (ii) 0.2325, 0.2325

Q7.4   (i) 4.8 (ii) 0.1, -0.1, 0.3, 0.1, -0.4 (iii) 0 (iv) Yes (v) 0.28

Q7.5   0.2646, 0.2646

Q7.6   (i) 6.4, 2.3, 1.517 (ii) 66.4, 2.3, 1.517 (ii) 416.4, 2.3, 1.517

Q7.7   (i) 0.265 (ii) cm (iv) 0.07 (v) cm^2 (vii) 0.0146 (viii) cm

Q7.8   (i) 6.6, 5.35, 7.75, 2.4 (ii) 17, 10.5, 25.25, 14.75 (iii) 56, 34, 72.5, 38.5

Q7.9   Web

Q7.10   (i) Web (ii) 2, 3, 10, 5, 4

Q7.11   (i) 4, 10, 7, 3 (ii) 0.167, 0.417, 0.292, 0.125 (iii) 1.001 (iv) Web

Q7.12   (i) Web (ii) 0.533, 0.057, 0.324, 0.086 (iii) 192, 21, 117, 31 (iv) Web

Q7.13   (i) 10, 4, 2, 2, 4,10 (ii) 0.6, 5.25, 8.5, 7.5, 4, 0.9 (iii) Web

Q7.14   (i) 4, 10, 7, 3 (ii) 14, 18, 22, 26 (iii) 56, 180, 154, 78 (iv) 468 (v) 19.5 (vi) 461.5 (vii) 19.23

Q7.15   (i) 1,2,3,4,5,6,5,4,3,2,1 (ii) Web (iii) 36

Q7.16   (i) 0.0278, 0.0556, 0.0833 etc. (ii) Web (iii) 1.00

Q7.17   0.05

Q7.18   (i) 0.0278, 0.0833, 0.1667 etc. (ii) 0.4167 (iii) 0.1667

Q7.19   (i) 0.243, 0.357, 0.271, 0.129 (ii) 0.129 (iv) 7, 11, 8, 4

Q7.20   1/52, 1/13, 3/13, 1/13, 1/4

Q7.21   9/16, 3/16, 3/16/, 1/16

Q7.22   7/24, 15/24, 1

Q7.23   1/26, 1/13, 4/13

Q7.24   1/13, 12/13, 1

Q7.25   1/36, 1/18, 1/6, 1/36

Q7.26   0.9, 0.1, 0.6561, 0.0001, 0.0036, 0.2916

Q7.27   0.0065

Q7.28   1/7, 2/7, 2/7, 2/7, 1

Q7.29   0.618

Q7.30 5.0×10^{-6}, 1.0×10^{-6}

Q7.31   0.8333

Q7.32 5.0×10^{-6}, $1.0 \times 10^{+6}$, 5.0, Yes

Q7.33   20

Q7.34   120, 1, 1, 0, 42, 10 100

Q7.35   24

Q7.36   1680

Q7.37   167 960

Chapter 8

Q8.1   0.1915, 0.2417, 0.5000, 0.6828, 0.9546, 0.9974, 1.000

Q8.2   1.5, 1.2, 1.2

Q8.3   0.0215, 0.4987, 0.9974, 0.0013, 0.9987

Q8.4   0.7734, 0.2734, 0.2734, 0.4920, 0.0723, 0.2778, 0.0062, 0.0375

Q8.5   1.25, -1.25, 1.96, 2.575, -1.96

Q8.6   1.96, 2.58, 1.96, 2.58

Q8.7   5, 3, 0, 1, -1, 2, -2, 0.4772, 0.6826, 0.9544, 0.0228

Q8.8   (i) 5, 3 (ii) 2, 8 (iii) -1 (iv) -1 (v) 9.935 (iv) 0.065 to 9.935 (v) -0.88 to 10.88

Q8.9   Web

Q8.10   19.08 to 26.92

Q8.11   1.0, 21.04 to 24.96

Q8.12   0.0228, 0.1359, 15.8225, 0.103, 14.69 to 15.31, 15.26

Q8.13 19.3 ± 2.6, 18.9 ± 1.1, Nothing

Q8.14   2.45, 1.94, 3.71, 3.14, 2.45, 2.23, 2.09, 2.01, 1.96, 2.45, 3.71, Web

Q8.15   5.65 to 6.39, 43.0 to 49.4, 10.01 to 17.19

Q8.16   553, 8.77, 3.92, 542 to 564

Q8.17   0.08 ppm

Q8.18 1.87 ± 0.21 (sd)

Q8.19 9.85 ± 0.04 (sd)

Q8.20   0.5905, 0.3281, 0.0729, 0.0081, 0.0004, 0.0000

Q8.21 5.9×10^{-6}, 1.4×10^{-4}, 0.233, 0.121, 0.028, 0.383, 0.617

Q8.22   (i) 0.262 (ii) 0.393, 0.246, 0.082, 0.015, 0.001, 6.4×10^{-5} (iii) 13, 20, 12, 4, 1, 0, 0

Q8.23 37 ± 6.1 (95%)

Q8.24   0.606, 0.303, 0.076, 0.014

Q8.25   (i) $CD_A = 0.99$, $CD_B = 1.33$ (ii) A (iii) Clumping

Chapter 9

Q9.1   No, <0.05, 0.05

Q9.2   F, T, F, F

Chapter 10

Q10.1   3.68, 3.77

Q10.2 $F_{STAT} = 3.99$, $F_{CRIT} = 5.99$, Not sig

Q10.3   (i) $F_{STAT} = 4.24$, $F_{CRIT} = 5.12$, Not sig

Q10.4   Two, One, Two

Q10.5   3.15 and 0.0078, 4.09 and 0.0018

Q10.6   1, 2, 1, 2

Q10.7   (ii) 0.05 (iii) 3.585×10^{-8}, 0.080×10^{-8} (iv) 2.60 (v) 1 (vi) 5 (vii) 2.02

Q10.8 $t_{STAT} = 1.83$, $t_{CRIT} = 2.06$, Not sig

Q10.9 $t_{STAT} = 2.54$, $t_{CRIT} = 2.45$, Sig, $p = 0.044$ (web)

Q10.10   Web

Q10.11 $r = 0.542$, $r_{CRIT} = 0.549$, Not sig, $p = 0.053$

Q10.12   Web

Q10.13 $z_{STAT} = 2.26$, $z_{CRIT} = 1.96$, Sig

Q10.14 $z = 1.68$, Not sig

Chapter 11

Q11.1   9.49, 6.63

Q11.2 $\chi^2 crit = 7.81$, No

Q11.3 $\chi^2 stat = 12$, $\chi^2 crit = 7.81$, Yes

Q11.4 $\chi^2 stat = 3.28$, $\chi^2 crit = 5.99$, No

Q11.5   A: $\chi^2 stat = 10.61$, B: $\chi^2 stat = 2.05$, $\chi^2 crit = 9.49$, Web

Q11.6   (ii) $\chi^2 stat = 3.42$ (iii) $\chi^2 stat = 4.05$, $\chi^2 crit = 3.84$

Chapter 12

Q12.1   (i) $W(-) = 28$, $W_L = 30$, Sig (ii) $W_L = 25$, Not sig

Q12.2 $W(-) = 3$, $W_L = 2$, Not sig

Q12.3 $U_B = 12$, $U_L = 12$, Sig, $p = 0.044$

Q12.4 $H = 7.43$, χ^2crit $= 7.81$, Not sig

Q12.5 $S = 9.53$, χ^2crit $= 7.81$, Sig

Q12.6 $S = 7.73$, χ^2crit $= 7.81$, Not sig (web)

Q12.7 $p = 0.0414$, Sig

Q12.8 $r_S = 0.303$, $r_S$crit $= 0.564$, Not sig (ii) $p = 0.214$

Chapter 13

Q13.1   Web

Q13.2   240.3

Q13.3 $SS_X = 0.0596$, $SS_y = 0.249$, $SS_{RESID} = 0.176$, $r = 0.542$

Q13.4 5.58 ± 0.16 (95% CI) g $100\,\text{mL}^{-1}$

Q13.5 $t_{STAT} = 3.54$, $t_{CRIT} = 2.78$, Sig

Q13.6   0.094

Q13.7   0.001 67, 0.002 55, 0.001 67, 0.014 42

Q13.8   3.89, 8.65, Sig

Q13.9   Web

Q13.10   Web

Q13.11   Media $p = 0.044$, Day $p = 0.0035$, Both sig

Q13.12   Temp $p = 0.252$, Press $p = 6 \times 10^{-6}$, Int $p = 0.003$

Q13.13   Web

Q13.14   0.061

Q13.15   11.92, 5.32

Index

Key page references are given in bold

α, Alpha - see Significance level
β, Beta - see Power of hypothesis test
χ, Chi - see Chi-squared
δ, Delta - see Deviation
μ, Mu - see Mean
π, Pi - 27
σ, Sigma - see Standard deviation
Σ, Capital sigma - 123
τ, Tau - see Time constant

Abscissa, 66, 306
Absorbance, 62, **66**, 80,**97**
Accuracy, 3
Analysis of variance, ANOVA, 261, 267, 272, 277, **292ff**
  One-way, 298
  Two-way, 299, 301
Angle, 29, 75, 134
Average, 124
Avogadro's number, 19

Base, 90
Bayes' rule, 154ff
Beer's law, 66, 80
Bin, 132
Binomial distribution, See Distributions, Mean,
    Probability, Standard deviation
Blocking, See Experiment design
Box and whisker plot, 130
Brackets, 40, 47, 51

Calibration, 286
Cancelling, 42, 50
Categories, 7, 132ff, 134, 245ff, 261

Central limit theorem, 181ff
Chi-squared
  statistic, 245ff, 269, 271, 316
  test, See Hypothesis testing
Class, 132, 137ff, 143ff
Coefficient of determination, 236
Coefficient of dispersion, **128**, 167, 200, 207ff, 278
Coefficient of variation, 191
Column graph, 134
Combinations, 163ff, 197
Concentration, **22ff**, 62, 66, 80, 85
Confidence deviation, **186**, 190, 194, 204, 305
Confidence interval, 167, 184, **186**, 191, 194, 203,
    225, 242, 289, 305
  in calibration, 289
  of binomial mean, 203
  of proportion, 203, 242
  of sample mean, 186, 225, 305
Confidence level, 187, 214, 225
Contingency, See Test for
Co-ordinates, 67ff
Correlation, **77**, 210, 219, **236ff**, 239, 274, **284ff**
  test for, 237, 274
Correlation coefficient, 236, 261, 274, **284ff**, 318
  Pearson's, 236, 261, 274, 284, 318
  Spearman's, 261, 274, 318
Coverage factor, 193
Critical value, 215ff, 221, 226, 246
Cumulative, See Frequency, Probability

Data, 1, 7
  continuous, 7, 111, 140
  discrete, 7, 111,
  frequency, 132ff
  interval, 7
  nominal, 7, 261

Data (*Continued*)
  ordinal, 7, 261
  ratio, 7
Data analysis tools, 79, 223, 239, 276, 290, 298, **314**
Decay, 5, 89, **97ff**, 107
Decimal places, 11
Decimal reduction time, 106
Deduction, 210ff
Degrees, 26ff
Degrees of freedom, 2, 187ff, 279
  ANOVA, 297
  chi-squared tests, 245, 247ff
  confidence interval, 186ff
  contingency, 252
  correlation, 238
  F-test, 221
  goodness of fit, 256
  t-test, 227
Denominator, 37, 42, 50ff
Density, 14
Deviation, 125, 226
Differential coefficient, 115ff
Differentiation, 116
Dilution, 24ff
Dispersion, 125, 128, 200, 207ff, 278
Distribution, 167
  binomial, 167, **196ff**, 200, 202, 206, 274
  continuous, 140
  frequency, 139
  normal, 167, **168ff**, 171, 181, 203, 245, 258, 315
  Poisson, 5, 167, 202, **205ff**, 207, 245, 256
  probability, 139
  standard normal, 171, 315
Dot plot, 132

Error, 2, 3, 81, 214, 219
  random, 3
  systematic, 3
Estimation, 13, 118, 141
Excel, See Appendix I - 313
Experiment design
  allocation, 305, **309ff**
  blocking, 270, 301, 305, **307**
  completely randomized, 298, 307
  cross-over, 308
  pairing, 225, 232, 264, 305, 308
  randomization, 298, 305
  randomized block, 301, 307

repeated measures, 305, 308
replicate measurements, 76, 122, 180,181ff, 190, 225, 268, 270, 286, 294, 298, 301, **305ff**
Exponent, 8, 89
Exponential, *e*, 5, 84ff, **90ff**, 97ff, 116
Extrapolation, 74, 81

Factorial, 160ff
Factors
  experimental, 209, 267, 270, 277, 292, 294, 296, 306
  mathematical, 41, 53, 98
  test for, 268, 270, 296
False positive/negative, 149ff
Fractions, 37, 50ff
Frequency, **131ff**, 133, 143, **246ff**
  density, 135, 169
  relative, 133, 139
  test for, 246
Friedman test, See Hypothesis testing
F-test, **220**, 297, 317
Function, 4, 31, 54, 314

Generation time, 89, 103ff, 116
Goodness of fit (test), 255ff
  normal distribution, 258
  Poisson distribution, 256
Gradient, See Slope
Growth, 5, 89, **97ff**

Half-life, 104
Histogram, 135, 140, 199
H-statistic, 269
Hypothesis testing, 211,**212ff**, 219, 278
  ANOVA, 297, 298ff
  chi-squared, 246, 257
  correlation, 237ff
  Friedman, 261, 270
  F-test, 220ff
  Kruskal-Wallis, 261, 267
  Mann-Whitney, 261, 265, 319
  proportion, 241ff
  Sign, 261, 273
  t-test, 226ff, 228, 230 232
  Wilcoxon, 261, **262**, 264, 318

Induction, 210ff
Interaction, 208, 252, 301, 308
  test for, 252, 301

Intercept, 65, **67ff**, **71**, 75, 77, 80ff, 281, 287
Interpolation, 74, 81
Interquartile range, 130, 261
Inverse function, 31, 54

Kruskal-Wallis test, See Hypothesis testing

Latin square, 311
Level, 209, 268, 277, 292, 298
Likelihood ratio, 157, 158
Linearization, 81ff
Logarithms, 59, 66, 85ff, **91ff**, 95
Loudness, 95

Mann-Whitney test, See Hypothesis testing
Mean, **124ff**, 137ff
  binomial, 200, 241
  normal, 169, 176, **179ff**
  Poisson, 207
  proportion, 241
  standard error of 181
  test for, 225ff, 228, 230
Mean of the sample variances, *MSV*, 293,
  295ff
Mean square, *MS*, 279ff, 295ff
Mechanisms, 167, 209ff
Median, 129, 261
  test for, 262, 266, 273
Michaelis-Menten, 84
Modelling, 4, 109, 111ff
Molar concentration, 22ff
Molar mass, 20ff
Molarity, 22
Mole, 19ff

Non-parametric, 219, 261
Normal, See Data, Distributions, Goodness of fit,
  Mean, Probability, Standard deviation
Numerator, 37, 42

Odds, 155ff
Order of magnitude, 13, 15
Ordinate, 66,
Outcomes, 148, 209, 277, 306

Pairing, See Experiment design, Test for
Parallel lines, 73
Parameter, 2,**123**
Parametric, 219, 261

Pearson, See Correlation
Permutations, 161ff
Perpendicular lines, 73
pH, 3, 96
Pie Chart, 134
Poisson, See Distributions, Goodness of fit, Mean,
  Probability, Standard deviation
Pooled standard deviation, 230, 243, 293
Population, 98, **122ff**, 180, 225, 278
Power, 8, 14ff, 56, 59, 82, 86ff, 90ff, 117, 192
  of hypothesis test, 215, 220, 234, 261, 272, 312
Precision, 3
Pressure, 14
Probability, 139, 139ff, **142**, 169, 171, 180
  area, 172ff
  binomial, 196ff
  combining, 145ff
  conditional, 149, 152
  cumulative, 141, 202
  density, 141, 168ff, 180
  joint, 152
  normal, 169ff
  Poisson, 205
  tree, 151ff
Proportion, 203, **241ff**, 248
  test for, 241
p-value, **214ff**, 227, 307, 310
  and significance level, 214, 220, 227
  in ANOVA, 298, 300, 302, 307, 310ff
  in Excel, 223, 234, 238, 276
Pythagoras, 29

Quadratic equation, 40, 56
Qualitative, 7
Quantitative, 7, 261
Quartile, 129

Radians, 26ff
Randomization, See Experiment design
Ranking, **129**, **263**, 266, 268, 270, 275
Reciprocal, 16, 37, 55
Regression (linear), **75ff**, 77, **281**, 306
Relative molar mass, 20
Repeated measures, 305, 308
Replicate measurements, See Experiment design
Residual, 239, 281ff, 306
Residual sum of squares, 282
Root mean square, 281
Rounding, 11

Sample, **122ff**, 180, 180ff, 184ff, 194, 226, 278
Scientific Method, 209, **210ff**
Scientific Notation, 8ff
SI units, 15
Sign test, See Hypothesis testing
Significance level, 187, **214**, 227, 234, 238
Significant figures, 10
Simultaneous equations, 60ff
Slope, 65, **67ff**, **71**, 75, 77, 80ff, 111ff, 114ff, 236, 281, 287
Solution (to equations), 54, 56ff
Solver (Excel), 62
Spearman's correlation coefficient, See Correlation
Speed, 14, 110
Square root, 48
S-statistic, 270
Standard deviation, **126ff**, 137ff, 180ff, 194
  binomial, 200
  normal, 169, **176ff**
  Poisson, 207
  proportion, 241
  relative, 191ff
Standard error of regression, 283, 288
Standard error of the mean, 181
Standard normal distribution, 171, 315
Standard uncertainty, 184, 190
Statistic, 2, **123**, 215ff, 221, 226, 228, 245, 263
Stem and leaf diagram, 131
Straight-line graph, **66ff**, 111, 281
Subjects, 209, 225, 277, 292, 305, **309**
Subscript, 123
Substitution, 40, 57
Sum of squares, SS, 126, 279ff, 284, 299

Tails, 187, 189, **213ff**, 221ff, 227, 238, 249, 262
Tangent (to a curve), 112, 115ff
Term, 36, 43ff
Test assumptions
  chi-squared, 245, 251
  correlation, 274
  F-test, 220
  non-parametric, 261
  t-test, 235
Test for
  association, 252
  contingency, 252ff, 255

correlation, 237, 274
factor effect, 268, 270, 296, 298
frequency, 246
goodness of fit, 255
interaction, 252, **301**
mean, 225, 228, 230
median, 262, 266, 273
pair differences, 232, 264
proportion, 203, 241, 248
variance, 220
Time constant, 89 ,107
Transmittance, **66**, 97
Treatment, 294, 298
Trendline, 79
Trigonometric functions, 28
True positive/negative, 149ff
True value, 179, 225, 241
Trueness, 3
t-test, See Hypothesis testing
t-value, 185 **187ff**, 203, 215, 316
Type I and II Errors, 214, 219, 309, 311

Uncertainty, 2, 167, **179ff**, 190ff, 286, 295
  calibration, 286ff
  combining, 192
  experimental, 190
  limiting, 194
  relative, 190, 192
  standard, 184, 190, 192, 194
Units, 14ff
U-statistic, 267

Variable, 1, 66
  dependent, 66, 306
  independent, 66, 306
Variance, **126ff**, 220, **278ff**, 292, 294ff
  test for, 220
Variance of Sample Means, VSM, 293, 295ff

Wilcoxon test, See Hypothesis testing
W-statistic, 263

XY graph, 5, 66

Yates' correction, 245, 248

z-value, **171ff**, 176ff, 203, 242, 315